Photovoltaic/Thermal (PV/T) Systems

Ali H. A. Al-Waeli • Hussein A. Kazem
Miqdam Tariq Chaichan • Kamaruzzaman Sopian

Photovoltaic/Thermal (PV/T) Systems

Principles, Design, and Applications

Springer

Ali H. A. Al-Waeli
Solar Energy Research Institute
Universiti Kebangsaan Malaysia
Bangi, Selangor, Malaysia

Hussein A. Kazem
Faculty of Engineering
Sohar University
Sohar, Oman

Miqdam Tariq Chaichan
Energy and Renewable Energies
Technology Center
University of Technology
Baghdad, Iraq

Kamaruzzaman Sopian
Solar Energy Research Institute
Universiti Kebangsaan Malaysia
Bangi, Selangor, Malaysia

ISBN 978-3-030-27823-6 ISBN 978-3-030-27824-3 (eBook)
https://doi.org/10.1007/978-3-030-27824-3

This Springer imprint is published by the registered company Springer Nature Switzerland AG
The registered company address is: Gewerbestrasse 11, 6330 Cham, Switzerland

This book is dedicated to my home country, Iraq, and to my family, Hussein, Jenan, Karrar, and Mohammed

I hope you are proud of me as I am proud of you.

Preface

In 2018, we decided to write a book which covers the topic of photovoltaic thermal (PV/T) collectors. The intentions of writing this book were to provide a reference and a textbook with detailed explanation of the field of PV/T that is useful to academic students, researchers, professionals, and even nonacademics. Hence, the decision was made to provide theory, literature review, and insights into the field. This book is intended to cover a broad range of topics related to PV/T technology which includes photovoltaics, solar thermal systems, and solar energy in general. We recognized the need for a book that summarizes and combines the vast research done for design and development of photovoltaic thermal systems. Hence, the work extends to technical, economic, environmental, and industrial aspects of PV/T technology. Calculations for different energy efficiency, exergy, and other parameters were provided to assist readers to grasp understanding of PV/T sizing, design, and evaluation. Given that PV/T is a hybrid technology which combines photovoltaic modules and solar thermal collectors, the first chapter brings an introduction to solar energy and theory of PV systems and flat-plate collectors (FPC) The chapter illustrates how research is conducted in either technology, and the main parameters affect PVs and FPCs. The second chapter discusses the subject of the book, PV/T collectors, in detail with mentions to rationale, chronology, and evaluation methods. The literature review in Chap. 2 presents conventional schemes of PV/T systems, while novel and innovative schemes are provided in Chap. 3 which mainly focuses on nanofluids' use for PV/T systems. This chapter represents an outlook into recent and advanced research conducted by various scientists and engineers in the field. The fourth chapter presents methods for cost assessment of PV and ST systems. A case study of economic assessment is provided in the chapter to assist the readers who conduct similar approach to assessing PV/T systems. Economic aspect is highly critical due to its impact on the future of PV/T applications which depends on the costs of these systems.

The fifth chapter of the book focuses on analyzing the performance of PV and PV/T systems across different environmental parameters. This is crucial for the end-user and businesses to assess feasibility of these systems and find optimum locations to install PV/T plants. Among the parameters are solar irradiance, ambient temperatures, dust, and wind speed, all of which leads to the need for predicting PV/T performance for long-term use. The applications of this technology are introduced and explained in Chap. 6 where various types of PV/T collectors (conventional and novel) are revisited and considered for commercialization. Few companies have sold/employed PV/T systems/products throughout the last two decades, and these systems/products are presented to offer the readers a glance into the utility and viability of this technology. The seventh chapter presents comprehensive conclusions and recommendations of this book. This chapter is made to provide ideas and novel concepts for future research by students, researchers, and other academics.

The topics we discussed in this book range from established and proven methods to new methods which have recently been presented in the literature. Some methods are too general and may not work in all cases, but researchers and/or students are expected to derive specific methods for their respective systems.

The main purpose of this book was to establish a frame of reference to describe and summarize the theory and findings for a new field in technology which does not have a global standard for design, installation, and evaluation. And so, we, the authors of this book, recommend putting forth initiatives to establishing a global standard for PV/T technology.

The chapters one, two and four present equations and approaches which students, researchers, and professionals can use to design optimum PV/T systems and evaluate their performance in terms of electrical, thermal, and overall efficiency, exergy, and peak energy yield. Furthermore we provide a critical review of recent and classical research studies which are classified according to the aspect discussed in each technology. Chapters one and five also discuss in detail the environmental effects and solutions presented to issues which result in unreliability of solar energy technologies.

Other aspects discussed in this book are the multidisciplinary dimensions associated with PV/T technology and solar energy in general, for instance, nanofluids and artificial neural networks. More focus into nanofluids was made given that more research has been conducted throughout the last 5 years in this topic. The use of ANN for performance prediction of PV/T in particular remains a new field of study and requires further time to establish a literature base.

The studies discussed in the book cover experimental and numerical simulations. Experiments are either conducted outdoor or indoor (under standard test conditions). Simulations are either conducted for electrical or thermal aspects using software such as TRNSYS, ANSYS fluent, Simulink MATLAB, HOMER, etc. The utility of experimental work is to investigate the dynamic behavior of PV/T under typical operating conditions and assess its potential through parameters such as payback period and cost of energy. In addition, experimental data are used to validate the numerical findings, using comparison methods to predict and measure parameters. Chapters 2 and 3 provide brief discussions on building integrated and grid-connected PV/T systems which are methods to establishing the technology for the commercial and residential users.

Numerous sources have been used in writing this book which include research published in WoS-indexed journals such as *Solar Energy*, *Renewable and Sustainable*

Energy Reviews, *Energy Conversion and Management*, *Case Studies in Thermal Engineering*, etc. All research articles, books, and websites which have been published by fellow scientists and engineers have given us a tremendous help in completing this book.

Other sources such as governmental and international reports and company records have also been helpful in providing description of the current status of solar energy technologies and PV/T applications.

Various authors have contributed to enriching the solar energy literature, and here we note the works of John A. Duffie and William A. Beckman from the Solar Energy Laboratory from the University of Wisconsin-Madison and also the works of H. C. Hottel and his colleagues at MIT and that of A. Whillier.

We, the authors, would like to acknowledge our respective institutes, namely, Solar Energy Research Institute (SERI) of Universiti Kebangsaan Malaysia (UKM), Malaysia; Faculty of Engineering of Sohar University, Oman; and Energy and Renewable Energies Technology Center (ERETC) of the University of Technology, Iraq.

The generous permissions which have been granted to us by various publishers for the use of their tables, drawings, and other materials in this book are highly appreciated. These materials have been very helpful in completing the book and further illustrate the studies discussed within it.

Finally, we would like to thank Springer Nature for providing the platform to publish this book and for making it available for academics and professionals and to the team of editors for their assistance and cooperation.

Bangi, Selangor, Malaysia — Ali H. A. Al-Waeli
Sohar, Oman — Hussein A. Kazem
Baghdad, Iraq — Miqdam Tariq Chaichan
Bangi, Selangor, Malaysia — Kamaruzzaman Sopian
June 2019

Contents

Abbreviations

ANN	Artificial Neural Networks
CFD	Computational Fluid Dynamics
CF	Capacity Factor
CNT	Carbon Nanotubes
COE	Cost of Energy
COP	Coefficient of Performance
CPC	Compound Parabolic Concentrator
CPV	Concentrated Photovoltaic
CSP	Concentrated Solar Power
DAQ	Data Acquisition device
EDS	Energy-Dispersive Spectroscopy
EDX	Energy-Dispersive X-ray
EPBT	Energy Payback Time
ETFE	Ethylene Tetrafluoroethylene
FESEM	Field Emission Scanning Electron Microscope
FiT	Feed-in-Tariff
FPC	Flat-Plate Collector
GA	Genetic Algorithm
GCPV	Grid-Connected Photovoltaic
GCPV/T	Grid-Connected Photovoltaic/Thermal
GIT	Grid Independence Test
GPBT	Greenhouse Gas Payback Time
HPWH	Heat Pump Water Heater
HRV	Heat Recovery Ventilation
ISC	Short-Circuit Current
LCA	Life Cycle Assessment
LCCA	Life Cycle Cost Analysis
LCOE	Levelized Cost of Energy
LCOH	Levelized Cost of Heat
LCPV/T	Low-Concentrated Photovoltaic/Thermal
MAE	Mean Absolute Error

MAPE	Mean Absolute Percentage Error
MC	The system total maintenance cost
MC_{or}	The maintenance cost of the *r*th component in the first year (USD)
MC_r	The maintenance cost (USD)
MLP	Multilayer Perceptron
MSE	Mean Square Error
MPPT	Maximum Power Point Trackers
MWCNT	Multi-Wall Carbon Nanotubes
OPOP	Optimal Power Operating Point
PBP	Payback Period
PCM	Phase Change Material
PLC	Programmable Logic Controller
POA	Plane-of-Array
PV/T	Photovoltaic Thermal
PVT. n-pcm. nf	PV/T with nano-PCM tank and nanofluids flowing through pipes
PVT. pcm. w	PV/T with PCM tank and water flowing through pipes
PVT. W	PV/T with water tank and water flows through pipes
Re	Reynolds Number
RMSE	Root Mean Square Error
ROI	Return on Investment
SAHP	Solar-Assisted Heat Pump
SAPV	Standalone Photovoltaic
SAPV/T	Standalone Photovoltaic Thermal
SiC	Silicon Carbide
SOFM	Self-Organizing Feature Map
STC	Standard Test Conditions
SVM	Support Vector Machine
TEM	Thermoelectric Module
TFMS	Thin Flat Metallic Sheet
TES	Thermal Energy Storage
UC_i	The cost per unit of the *ith* component (USD/unit)
VOC	Open-Circuit Voltage
XRD	X-Ray Diffraction
YS_d	The daily/monthly specific yield
YF	Final Yield
YR	Reference Yield

List of Symbols

A_c	The collector area (m^2)
$A_{PV/T}$	The surface area of PV/T
c1,c2 and c3	Empirical coefficients allowing different power parameters to vary linearly with dc-voltage input
CA_i	The capacity of the *ith* component of SAPV, SAPVT nanofluid, and nano-PCM
$C_{capital}$	The capital cost of a project
$C_{O\&M}$	The yearly operation and maintenance costs
Cp	Heat capacity of fluid
Cp_{PVT}	The heat capacity of PV/T
Cp_{wax}	The heat capacity of wax and equal to 2100 J/kg °C
$C_{replacement}$	The cost of all equipment replacement and repair
$C_{salvage}$	The net worth of the system at the final year of project lifetime
D_{coil}	The inner diameter of the coil
E_{AC}	Alternating current energy
E_{DC}	Direct current energy
G	The incident solar irradiance (W/m^2)
G_{stc}	Solar irradiance (1000 W/m^2) at standard test conditions
h	Air heat transfer coefficients ($W \cdot m^{-2} \cdot K$)
h_{fluid}	Heat transfer coefficient of fluid inside coil
ICI	Total constant cost, including the cost of installation and civil works (USD)
IC_k	The initial cost of the *k*th component (USD)
IC_r	The initial cost of the *r*th component (USD)
K	Value of 1 and 2, which are equivalent to the inverter and pump, respectively
k_r	A constant refers to the maintenance cost as a percentage of the initial cost of the *r*th component
K_{fluid}	Thermal conductivity of water
K_{PVT}	Thermal conductivity of PVT
$\dot{m}$	Mass flow rate (kg/s)
N	Number of years

Nf	Nanofluid
N_r	The number of components replaced over the lifetime of the system
$P_{in}(t)$	The instantaneous input power
P_{Inv}	The inverter power
$P_{loss}(t)$	The instantaneous power losses
P_{PV}	The PV module power
P_{peak}	The PV peak power
P_R	Rated power
Q_i	The solar heat that equal to αG
Q_{coil}	Represent the heat transfer to the coil system
R	The equivalent to SAPVT components
RC_k	The replacement cost of the *k*th component (USD)
R_{PV}	Performance factor
t_1	The hour, day, month
t_2	The minute, hour, day
T	Temperature
T_a	Ambient temperature
T_c	The cell temperature
$T_{fluid, in}$	Input fluid temperature (approximately 25 °C)
T_{sk}	Represent sky temperature
T_{STC}	The temperature of (25°C) at standard test conditions (STC)
U_{fluid}	Velocity of fluid
U_{wind}	Wind velocity
W	Water
W_R	The total uncertainty of the used devices
z	Length gradient

Greek Letter

α_T	The temperature coefficient of PV
η_{inv}	Inventor efficiency
η_{wire}	Wires efficiency
μ_{fluid}	Viscosity of fluid
$^\circ$	Degrees
ρ_{PVT}	Density of PVT and equal to 2330 kg/m^3
ρ_{wax}	Density of wax
σ	Represent Stefan-Boltzmann constant (5.670367 × 10^{-8} W·m^{-2}·K^{-4})
τ	PV transmittance

Chapter 1
Introduction

1.1 Background

The continuous rise in global energy demand is heavily linked to the growth in population and improved economic and technological situations in most parts of the world. This demand is primarily met using coal, oil, and gas, or else known as "fossil fuels" resources [1]. The term "fossil" refers to these fuels being formed from remains of living organisms over the course of millions of years, while the term "fuels" is a description of their nature as they produce heat energy when they are burned. These resources can be used to generate the energy required for a variety of industrial processes such as heavy machinery, furnaces, and monitoring systems. The main issue of fossil fuels is that they produce carbon dioxide (CO_2) gases when they are burned. CO_2 is classified as a greenhouse gas; hence increase of CO_2 production leads to increase in global warming [2]. Among the three resources, coal is the highest producer of CO_2, while natural gas is the lowest. In addition, fossil fuels are classified as nonrenewable energy sources, meaning they can replenish or run out. This can occur if the rate of consumption of these resources is higher than the rate of formation. This partially leads to continuous fluctuations in fossil fuel prices and rise in greenhouse gases (GHG) at the ozone layer [3]. Alternative energy resources which does not replenish were discovered and integrated to generate energy. These resources include solar radiation, wind, geothermal, tidal, hydro, and bioenergies and are referred to as modern renewable energies. These resources have been widely welcomed as they are less likely to produce (GHG) emissions and will not run out. Renewable energy consumption has been continuously rising since the 1950s which is motivated by the need for an alternative energy resource and reduction of carbon footprint which is a goal for many countries that abide by the UNFCCC global treaties, mainly the "Kyoto Protocol" and the "Paris Agreement." According to the REN21 report 2017 [4], the modern renewable share of total final energy consumption in 2016 is estimated to be around 10.4%, while fossil fuel forms around 79.5% which is illustrated in Fig. 1.1.

A. H. A. Al-Waeli et al., *Photovoltaic/Thermal (PV/T) Systems*,
https://doi.org/10.1007/978-3-030-27824-3_1

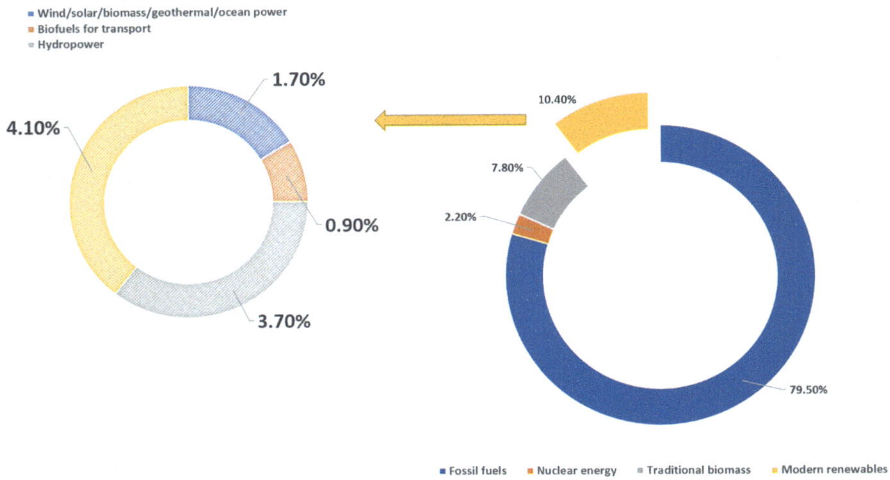

Fig. 1.1 Modern renewables share of total final energy consumption in 2016

Moreover, 5.4% average growth rate of modern renewables is observed across the past decade. The report labelled the year 2017 as a "record-breaking one" for renewable energies.

In addition, REN21 reports uneven growth in renewable energy technologies, with vast majority of installed capacity being that of solar PV, wind power, and hydro. This is because the output, cost-effectiveness, and benefits of each technology vary tremendously, which calls for the need of research and development (R&D) concerning each technology, separately. The highest technology in terms of installation capacity for the year 2017 is the solar PV which is the central focus of this book. To view the scope of renewable energies and its current status, Fig. 1.2. shows the renewable energy consumption by technology in 2017, according to the IEA [5].

Solar radiation is a renewable source stemming from the Sun, and if utilized to create thermal or electrical energies, it is referred to as "solar energy." This energy is utilized with two main technologies which are the solar cell and the solar thermal collector. The solar cell is also referred to as "solar photovoltaic – PV." According to the International Energy Agency (IEA) [6], the cumulative solar PV capacity reached 398 GW which represents 2% of the global power output. While, cumulative installation capacity of solar thermal has reached an estimated 472 GWth by end of 2017. In 2017, solar PV market has been the forefront of power generating technologies being the top source of new power capacity for different markets such as China, Japan, the USA, and India. The renewable energy consumption of solar PV in major markets is shown in Fig. 1.3 [5].

China seemingly is the global leader in solar PV consumption which is around 9.19% and 82.7% higher than the European Union and the USA in 2017. Also, it

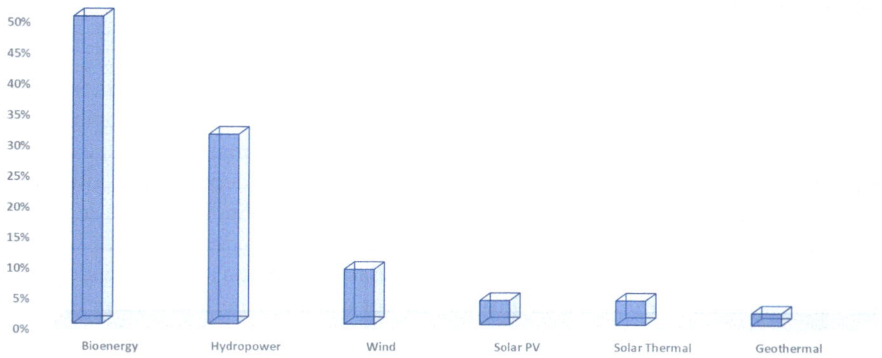

Fig. 1.2 Renewable energy consumption by technology in 2017

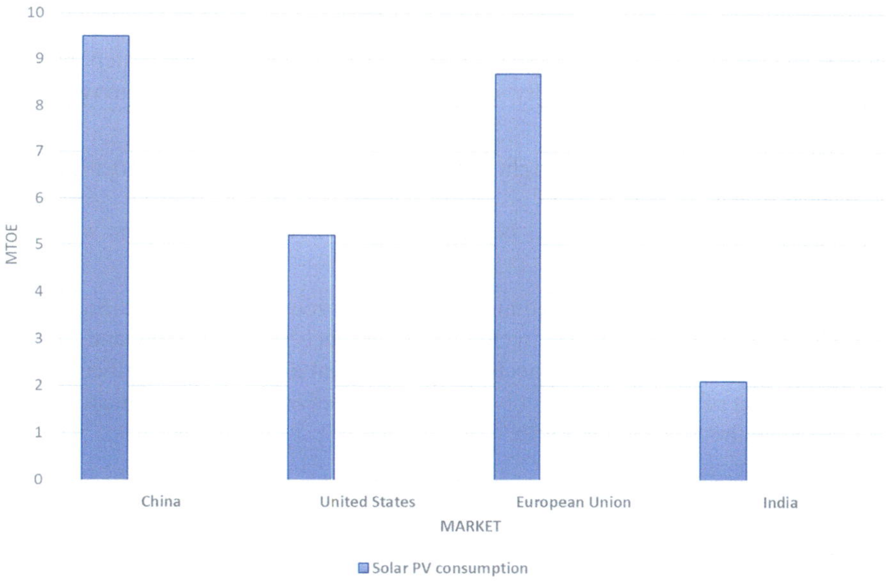

Fig. 1.3 Solar PV consumption in major markets in 2017

produces about 63% of the world's solar PV units. This chapter will introduce the fundamental concepts of solar energy, solar PV, and solar thermal collectors. Moreover, various research works will be introduced to form the literature and reviewed to provide a general understanding of R&D scene in this field.

1.2 Solar Energy

Solar energy is a term that refers to the combined radiant light and heat from the Sun which is harnessed using different energy conversion and transfer technologies such as the photovoltaic cell and solar thermal collector, etc. The main advantage of this technology is that its source, the Sun, is endless and cannot deplete. Moreover, the Sun shines everywhere which means that solar energy technologies are a viable option for everyone. The interesting element these technologies introduce is their relatively lightweight and viability for residential installation. Simply put it, not everyone can install or invest in hydro power, nor use a geothermal system in their backyard and certainly not tidal power. However, Installation of a wind turbine is possible, but size and weight limitations may affect the consumer's installation capacity along with the drawback of it not being cost-effective in low wind velocity locations. Solar energy is a suitable energy source for residential users because it can be installed on rooftops, moved around, and even set up for different configurations. Although the main disadvantage is that these technologies do not function at night due to absence of solar irradiance [7]. It is important to note as well that solar energy may vary in terms of its intensity from one place to another, which will be discussed in this chapter. Generally speaking, solar energy is divided into two processes which are:

1. The conversion of its visible light into electricity, which is referred to as photoelectric effect.
2. The storage and transfer of its heat component for heating purposes, which is referred to as solar thermal.

These two types are introduced and examined thoroughly throughout Sects. 1.3 and 1.4. Moreover, solar energy technologies produce less emissions and, in the case of photovoltaic, are free of mechanical movement and noise. To further understand the effect of solar radiation, it is important to understand the dynamic relationship between the Sun and the Earth.

1.2.1 Sun-Earth Relationship

The diameter of the Sun is around 1.391016 million km which makes it around 109 times the size of the Earth. Earth is the third planet in the solar system, which is its order from the Sun. The Earth rotates around the Sun every 365.256 days. Table 1.1 shows a comparison between various characteristics of the Sun and Earth to properly present them in simple terms [8–14].

The Sun emits solar radiation, which is the combination of its visible and near-visible (ultraviolet and near-infrared) radiations. The regions of the solar radiation exist throughout different wavelength ranges. These ranges are part of the broad band range of 0.2 to 4.0 microns. Figure 1.4 shows the regions of solar radiation.

Table 1.1 Comparison between the Sun and planet Earth [8–14]

Element of comparison	Sun	Earth
Classification	Star	Planet
Age	4.603 billion years	4.543 billion years
Formed of (mostly)	Hydrogen and helium	Iron, oxygen, silicon, magnesium, sulfur, nickel, calcium, and aluminum
Shape	Spherical	Oblate spheroidal
Surface area (Km^2)	6.09×10^{12}	510,072,000
Radius (km)	695,946	6371
Distance to the other (km)	149.6 million	149.6 million
Surface temperature (K)	5778	288
Layers	6	4

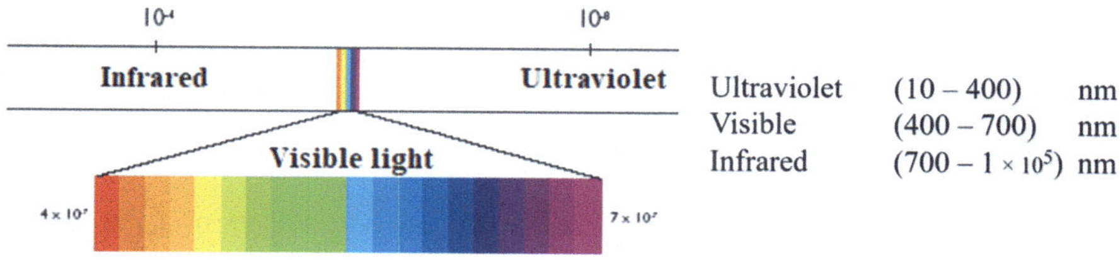

Fig. 1.4 Solar radiation regions

The solar radiation can be classified into two types, depending on its location with respect to the Earth. If the incident solar radiation is outside the earth's atmosphere, then it is referred to as extraterrestrial radiation. Terrestrial radiation, on the other hand, is the solar radiation absorbed and then emitted by the Earth.

Extraterrestrial

This radiation is expressed in terms of power per unit area (w/m^2) on a plane normal to the Sun. The amount of extraterrestrial radiation is dependent on the Earth-Sun distance which varies throughout the year, as Earth orbits around the Sun. Figure 1.5 shows Earth-Sun positioning in space. Although on average extraterrestrial irradiance is around 1367 W/m^2 which is also known the "the solar constant." This value varies by ±3% due to varying distance between the Earth and the Sun.

The value of extraterrestrial radiation can be calculated using Eq. (1.1) which accounts for the variation in the value [15].

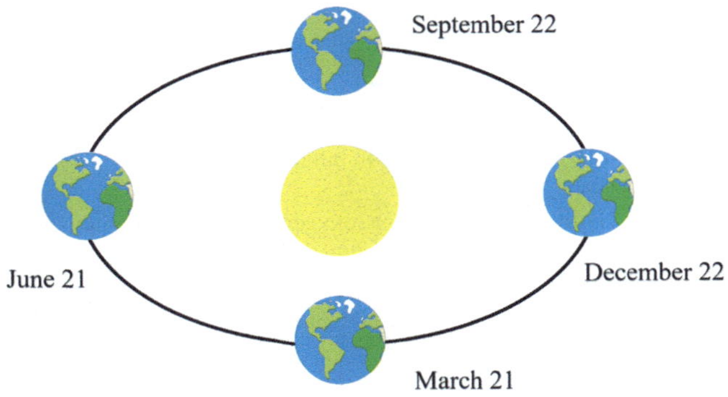

Fig. 1.5 Earth orbits around the Sun

$$E_a = E_{sc} \times \left(\frac{R_{av}}{R}\right)^2 \tag{1.1}$$

where E_a and E_{sc} represent the extraterrestrial radiation and solar constant, respectively. While, R_{av} and R represent the mean and actual Sun-Earth distances, respectively [15].

$$\left(\frac{R_{av}}{R}\right) = 1.00011 + 0.034221\cos(b) + 0.00128\sin(b) + 0.000719\cos(2b) + 0.000077\sin(2b) \tag{1.2}$$

where $b = 2\pi \frac{DOY}{365} radians$ (DOY is an integer value representing the day of the year).

The uncertainty of predicting solar radiation incident on a plane surface or solar collector is mostly a product of the Earth's atmosphere (which includes moisture, particulates, aerosols, clouds, etc.).

The nature of the solar irradiance within the Earth's atmosphere can vary depending on the obstacles and mediums the radiation may go through. Figure 1.6 (a) and (b) illustrates and explains the different types of solar irradiance.

Figure 1.6 explains extraterrestrial radiation, total solar radiation, beam radiation, and diffuse radiation. The extraterrestrial radiation is the solar radiation outside the Earth's atmosphere. The radiation that enters the atmosphere goes through different cases. Some are absorbed by clouds, while remaining is reflected by them, referred to as reflected sky. Some are absorbed by the atmosphere, while the rest is reflected into outer space. Huge portion of solar radiation is absorbed by the Earth's surface, while some of it is reflected into the absorber (known as reflected radiation) or back into outer space. Some are scattered, by the atmosphere, by the time they reach the collector. Those are known as diffuse radiation. Finally, the absorbed solar radiation which has not been scattered by the atmosphere is known as direct or beam

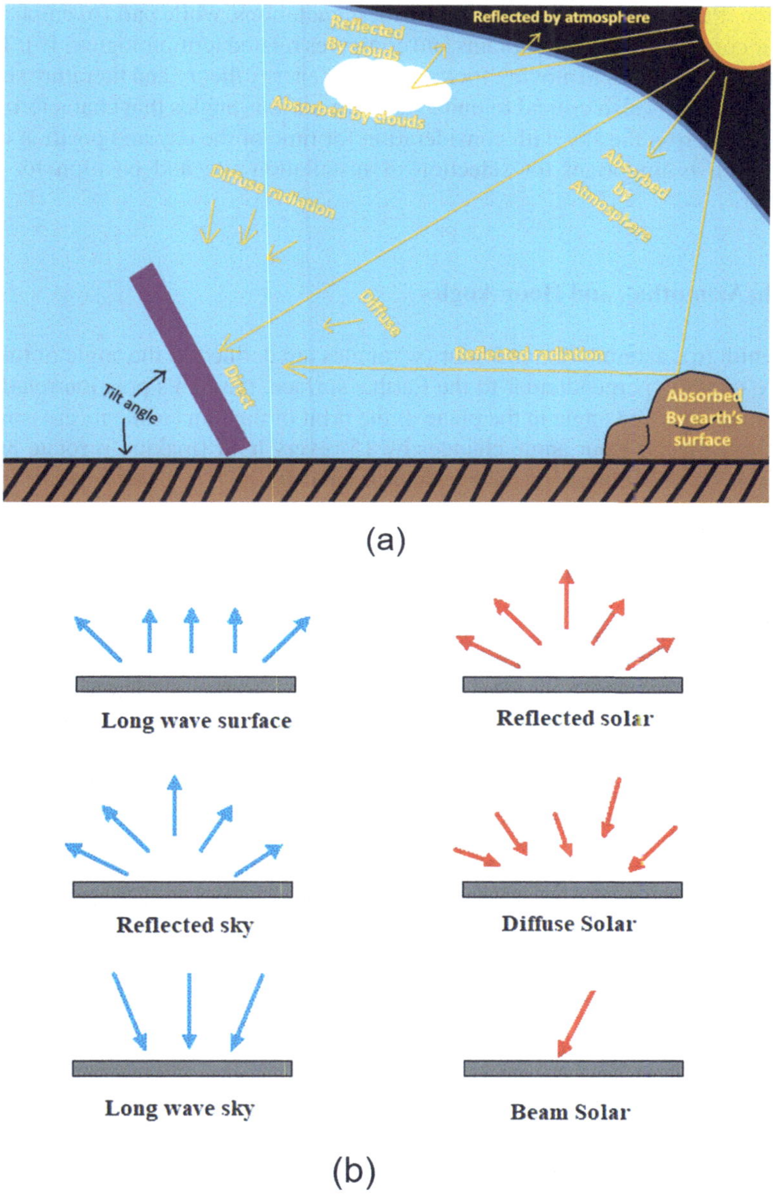

Fig. 1.6 Types of solar radiation (**a**) process (**b**) concept

radiation. Part (a) shows the trajectory of these radiations, while part (b) emphasizes on their concept. Table 1.2. explains and compares related terminologies [16]. These definitions are critical to moving forward in solar energy theory and literature survey.

Moreover, it is also critical to understand the various angles that characterize the Sun's path across the sky with consideration for time of the day and position of the Sun, which is important for selection of installation site and position for solar collector.

Zenith, Azimuthal, and Hour Angles

The zenith (q), azimuth (a), and hour (w) angles are defined as the angle of the Sun relative to a line perpendicular to the Earth's surface, the Sun's position relative to north-south axis and angle in the plane of the orbit of the Sun (while its crossing the sky), respectively. Hour angle changes by 15° every hour (making it rotate 360° a day). The value of hour angle during solar noon is 0°.

Solar and Local Standard Time

Description of the Sun position in local standard time is important. However, one must differentiate between solar time and local standard time. The former is the time for the entire time zone, while the latter relates position of the Sun with respect to observer (which is dependent on the longitude in which solar time is calculated).

To calculate the true solar time, the local time must be added to the offset time (offset time is taken in minutes). Equation 1.3 shows the time offset equation [17].

$$T_{\text{offset}} = \text{EOT} - 4L + 60T_{\text{Zone}} \tag{1.3}$$

where EOT and L are the equation of time and longitude, respectively. T_{zone} on the other hand is the number of hours of the local time zone.

Table 1.2 Terminology [16]

Term	Definition	Symbol	Unit
Solar irradiance	The rate at which radiant energy is incident on a surface per unit area of surface	G	w/m^2
Solar irradiation	The incident energy per unit area on a surface, found by integration of irradiance over specified time (hour/day)	H, I	J/m^2
Insolation	Solar energy irradiation (H, over day; I, over an hour)	H, I	J/m^2
Radiosity	The rate at which radiant energy leaves a surface per unit area by combined emission, reflection, and transmission	J_e / M_e	W/m^2
Emissive power	The rate at which radiant energy leaves a surface per unit area by emission only		W/m^2

Equation of Time

The equation of time (EOT) expresses the time difference between solar time and local standard time which is a product of the change in the Sun's position. EOT is expressed in Eq. 1.4 [17].

$$\begin{aligned}\mathrm{EOT} = 2.292(0.0075 + 0.1868\cos(\beta) - 3.2077\sin(\beta) \\ -1.4615\cos(2\beta) - 4.089\sin(2\beta)\end{aligned} \tag{1.4}$$

where

$$(\beta) = 2\pi(d-1)/365 \text{ in radians} \tag{1.5}$$

Sunrise and Sunset Times

When the Sun is at the horizon, it can either be a phenomenon of the sunrise or the sunset. Either way the cosine of the zenith angle is zero.

1.2.2 Measurement of Solar Irradiance

Measurement of solar irradiance is critical for sizing solar energy technologies such as photovoltaics and solar thermal collectors. The decision-making process for installation of an absorber should be strongly based on the solar profile of the installation site, or nearby location. The instruments used to measure the solar irradiance can be classified into two types, the pyrheliometer and the pyranometer. Figure 1.7 shows the types of solar radiation data produced from the two instruments of measurement.

1. Pyrheliometer is an instrument using a collimated detector for measuring solar radiation from the Sun and from a small portion of the sky around the Sun (i.e., beam radiation) at normal incidence. Commonly, the device is connected to a Sun tracker to be able to measure the direct solar radiation.
2. Pyranometer is an instrument that measures the broadband solar irradiance on a planer surface. The pyranometer can measure the diffuse radiation if it is shaded from the beam radiation by a shade ring or disc.

After collection of solar irradiance data, the researcher can use them in several forms and for a variety of purposes. Beam and diffuse solar radiation on a horizontal surface and by hours are the most detailed information available. Fig. 1.7 shows that the integration of irradiance over time is used to calculate irradiation, and the irradiation over time is used to calculate average irradiation. The pyranometer is

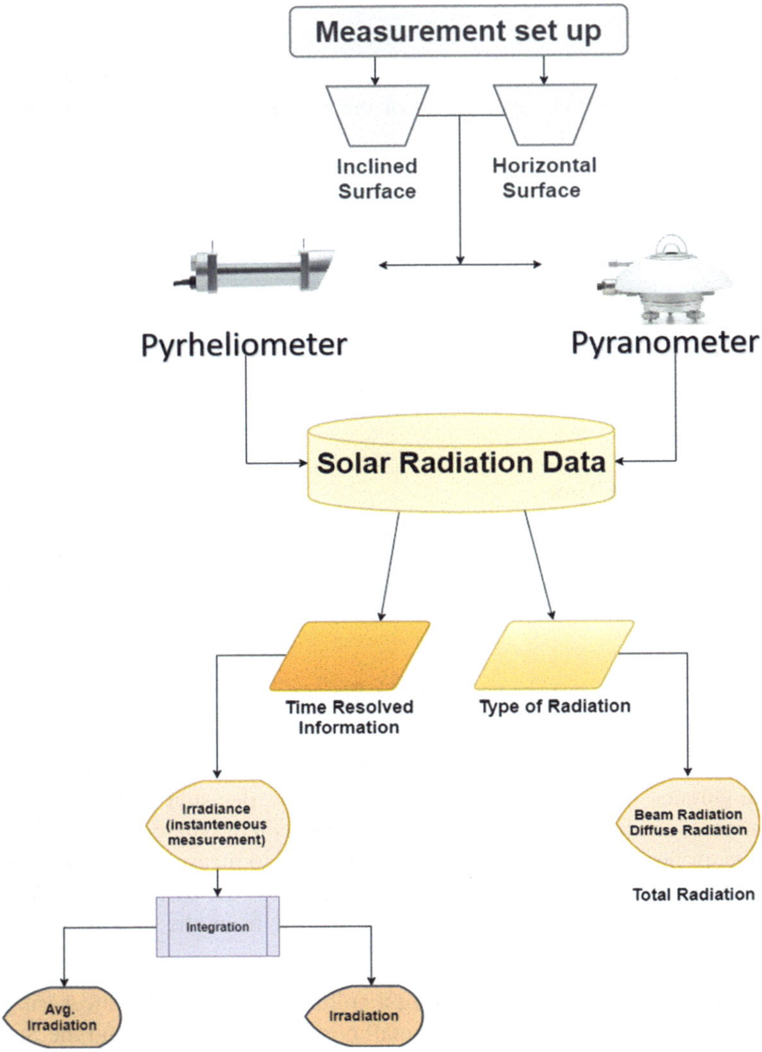

Fig. 1.7 Difference between measuring equipment

composed of a detector which is covered with a glass dome (usually an inner and an outer dome), sun shield, adjustable feet, and housing. Pyrheliometer looks more like a thin-tall cylinder. It is composed of housing, detector, smart interface, drying cartridge, aperture rings, quarts window, and a rain shield. Usually these two instruments are installed within the same experimental setup. Both instruments are supplied with connectors which can be plugged into a data acquisition system (DAQ) for automatic data collection and storage. It is noteworthy to mention that the position and tilt of these instruments are very crucial to capturing accurate readings. If these systems are used for experiments involving solar technologies, it is best to place them very close to the technology being investigated. Knowing accurately the amount of solar irradiance striking the collector and the output of the collector helps in finding the correlation and understanding the behavior of the technology. This can be useful for different types of simulations of solar processes. It is common to record daily data and estimate hourly radiation from it. Figure 1.8 (a) and (b) shows the global horizontal irradiation (GHI) and direct normal irradiation (DNI), respectively [18].

Various observations and conclusions can be made from the information provided in Fig. 1.8. These observations are:

- The Middle-East, Africa, Australia, and parts of South America have higher daily and yearly sum of global horizontal irradiation (GHI).
- The south of Africa, Australia, and parts of North America have higher daily and yearly sum of direct normal irradiation (DNI).
- The solar resource maps for GHI and DNI allow for examination of solar potential everywhere in the globe.

1.2.3 Solar Energy Industry

Global solar market increased around 55% from 2004 to 2005 to reach a whopping US$ 11.8 billion. By 2006, the solar installation was around 2.7 GW. Solar photovoltaic installation has grown exponentially throughout the period between 1992 and 2017 [19]. The support for investment in solar by some governments with programs like feed-in tariffs (FIT) was pivotal to this growth. This highlights the importance of energy policy and research in such field. Also, the declining prices of solar as a consequence of experience curve effects such as the improvement in the technology (itself) and economies of scale. This highlights the importance of research and development in the field. The cumulative photovoltaic capacity reached 402.5 GW by the end of 2017, and this means it can supply around 2.1% of global electricity demand [20], while cumulative solar thermal installations reach 472 GWth. However, solar thermal market continues to slow down for fourth year in a row. Development of the technology's efficiency coupled with economies of scale leads to flourishment of the industry. Figure 1.9 shows development of photovoltaic efficiency over time.

As Fig. 1.9 illustrates, the efficiency of the photovoltaic technology has increased immensely since its first conception. However, the solar cell is still under 50% efficiency, and commercially the highest achieved efficiencies do not exceed 23%. Many reasons led to the development of the solar PV industry, and some are summarized in Fig. 1.10. The same is made for solar thermal in Fig. 1.11.

In-depth study of photovoltaic and solar thermal systems is provided in the next two sections, 1.3 and 1.4, of this chapter. The theory will go through the concept and entire system of photovoltaics and solar thermal collectors. However, literature will only focus on research in PV system sizing and optimum configuration. As for solar thermal, the focus is mainly on flat-plate solar collectors.

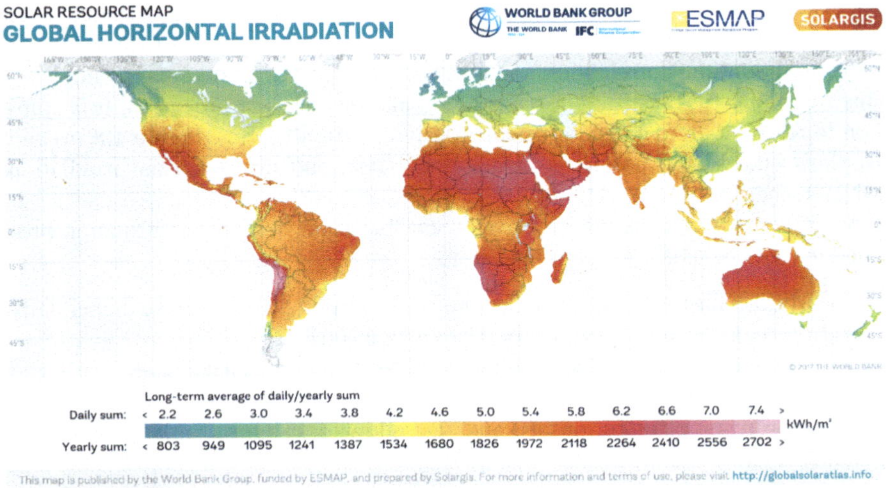

(a)

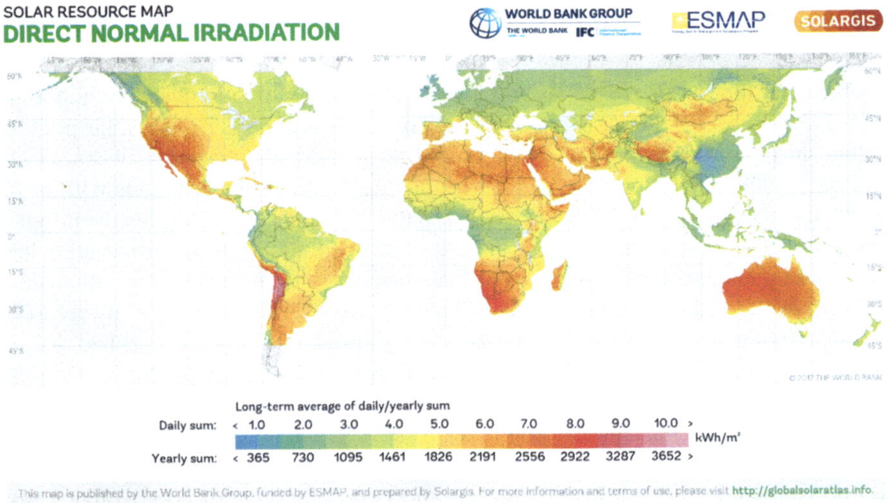

(b)

Fig. 1.8 (**a**) GHI, (**b**) DNI across the map

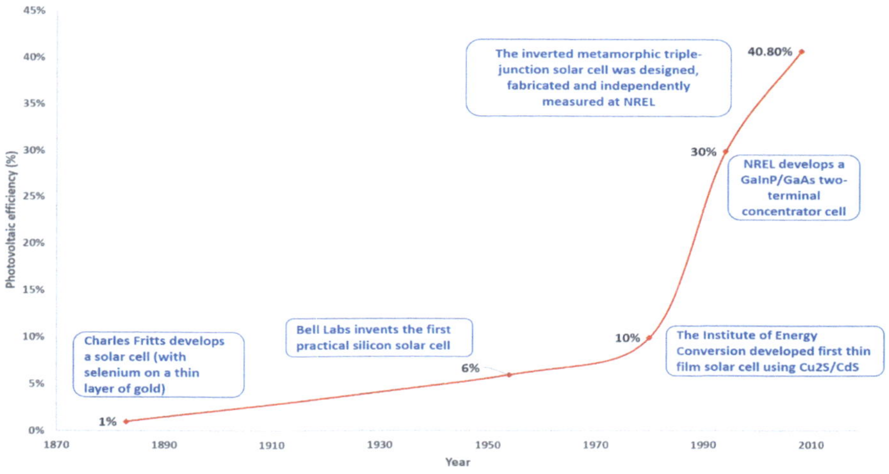

Fig. 1.9 Chronology of photovoltaic efficiency development

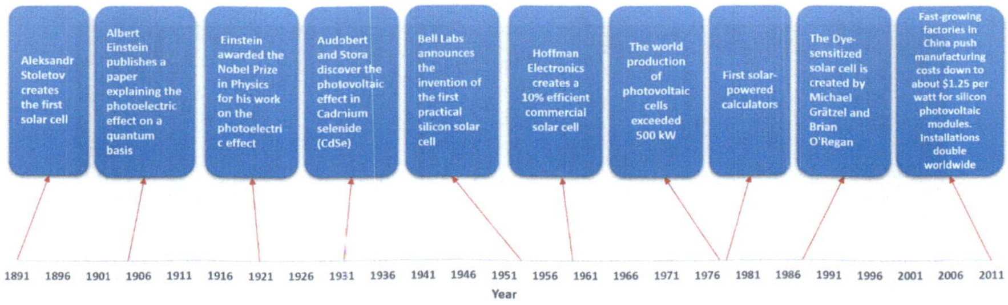

Fig. 1.10 Solar photovoltaic pivotal moments

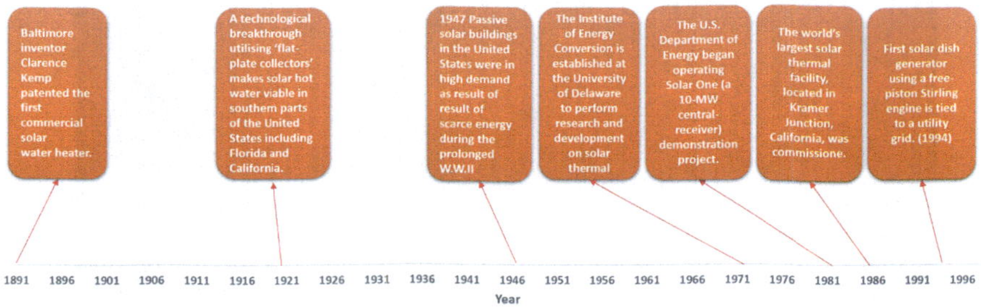

Fig. 1.11 Solar thermal pivotal moments

1.3 Photovoltaic Systems

The solar cell or photovoltaic (PV) is a semiconductor device that converts sunlight into electricity. This cell is made of wafers of highly purified silicon that is doped with impurities which leads to creating electron/hole abundance in the lattice structure. Figure 1.12 shows a solar cell. The silicon semiconductor is considered a p-n junction. It has a P-type layer and an N-type layer. In order to understand how the solar cell works, it is necessary to explain the P-type and N-type.

1.3.1 P-Type

Silicon semiconductors atom have four (valence) electrons in its outermost orbit (2, 8, 4). These electrons are shared, allowing it to form a covalent bond with four other silicon atoms. It can be doped with boron (B) which has three valence electrons (outer most). Hence an electron can be taken by the boron, leaving a hole, in its

Fig. 1.12 Silicon solar cell

place, in the valence band of silicon. This will allow the electrons in the valence band to move freely.

The electrons will try filling the holes, leaving their place, or leaving holes in their place. This will appear as if the hole is moving in the opposite direction, as positive charge across the semiconductor. Figure 1.13 shows P-doping with boron.

Because hole movement is a shortage of electrons, the doped crystal becomes a positive pole. Hence, it is considered P-type semiconductor. As for the dopant, boron, it will include an electron which makes it negatively charged. Hence, it is referred to as an acceptor.

1.3.2 N-type

Silicon can also be doped with phosphorous (P) impurities which has five outer orbit electrons. The fifth electron is free to move and is the charge carrier. The electron will leave the phosphorous impurity and is now added to the silicon atom. Hence, phosphorous (P) is considered a donor. This electron requires less energy than the other electrons to move from valence band to conduction band. Figure 1.14. shows N-doping with phosphorous (P). Silicon atoms are now N-type semiconductors as the majority carrier is negatively charged electrons, while phosphorous becomes positively charged.

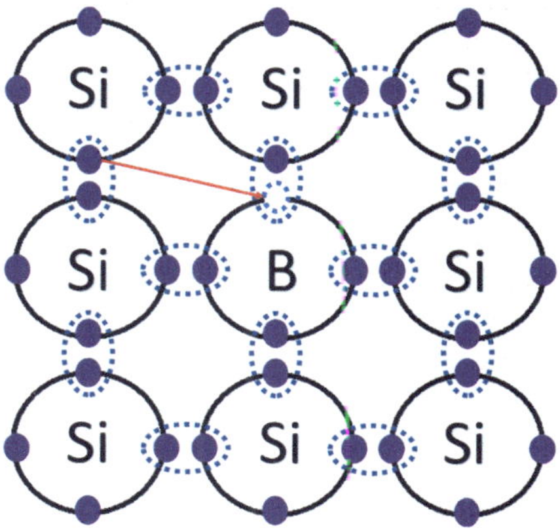

Fig. 1.13 P-doping of Si-semiconductor with B-boron

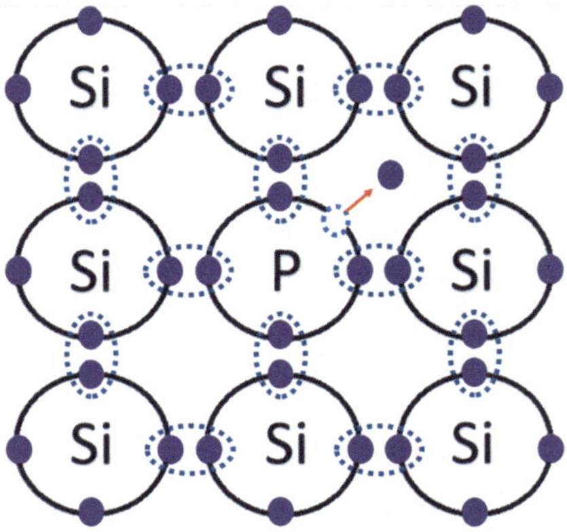

Fig. 1.14 N-doping of Si-semiconductor with phosphorous (P)

1.3.3 p-n Junction

A p-n junction represents an electrical component known as diode which has a function known as forward biased condition which allows an electric current to flow in one direction and a reverse bias condition which blocks the current in opposite direction.

Both P-type and N-type are electrically neutral. However, when the two layers are added together diffusion occurs. The process of diffusion occurs when electrons move from the N-type into the P-type, leaving holes behind, while the opposite happens to the P-type. This creates the p-n junction. Diffusion will continue to occur till electrons have large enough electrical charge to repel more carriers from crossing the junction. This process is illustrated in Fig. 1.15.

Once the state of equilibrium is achieved, a potential barrier at the junctions will be produced. Potential barrier will oppose the flow of both electrons and holes from crossing the junction. Hence, donor will repel holes, and acceptor will repel electrons. So free charges cannot rest in a position and are depleted. This is referred to as the depletion layer. When photons from sunlight are absorbed by silicon cell, electrons are free to move across the depletion layer. The electrons will move from the P-type into the N-type but will not return. Given that now the N-type has an excess of electrons, it is the negative side, while P-type is the positive. If electrical conductors are attached to the positive and negative sides, forming an electrical circuit, the electrons can be captured in the form of an electric current. The current produced is direct current (DC) [21].

1.3.4 PV Cell Components

The front of PV is comprised of metallic strips which collects electrons. They are the positive connection. The back of PV is made of aluminum/molybdenum metal, and it is the negative connection. Components of PV cells are shown in Fig. 1.16 [22].

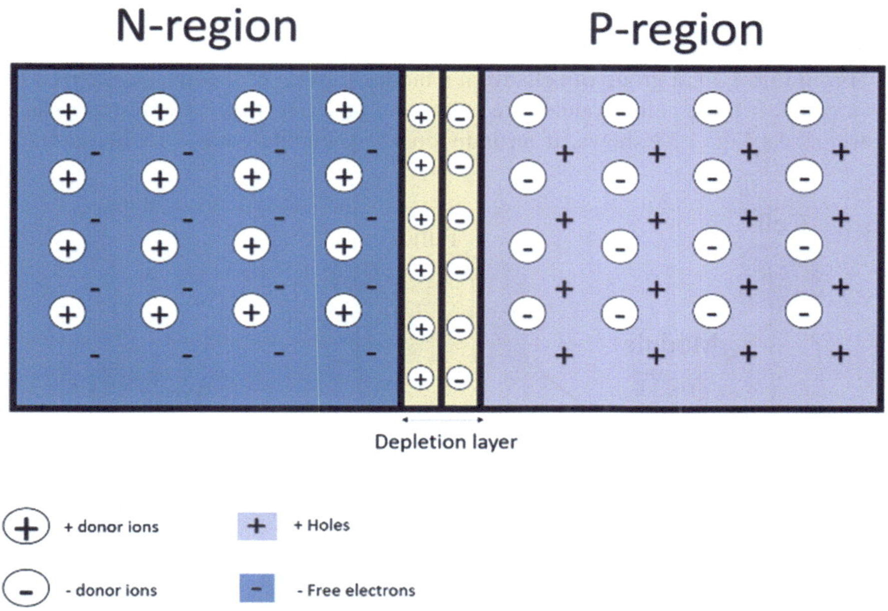

Fig. 1.15 p-n junction

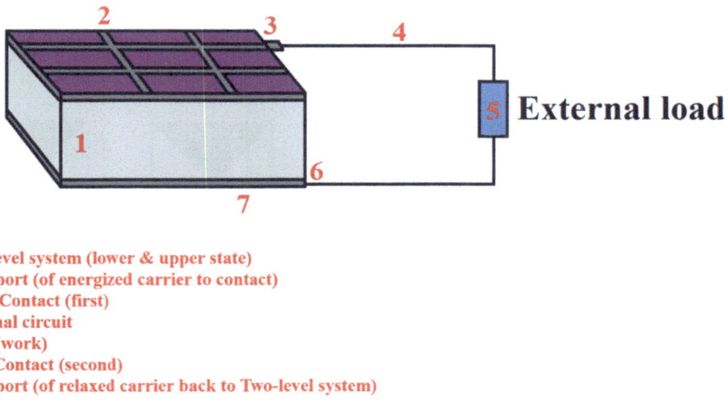

Fig. 1.16 Component of the PV cell

It is noteworthy to mention that there are two types of contacts, which are contacts for excited carriers and contacts for relaxed carriers. PV cells connected, in series, and framed are referred to as "PV module." A number of these modules can be wired in series, parallel, or combination of both to produce a "PV array." The wiring is done in order to either increase the voltage or current delivered to the load/battery. The manner of which PV modules are connected can be referred to as configuration. Figure 1.17 illustrates the concept of cell, module, panel, and array, while Fig. 1.18 illustrates the connections of PV modules.

From Fig. 1.17, a group of cells from a module, group of modules is referred to as a panel. If these cells/modules are connected in series, they are referred to as a "string." As Fig. 1.18 shows, if an individual solar module has 16 volts and 2.5

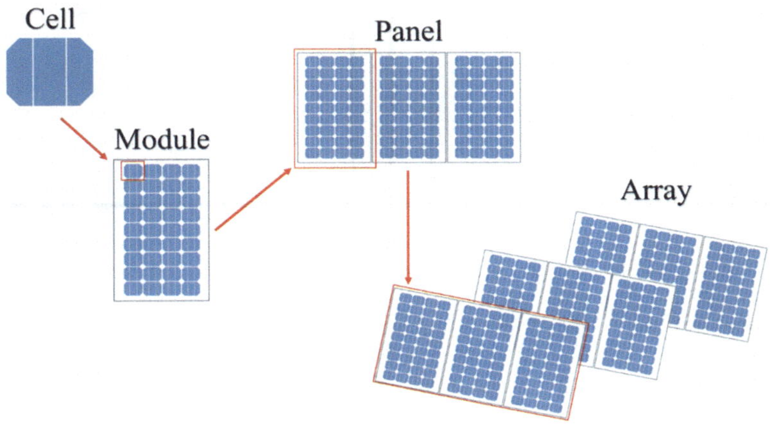

Fig. 1.17 Cell, module, panel, and array

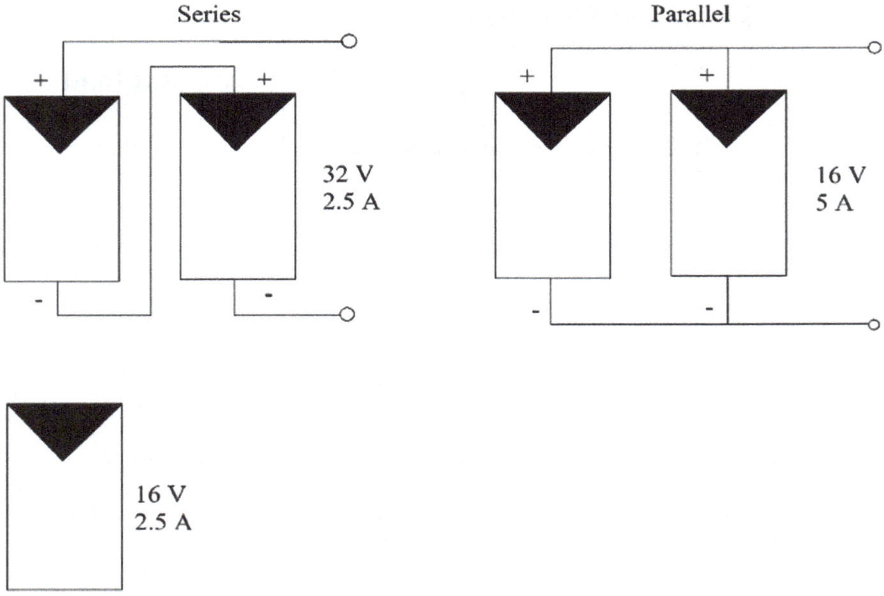

Fig. 1.18 Connection of the PV modules

Amps, connecting the solar modules in series leads to increase in its voltage (as the current flows in one direction). Hence, if two solar modules are connected in series, the current remains the same (e.g., 2.5 A), while the voltage is raised by the addition of the two (e.g., 32 V). The opposite happens for parallel connection of solar modules. Two modules connected in parallel will have the same voltage across (e.g., 16 V); however the current will be the addition of the two currents (e.g., 5 A). This is because the current crosses more than one path. So, in conclusion, if the modules are connected in series, the voltages will be added. If the modules are connected in parallel, then the currents are added. Connection procedure of the PV modules can be further illustrated in Fig. 1.19.

1.3.5 PV Equivalent Electrical Circuit

The solar cell can be represented with an equivalent electric, which is shown in Fig. 1.20. The circuit is composed of a current source, diode, parallel resistance, series resistance, and the electric load. The parallel resistance is referred to as R_{Shunt}, while the series is R_{Series}. The ideal circuit is represented by the current source and the diode. However, introducing both R_{series} and R_{shunt} is more accurate for the practical PV device [23].

The voltage across the circuit is shown in Fig. 1.20. This voltage is dependent on other factors such as solar irradiance and cell temperature. The maximum voltage of the solar cell could only be reached at zero current, and it is referred to as "open-

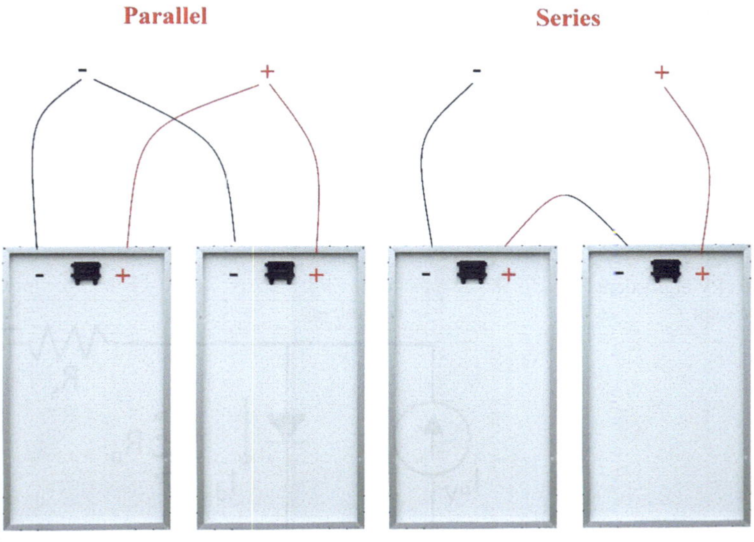

Fig. 1.19 PV module connections

circuit voltage" or V_{OC}. This voltage corresponds to the amount of forward bias on the solar cell. Equation (1.6) shows V_{OC} which can only be made by setting the current zero [24].

$$V_{OC} = \frac{nkT}{q} \ln\left(\frac{I_L}{I_o} + 1\right) \tag{1.6}$$

where I_L and I_O are the light-generated current and dark saturation current, respectively. K is the Boltzmann constant, and T is the absolute temperature, while q and n are the absolute value of electric charge and the ideality factor, respectively.

If the voltage across the solar cell amounts to zero (e.g., solar cell is short-circuited), then the current is at its highest. This current is known as the "short-circuit current" or I_{SC}. This current occurs due to generation and collection of light-generated carriers. The light-generated current and short-circuit current are identical in ideal solar cell, with moderate resistive loss. For an ideal circuit, the current is provided in Eq. 1.7 [24].

$$I = I_o\left[\exp\left(\frac{qV}{nkT}\right) - 1\right] - I_L \tag{1.7}$$

where I is the net current flowing through the diode. Equation 1.8 shows the short-circuit current.

$$I_{sc} = I(V = 0) = I_L \tag{1.8}$$

The short-circuit current is a reverse bias current because electrons flow toward the cathode, and the holes flow to the anode. When light is absorbed by the cell, a photocurrent that corresponds to the reverse current will be produced. Once forward bias is applied, this will generate a current in the opposite direction with respect to the light-generated current (I_L) and will compensate it.

The power of the solar cell is calculated in Eq. 1.9 [24].

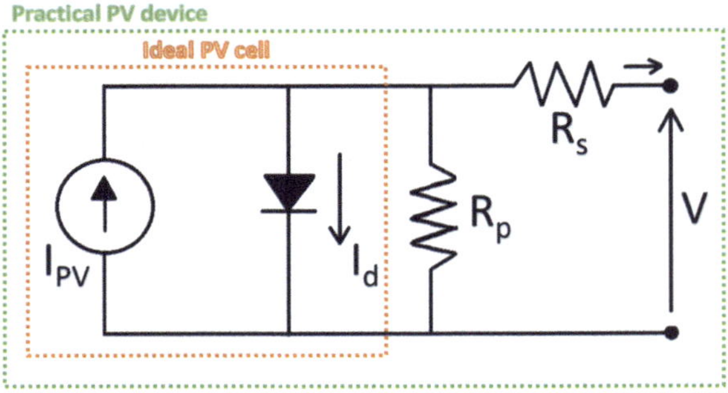

Fig. 1.20 PV equivalent electric circuit

$$P = I_{mp} \times V_{mp} \tag{1.9}$$

The unit of the electric power is watts (W_p), and it is the product of voltage times ampere. To have better understanding of the short-circuit current, open-circuit voltage, and power, Fig. 1.21 shows the IV and power curves.

The following points are:

1. The highest voltage produced by the solar cell is referred to as maximum power voltage (V_{mp}), while the highest voltage possible is referred to as open-circuit voltage (V_{OC}).
2. The highest current produced by the solar cell is referred to as maximum power current (Imp), while the highest current possible is referred to as short-circuit current (Isc).
3. Given that current remains constant during the progression of voltage, until a certain point then it starts to drop, the output power from solar cell is significantly affected by its voltage. Reduction in voltage leads to drop in power output.
4. The point at the I-V curve with the highest power, as a product of voltage and current, is referred to as the maximum power point (MPP). Hence, it provides the highest power output (P_{max}). The current and voltage at that point are (I_{mp}) and (V_{mp}), respectively.

Given that the power is zero at the open-circuit voltage and short-circuit current, fill factor (FF) is used to find the maximum power produced. It can be defined as the ratio of maximum power of PV to the product of its V_{OC} and I_{SC}. Figure 1.22 shows the fill factor. It is graphically defined as the "squareness" of solar cell, as in largest rectangular to fit in I-V curve. Equations 1.10 and 1.11 describe the fill factor and maximum power, respectively [24].

Fig. 1.21 I-V and power curve

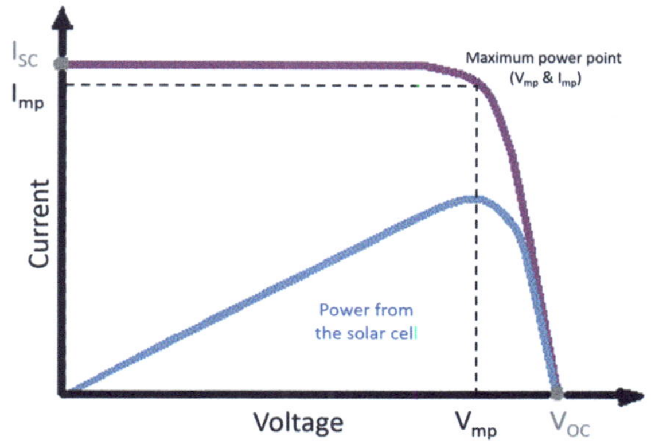

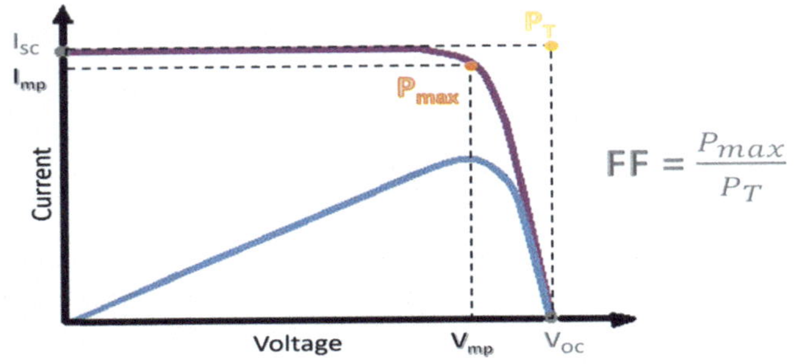

Fig. 1.22 Fill factor of the solar cell

$$FF = \frac{I_{mp} V_{mp}}{I_{sc} V_{oc}} \tag{1.10}$$

$$P_{max} = V_{OC} I_{SC} FF \tag{1.11}$$

Finally, the efficiency (η), or power conversion efficiency (PCE), of the solar cell device which is the same as any other device (output divided by input) is illustrated in Eq. 1.12. The input to the solar cell is the incident light power at AM1.5 solar spectrum [24].

$$\eta = \frac{V_{OC} \times I_{SC} \times FF}{P_{inc}} \tag{1.12}$$

The efficiency of the solar cell is a major area of research. It is necessary to develop and raise the efficiency, under laboratory conditions, and maintain it under practical or outdoor conditions. This book presents the photovoltaic thermal (PV/T) technology which aims to maintain this efficiency by cooling the photovoltaic module. Photovoltaic systems can be classified based on their configuration into two categories: stand-alone PV systems and grid-connected PV systems.

1.3.6 Stand-Alone PV Systems

Stand-alone PV configuration is when the PV system is not connected to the grid. This configuration is suitable for rural and isolated areas where electricity is needed but national utility grid is nonexistent. It may be costly to build underground cables or overhead lines for hundreds of kilometers for low demand. A better solution is to be autonomous by installing a stand-alone PV system. These systems are commonly composed of photovoltaic panels (or arrays) connected to charge controllers which are connected to battery banks and the load. If load is AC, then an inverter is added to the system. Figure 1.23 shows a typical stand-alone PV system [25].

The PV module is the power source of this system. It draws its power from sunlight. Hence, it can only provide power during daytime. This is why battery banks are usually included in such systems. The battery banks stores produced power during the day, so that it can supply the load/demand at night. The charge controller is a device that is responsible for voltage regulation and system safety. If the battery is fully charged, the device can isolate it from further charging (separation from solar cell). If the battery is fully consumed, or almost empty, then it will isolate it from further discharging (separation from the load). The system can power both AC and DC loads if it has an inverter. The inverter can convert direct current (DC) waveform into alternating current (AC) waveform. Most residential appliances are AC loads (e.g., television, air conditioner, fridge, hair dryer, computer, fans, and lights). DC loads can also include portable lights, small fans, and remote controls.

1.3.7 Grid-Connected PV System

If the PV system is connected to the utility grid, it is referred to as "grid-connected" or "grid-tie." These systems are commonly composed of solar panels (or array), switch board, smart inverter, electric meter, and the grid [26]. Figure 1.24 shows the components of the grid-connected PV system. Grid-Connected PV's can either be used to fully supply its power to the grid or supply the local load demand and give excess power to the grid. Hence, there are four scenarios for grid-connected PV systems:

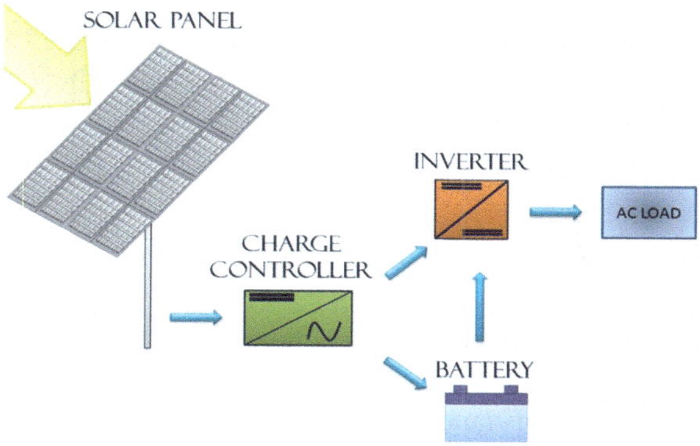

Fig. 1.23 Typical stand-alone PV system

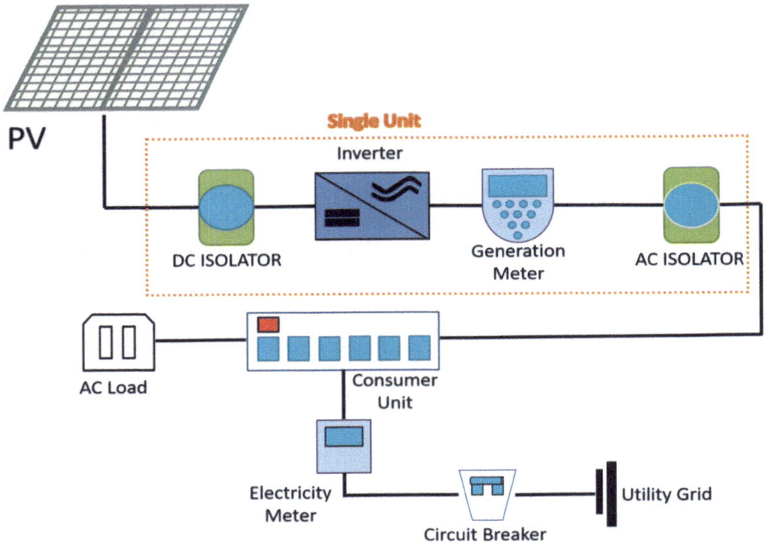

Fig. 1.24 Grid-connected PV system

1. Solar PV power equal load requirements: The PV fully meets the demand of local load with no excess power. Hence, the power of the load is supplied by solar PV.
2. Solar PV power is greater than load requirements: PV fully meets the demand of local load with excess power being supplied to the grid. Hence, the power of solar PV is supplying the load and exporting to the grid.
3. Solar PV power is lesser than load requirements: The PV cannot meet the demand of local load; the grid supplies the excess load demands. Hence, the power of PV and grid is joined to supply the load.
4. Solar PV is zero. This is during nighttime where PV cannot produce power. The utility grid fully powers the load demands which are known as import from the grid.

1.3.8 PV System Literature Review

The research in PV systems covers a broad range of topics including optimum tilt angle, sizing techniques, cost-effectiveness, effect of environmental conditions, and applications of stand-alone PV systems. This section showcases recent and historic works in PV system research [27–52].

Optimum Tilt Angle and MPPT Tracking

The position of the photovoltaic module is crucial for absorbing maximum incident solar irradiance; Fig. 1.6 (a) shows the tilt angle of the PV module. Hence, researchers investigate the optimum tilt angle and employ tracking systems for different locations for better PV productivity.

Hussein et al. [27] provided an optimization method to achieve the optimum tilt angle for PV systems in Sohar, Oman. The method was numerical and developed in MATLAB environment accounting for hourly meteorological data and load profile (demand). The flowchart and initial assumption are both illustrated in the article. The findings show that it is appropriate for adjustment of PV array twice a year for Sohar Zone. The tilt of the array should be around 49 degrees between 21st September and 21st March and be horizontal between 21st March and 21st September. These two adjustments lead to gain of collected energy by 24.6%.

Belhachat and Larbes [28] presented a review that discusses advancements of global maximum power point tracking GMPPT for photovoltaic systems which are affected by partial shading conditions (PSC). The aim of the review is to offer users the appropriate choice suitable to their system. The MPPT approaches are classified by the author to optimization, mathematical model-based, hybrid, and other methods of GMPPT. The effect of partial shading condition is initially presented, providing the problem of which GMPPT can solve. The optimization-based methods are PSO, LPSO, ABC, ACO, SA, BA, FFA, FWA, etc.; based algorithms are discussed and compared in the study. Moreover, the hybrid MPPT approaches are ANFIS-PI, INC-FFA, SA-PSO, Jaya-DE, P&ONN, etc. The mathematical models are Modified Beta Algorithm, modified Fibonacci search algorithm, modified extremum seeking control, etc. Other methods include transient evolution of series capacitor, curve fitting, and active bypassing of the shaded cells. The gathering of information and literature on this topic is remarkable, and comparative studies are important to review advances in this field.

Kacira et al. [29] used a mathematical model for estimating total solar irradiance on the tilted surface of PV and determination of optimum tilt angle for Sanliurfa, Turkey, by searching tilt angles that corresponds to maximum total irradiance for studied period. Furthermore, the study looks into the effect of two-axis solar tracking on energy gain of installed PV relative to a fixed PV. The system was set up on roof of a research facility located in Sanliurfa. The panel is composed of two single-crystalline PV modules (120 Wp rated each). The first was mounted at a fixed tilt angle, while the second was coupled with a two-axis solar tracking. The authors found that optimum tilt angle varies much throughout the year, where it reaches a 13° minimum and a 61° maximum during June and December respectively. The results obtained show a gain in daily average solar irradiation and generated power of around 29.3% and 34.6%, respectively, for the two-axis solar tracking as appose to its fixed (14° tilted) counterpart.

Sizing Techniques

Different methods are investigated throughout the literature to size the different components of the PV system and find optimum PV configuration. The investigations covered the use of intuitive, numerical, and experimental approaches.

Mellit [30] conducted a literature review on the role of AI techniques in sizing of PV systems of different configurations such as stand-alone, grid-connected, and hybrid PV-Wind. The role of AI encompasses a broad range of functions such as modeling, optimization, identification, prediction, forecasting, and control. The review explains different branches of AI such as fuzzy logic, artificial neural networks (ANN), and genetic algorithm. Conventional methods of PV system sizing are also presented from different works in the literature for the different configurations targeted by the study. The authors claim that the main issue associated with conventional methods is that it is highly dependent on metrological data, and so it is hard to use conventional sizing for remote area where these data does not exist, or for long-term operation. The solution to this is the implementation of AI techniques. Authors claim that AI methods are based on conventional hybrid technique for PV system sizing.

Bert Herteleer et al. [31] developed an intuitive method via software tool – spreadsheet for PV system sizing for office applications. The aim of the spreadsheet is to help non-expert users size the PV system. The model is described through a series of equations aimed to sizing the storage (batteries), inverter, and PV panels. Both on- and off-grid PV systems are discussed in the study. The authors performed an analysis of predicted annual irradiation on a tilted, oriented plane. The authors performed a comparison between the spreadsheet and HOMER software for the stand-alone system sizing of two locations in Africa. Irradiation predictions varied between the spreadsheet and HOMER by −0.5% and + 8%. The tool allows the users to achieve rapid system sizing and cost estimation. Factors calculated include payback period, net present value, return on investment, performance ratio, etc. Although the tool is designed for Africa, the authors claim it can be used in "any country within the "sun belt" of the world."

Zanesco et al. [32] presented an experimental evaluation of an analytical method for optimum sizing of photovoltaic (PV) system. It takes into consideration the size of the system (PV and batteries) to fit the load demand and behavior of the solar radiation with respect to loss of load probability (LLP). The analytical calculations lead to conceptualization of the system. The experimental aspect uses a stand-alone PV system consisting of two 50 Watts PV's, two lead-acid batteries, timer system, and a charge controller, while the load is made up of four 40 watts fluorescent lamps. The study concludes that the system successfully supplies the load demand. The highest daily energy readings (Wh/day) were in January and December with around 275 Wh/day. In June the energy of the system was around 207 Wh/day.

Bruno Burger and Ricardo Ruther [33] proposed a different inverter sizing strategy for PV plants by investigating different solar radiation data recording times. Authors compared instant measurements of 10 seconds (high resolution) to average hourly irradiation values and found sizable differences in optimal invertor sizing.

The authors used solar irradiation data from two different cities located in Germany and Brazil. The study shows, how for annual inverter efficiency, higher averages (hourly) leads to higher calculated losses which is attributed to inverter under sizing, as claimed by the authors. The issue with average hourly measurement is that it filters peak values which might lead to inaccurate performance estimation. The authors broke down the solar energy distribution over the two cities by range of solar irradiance (amount of w/m^2), percentage of daytime, and total energy for data sampling rate of 10 seconds, 1 minute, and 1 hour. Calculation of inverter yearly efficiency and power losses is necessary to draw on the differences between the sampling rates. This shows how optimum ratio between nominal power of solar panel and inverter is necessary for correct design of PV plants. The study concludes that effect of PV tilt and orientation on the PV-Inverter power ratio varies according to time resolution of solar radiation data. In real scenarios, the tilt and azimuth are more sensitive, and it is difficult to account for this sensitivity with hourly averages.

Cost-Effectiveness

The cost-effectiveness of the PV panel is very important for the technology to compete with rival energy source devices like the diesel generator or simply the utility grid. Different parameters must be provided such as life cycle cost, cost of energy, and payback period. These parameters are explained in Chap. 4.

Muyiwa S. Adaramola [34] conducted an experimental and theoretical study on a 2.07 kW grid-connected PV system installed on a rooftop in Norway. The theoretical aspect is of the economic analysis which covers internal rate of return on investment, annual and monthly costs of energy (produced), and premiums over levelized costs of energy. The economic life of the project is 25 years. The system installed consists of nine 220–240 Wp PV modules which were tilted 37° and oriented southward, a 4.4 kW inverter which makes it oversized, because PV-Inverter ratio is 0.5. Data collection was carried out for 1 year, from 2013 to 2014, measuring different parameters (total module in-plane irradiation, array output power, and system final energy output power). The evaluation of the systems' technical performance was made by calculating indices such as final energy output, different energy yields, capacity factor, efficiency, system losses, and performance ratio.

The study displays seasonal array and system losses, which are relevant to economics of the system. After feed-in tariff, US$ 0.356/kWh, for a cost of energy of US$ 0.246/kWh, the premiums are US$ 0.110/kWh. The study concludes that system performance during spring and summer seasons is encouraging. Furthermore, FiT and initial investment financial support of up to 40% the installation costs, are attractive incentives which can lead to an increase of PV system installation in Norway.

Jeong et al. [35] assessed a rooftop PV system over a military facility. The assessment focuses on life cycle economic and environmental aspects. Three factors were taken into consideration which are installation area, slope of installed panel, and orientation of the gable roof, which amounts to 12 scenarios being established for the PV system. Moreover, three prototypes were considered, named

P1, P2, and P3 with orientation south-north, southeast-northeast, and east-west, respectively. The simulation of the work was conducted using "RETScreen" software. The work of this study is divided into three steps: selection of target military facility and PV system, calculation of PV generation using PV scenario, and, finally, the assessment takes place. The findings show that absolute investment value (NPV25) for all scenarios can improve within range of 5.58% to 15.82%. Moreover, relative investment value (SIR25) for all scenarios was positive. Finally, differences in weather conditions showed differences in PV generations, which are consistent with the literature and theory.

Makbul A.M. Ramli [36], hybrid PV-diesel system is proposed and analyzed using HOMER software for economic and environmental benefits, for Makkah, Saudi Arabia. The system employs flywheel for storage of excess PV energy. In addition, a battery type (Surrette 4 KS25P) is also proposed. The load is 32,962 MWh/d with a peak of 2213 MW. The use of flywheel is the major focus of analysis and how it impacts different costs (e.g., energy and net present costs) associated with certain configurations of the proposed system. For environmental study, the fuel consumption along with emissions of carbon is also studied and compared. The monthly load profile for Makkah is displayed. Moreover, the sizing and details of each component are illustrated as well. The study displays three configurations: (i) diesel/flywheel hybrid system, (ii) PV/diesel/flywheel hybrid system, and (iii) PV/diesel/battery/flywheel hybrid system. The lowest COE and CO2 emissions are attributed to third (iii) proposed system. Two array sizes were proposed for the third system, which are 1.1 GW and 2.2 GW. The latter showed lowest COE with around 33% of renewable penetration. Authors conclude that addition of flywheel to the hybrid system makes it "more economical."

In addition, it helps in decreasing the CO_2 emission and fuel consumption. The cost of energy for systems (i), (ii), and (iii) are 0.451, 0.370, and 0.369 US$/kWh, respectively. When the flywheels are not used, the diesel generator is favored, where the cost of energy becomes 0.463, 0.389, and 0.388 US$/kWh.

Saib et al. [37] presented an approach for optimization of a hybrid grid-connected PV and wind with energy storage with consideration for voltage fluctuations by utility grid. Minimization of the hybrid-system size is done by performing a techno-economic analysis first with consideration of the benefit-cost. The system is composed of PV system with converter, wind turbine with inverter, battery bank with converter, and inverter connected to grid. The batteries used are of the lithium-ion type and are intended to manage electricity sell to grid and reduction of voltage fluctuations. FC-VACPSO method is used for optimization. The authors performed a comparative study with numerical simulation between standard PSO and PSO-based methods for optimum sizing for technical and cost purposes. The simulation extended to sensitivity analysis for variable battery types and costs. Hence, two cases were compared: constant-state and variable-state battery state-of-charge (SoC). The benefit cost for different methods in MUS$ is 51.3, 52, 53, and 55 for the PSO, VCPSO, VACPSO, and FC-VACPSO methods, respectively. While for case 1, the benefit cost in MUS$ is 77, 27, 62.8, and 26.6 for Li-ion (1500∗5), Ni-Cd (1200∗5), VRB (600∗5), and lead acid (300∗5), respectively. As for the second case, the cost benefit in MUS$ for the aforementioned battery types is around 79, 26.4, 63.6, and 52, respectively. The study concludes that improved FC-VACPSO method gives rise to a better convergence.

Rajput et al. [38] presents a life cycle assessment of a 3.2 kW cadmium telluride (CdTe) PV system in an outdoor experimental setting, to view performance and life cycle for India's climate. The system is composed of 40 PV modules containing 4 parallel connections of 10 PV panels in series, an aluminum structure, weather station, inverter with built-in MPPT, and a data logger. The collected data are analyzed under same environment conditions of solar irradiation, ambient temperature, etc., with energy metrics, life cycle assessment, carbon credit earnings, and unit cost of electricity. Life cycle assessment equations and concept are illustrated throughout the article. Moreover, a description of the embodied energy of different PV components is provided. The results obtained show an energy payback time (EPBT), energy production factor (EPF), life cycle conversion efficiency (LCCE), and unit cost of electricity, for a 5% interest rate and 30-year life space, of 3.6 years, 0.27, 0.0018, and 9.85 INR/kWh, respectively.

Effect of Environmental Condition

Environmental conditions such as ambient temperature, solar irradiance, humidity, and dust are all influencers on PV systems. Conducting research on PV systems for a particular location helps in understanding the viability of these systems in that location, and it also prepares researchers to find solutions or alternative renewable energies systems in case it is problematic to implement those PV systems. Understanding the behavior of PV systems under specific conditions allows us to impose their advantages and strive to eliminate the disadvantages or at the very least, explain them.

Fesharaki et al. [39] discussed and simulated the relationship between solar radiation temperature and PV conversion efficiency under cloudy climate. Although simulation software or specifics of the condition are not illustrated in the study, the associated equations and efficiency variations with different temperatures across time are presented. For current values, the authors assume the equation:

$$I(t) = 1000 + 500\sin(t/5) \tag{1.13}$$

Efficiency was found to drop with increase in PV module temperature. Moreover, the frequent variation of Sun irradiance leads to efficiency variation. The findings are consistent with the literature. Jiang et al. [40] presented a numerical simulation of PV cell/module system and converter power stage for I-V and P-V characteristics curves of PV module under uniform shading conditions (USC) and PSC. The simulation is hybrid in nature, as in MATLAB/Simulink and Pspice were used. The produced simulation model is intended to investigate different parameters that influence PV modules. It is also suitable to simulate PV array, several PV panels connected in series/parallel whether they are homogeneous or heterogeneous. To simulate the PSC conditions, the authors used a Pspice model, adjusting values of current sources allowed for simulating PSC. The authors described the difference between such model and use of either MATLAB-based code or P-Spice; MATLAB is inconvenient dealing with electronic power stage elements, and Pspice is less accurate given it cannot account for surrounding conditions information. Hence, the cell temperature and insolation effects on photocurrent of the solar cell are accounted for using a mathematical model in MATLAB/Simulink. Cadence® software SLPS was used to combine the Pspice model and the Simulink functions. For I-V and P-V curves, the cell was subjected to insolation of 400 W/m^2 to 1000 W/m^2 with increments of 200 W/m^2. Under PSC, the P-V curves exhibit multiple peaks. When one cell is shaded, the output is higher than if two cells were shaded. For the case of connected PV to a power converter, the voltage and current of PV are determined by load voltage and duty cycle of power converter.

Jiang et al. [41] conducted an experimental investigation of effect of type and material of the solar cell itself on efficiency degradation caused by dust accumulation. The authors distinguish this work by claiming it considers the influence of PV properties, while most research focuses on dust properties alone. The experiments were made indoors, using a solar simulator and a test chamber that contains fan, pyranometer, and thermocouple. A particle counter and a dust generator were employed, as well. The PV panel types were mono- and polycrystalline and amorphous silicon. The polycrystalline was covered with epoxy, while other modules covered with glass. The fine test dust was employed in the experiments as well. The study results show significant impact of dust on PV output power. As dust deposition density increases from 0 to 22 g m^{-2}, it caused a degradation of efficiency from 0 to 26%. Authors claim that the relationship between dust deposition density and efficiency reduction is linear. However, not much difference can be attributed to the type of the cell. Among the three modules, polycrystalline (PV) module covered with epoxy experiences faster degradation rates than modules with glass surface, for the same dust concentration.

Mekhilef et al. [42] reviewed simultaneously the effect of dust accumulation, humidity level, and air velocity. All of which are studied separately, then their relationship was clarified. The findings of the review are in threefold: one, dust deposition on PV panel surface leads to drop in its efficiency; two, humidity leads to decrease of PV efficiency; and finally, increased air velocity leads to reduction of panel temperature and hence an increase in efficiency. Higher wind can be both beneficial and harmful to the performance of PV, where higher wind velocity leads to decreased relative humidity around solar panel, while it causes for dust in the environment to spread over PV surface and hence more shading which leads to reduced efficiency. As for relative humidity, the study classifies effects into two types: effect of water vapor particles on solar irradiance and effect of humidity ingression on PV cell enclosure. The authors recommend not to study these factors separately but to consider them all at once.

Chandra et al. [43] offered an experimental investigation into the effect of wind on systems installed in hot and dry climate zones. The authors justify their study with the premise that PV performance is site dependent, as it affected by environment conditions such as irradiance, temperature, and wind.

Two polycrystalline type PV modules of same specifications were used for the experiments for selected months of the year (March, May, September, and December) of different seasons. The setup is done indoors, with artificial solar irradiation and artificial wind, using halogen lights and cooling fans, respectively. One module was cooled with the fan, while the other was not. For the cooled PV module performance, ratio is improved by 7.14% with net energy gain of 7.69% in considered time. For temperature and solar irradiation, a temperature sensor and solar power meter were used, respectively. The energy for both modules was calculated theoretically, and comparisons were made with experimental measurements. The study shows correlation coefficient and root mean square percentage deviation between measurements and calculations. The study concludes that it is important to account for wind, especially in windy locations, before sizing of PV system. The authors recommend considering average wind speeds at any installation location, prior to estimation of PV performance and system size to reduce cost and payback of system. The study findings show two major points, firstly the importance of wind to minimizing efficiency losses due to high temperatures and, secondly, the utility of artificial cooling mechanisms.

Dubey et al. [44] conducted a review centered around the temperature-dependent nature of PV cells and how it affects its efficiency which impacts PV production worldwide. The paper offers different equations and correlations for cell temperature that involve environmental variables and numerical parameters that are either material or system dependent. The study discusses the electric performance and installation of one-sun commercial silicon cells. The authors observe the correlation between latitude and performance ratio (PR), where PR drops with latitude as a result of high temperature. In addition, how altitudes are correlated to high PR, due to low temperatures in those regions. The study offers different PV efficiency correlation equations as function of temperature. Furthermore, PV potential is discussed with respect to annual total irradiation and PV energy generation by crystalline silicon.

Hybrid PV Systems

Hybrid PV designs are those implementing more than one type of energy sources or devices coupled with the PV device. Famous examples of hybrid PV systems include the hybrid wind turbine PV system and hybrid PV-Genset system. Both are introduced in this section.

This configuration can be more cost-efficient than separate PV system, in some countries or locations. Hence, it is worthwhile to investigate those types of systems.

Shiroudi et al. [45] used HOMER software to perform system optimization for a hybrid PV-Wind system with battery storage in Taleghan, Iran. The data collection includes the main inputs to HOMER system design which are solar irradiance data, wind speed, and the electrical load demand. The study presents the techno-economic specifications of the system components. The simulation was set for a project of 20 years lifespan and daily load of 5.5 kWh. The optimum system configuration consists of a PV, two wind turbines, inverter, and eight 12 V batteries with ratings of 0.8 kW, 0.4 kW each, 2.5 kW, and 200 Ah, respectively. The initial capital, net present, and energy costs are 22,998 US$, 24,623 US$, and 1.655 US$/kWh, respectively.

Baghaee et al. [46] proposed a novel multi-objective optimization algorithm and implemented on a hybrid Wind-PV microgrid system coupled with a hydrogen storage with a lifetime of 20 years. The simulation of 1 year with 1 hour time step was conducted. Outage probability of three main system components (PV, wind turbine generator, and converter) was taken into consideration. The algorithm minimizes the annualized cost of the system, loss of load expected, and loss of energy expected. Hence, the aim is to conduct a reliability–/cost-based optimal design. Variables of optimization are the number of PV arrays (and angle), wind turbine generators, hydrogen tanks, electrolyzers, fuel cells, and DC/AC converters. Moreover, study presents an approximate method for reliability evaluation of hybrid system, in order to reduce computation time. MATLAB software was used to develop the software. The effectiveness of the algorithm in optimal sizing over conventional single-objective methods is claimed by the authors. The study concludes that system cost is directly dependent on the reliability of its components.

Prévost [47] presented in his thesis a case study of optimal sizing of a hybrid PV-biomass power plant, focusing on gasification, for a village in Indonesia. Load estimation technique is used, and 13.3 MWh/day is assumed as the daily load. Different scenarios are considered for the optimal design, and hence numerical simulation using AKIMOS software was performed. The scenario in which gasification runs at half load during day, while batteries (charged by PV during the day) supply during the night is the best one. Hence, optimum system is composed of 1600 kWpeak PV, 450 kW gasification, and 1275 kWh storage capacity. The levelized cost of energy is quite low and suitable with around US$ 141/MWh. The system is compared to a hybrid PV-Genset and proved to have lowered levelized cost of energy.

PV Applications

There are many applications for PV systems in industrial, commercial, and residential settings. To utilize a PV system better for those applications, it is necessary to investigate the behavior and optimum configuration of the system. Among those applications are PV water pumping systems (PVWPS), solar powered electric vehicle, hybrid cars, etc. As mentioned earlier, PV-powered calculators are famous example of PV applications.

Bright et al. [48] presented a methodology for a PV system that supports a satellite-derived PV power nowcasting. The paper investigates the upscaling-only PV fleet methodologies and a conventional satellite and compares them to hybrid cases which are newly developed. The four cases are satellite-only approach, upscaling-only approach, correction approach, and hybrid approach. When using single reference PV for estimation of different targets in Canberra, Australia, the fourth case improves the performance of PV power estimates by 26%, 25%, 14%, and 14% for MBE, rMBE, RMSE, and rRMSE, respectively. The relative improvements are similar when using 30 reference PV systems, which are around 72.22%, 73.04%, 40.96%, and 41.29%, respectively. The exploration of these different approaches in real-time estimates of power production for different target systems is the main contribution of this study.

Alwaeli et al. [49] presented a techno-environmental evaluation of a PV water pumping system (PVWPS) using HOMER and REPS.OM software. The study is made for a PVWPS in Sohar, Oman, and presents a diesel generator system for comparison purposes. The article displays the sizing equations and system components. The techno-economic criteria for the designed PVWPS are illustrated as well. The pump load demand is around 2.2 kWh/day. The system is composed of 0.9 kW PV array (using six 150 W PV modules), 450 W – water pump, 1 kW inverter, and 200 Ah battery storage. The competing diesel system is simply composed of the diesel generator. The cost of energy of the system is around US$ 0.309/kWh, while diesel generator had cost of energy of US$ 0.79/kWh. For the diesel generator, carbon dioxide emissions were around 924 kg/year, while for PV system the CO_2 emissions are zero.

COLAK et al. [50] designed a model of an electric vehicle charging station which uses a PV system as its power source. The model forecasts the systems total power output for different conditions in Ankara, Turkey. The methodology of the paper starts with determining PV cell parameters then forming PV array (of cells). The model uses actual irradiation and temperature data from measurements. The system is made up of 310 solar modules with 30 V_{mpp} each, a buck converter, an energy storage system battery of 120 V and 1000 Ah, back converters, and PI controllers. Authors encourage using worst case for solar irradiation and temperature to model the output of PV. The study considers different battery types according to efficiency, energy, and power density. The article displays the PV model design schematic, software/program flowchart, and IV, PV and PI curves for PV system.

Tiwari et al. [51] conducted a performance investigation of a PVWPS with helical rotor type pump under outdoors condition. Analysis of solar radiation and total head effect on water output is presented and coupled with optimization of PV array configuration. The pumping heads proposed are 4-bar, 6-bar, 8-bar, and 10-bar. The system was composed of PV array (1.6 kW) with two-axial manual tracking. The system was installed with an artificial well. A submersible pump with helical rotor was implemented. A pressure control valve was used to regulate the required head. For the experiments, the measurements included the instantaneous flow rate, electric current and voltage, and the solar radiation intensity. The highest values of solar irradiance in the installation location are during the month of May, while the lowest are in August. This gives a range of 4–7 kWh/m^2/day. Different parameters related to pump performance were investigated to analyze the data. For data analysis, estimation of electrical energy required to run the pump, generated hydraulic energy, received incident energy, and pump and system efficiency of different heads using optimum PV configuration was done. The authors conclude that best efficiency is achieved when total head is 10-bar with a helical rotor submersible pump. The study recommends sizing PV systems in Nagpur, India, for an irradiance range of 400–800 W/m^2.

Biyik et al. [52] conducted a comprehensive review on building integrated PV (BIPV) systems which discusses various aspects of the research conducted in the field. The study covers BIPV and BIPV/T along with applications, experimental and numerical studies, cell design studies, and grid-integration studies and policies. The study emphasizes on the improvements in system efficiency through ventilation which is a consequence of reduced panel temperature and advanced thin film technologies which are suitable for BIPV. The authors note that BIPV holds a small share in PV market due to common misconception regarding system costs. Among strong solutions for BIPV systems is the multifunctional PV façade which exhibits active and passive benefits for buildings. The authors observed that more work is being conducted to study energetic aspects as appose to exergetic studies which are very low in numbers. Another noteworthy observation is that the two most commonly used software in simulation of BIPV are TRNSYS and EnergyPlus software.

Summary of PV Literature

A detailed summary of the literature is provided in Table 1.3, illustrating year of publication, study area, research area and approach, and some findings.

1.4 Solar Thermal Systems

Active solar heating systems utilize solar energy by absorbing sunlight and transferring the heat into a fluid medium such as water or air which is transferred directly into interior space or stored for later use in a proper storage unit. The device that capitalizes on solar energy to collect heat is commonly referred to as solar thermal collector. Although it commonly refers to solar water heaters (SWH), it can also mean solar parabolic troughs, solar towers, and solar air heaters. Figure 1.25a, b, and c shows the solar water heater, solar air heater, and solar parabolic troughs, respectively.

The main difference between solar water and solar air heating is the type of fluid which causes massive difference in the output and process. Cold air enters from the building, through pipes, into the collector to be heated then injected back into the building. Although solar air heaters pose an interesting solution and they do not freeze in extremely cold climates, unlike water heaters, it is still behind in terms of popularity. This is mainly because more success is attributed to solar water heating which is considered most cost-effective. Modern systems use nanofluids and other highly conductive material to increase the systems thermal efficiency. Both technologies are generally used for low temperature ranges, around 100 °C. On the other hand, solar parabolic troughs are straight in one dimension and curved, as a parabola, in the other two, lined with a polished metal mirror. The mirror parallel to its plane focuses the incident sunline into a focal line which forms the absorber tube. The efficiency of this type of collector may reach as high as 73%, according

Table 1.3 Summary of research conducted in the literature

Author(s)	Year	Location	Research approach	Research area	Findings
Kacira et al.	2014	Turkey	Mathematical and experimental	Optimum tilt angle and MPPT tracking	Minimum tilt of 15° in June and maximum of 61° in December. Tracking led to power generation increase of 34.6%
Belhachat and Larbes	2018	Algeria	Review	MPPT tracking	Classified models in literature to optimization, mathematical model-based, hybrid, and other methods of GMPPT
Hussein et al.	2013	Sohar, Oman	Numerical, using MATLAB	Optimum tilt angle	Two yearly adjustments in tilt: 49° during 21st September–21st March and horizontal during 21st March–21st September. This led to gain of energy by 24.6%
Bert Herteleer	2012	Africa	Intuitive method, using Excel sheet	PV system sizing of stand-alone PV and grid-connected	Irradiation predictions varied between the spreadsheet and HOMER by −0.5% and + 8%. Tool developed can be used for Africa and other places within the Sunbelt
Mellit	2007	Turkey	Review	AI sizing techniques for PV performance prediction	Presented different branches of AI and their utility for PV performance prediction, especially for remote areas
Zanesco et al.	2004	Porto Alegre, RS, Brazil	Experimental and analytical method	Optimum sizing of a stand-alone PV system	Highest daily energy produced is around 275 Wh/day in January and December System consist of two 50 W PV's, two lead acid batteries, timer system, and a charge controller

(continued)

Table 1.3 (continued)

Author(s)	Year	Location	Research approach	Research area	Findings
Bruno Burger and Ricardo Ruther	2006	Germany and Brazil	Experimental and theoretical	Optimal sizing of grid-connected PV (C-silicon) configuration	More accuracy in solar irradiation data recordings leads to better inverter sizing and less power losses for PV plants
Muyiwa S. Adaramola	2015	Norway	Experimental and theoretical	Cost analysis of grid-connected	The cost of energy of the system is around $0.246/kWh after FiT of $0.356/kWh
Jeong et al.	2015	Korea	Numerical, using RETScreen	PV application in military purposes. Life cycle and environmental assessment	Improvement in absolute investment value (NPV25) for all scenarios ranging between 5.58% and 15.82%. Positive relative investment values
Makbul A.M. Ramli	2015	Saudi Arabia	Numerical simulation, using HOMER	Hybrid PV systems, economic and environmental	PV/diesel/battery/flywheel hybrid system had lowest CoE and CO_2 emissions among proposed configurations
Saib et al.	2018	Algeria	Simulation, using MATLAB	Hybrid PV systems, grid-connected PV techno-economic	The study concludes that improved FC-VACPSO method gives rise to a better convergence
Rajput et al.	2018	India	Mathematical, intuitive, and experimental	Life cycle assessment of grid-connected (CdTe) PV system	The system is found feasible in India's climate with EPBT, EPF, LCCE, and CoE of 3.6 years, 0.27, 0.0018, and 9.85 INR/kWh, respectively
Fesharaki et al.	2011	Iran	Numerical simulation	Effect of temperature and solar irradiation	Efficiency drop correspond to increase of PV module temperature. Variation of efficiency also due to variation of solar irradiance

(continued)

Table 1.3 (continued)

Author(s)	Year	Location	Research approach	Research area	Findings
Jiang et al.	2011	USA	Numerical simulation, using (Pspice, Simulink and MATLAB)	Effect of solar irradiance and shading on PV cell/module	Under PSC, the P-V curves exhibits multiple peaks. Increasing number of shaded cells was found to reduce the output of the system
Jiang et al.	2011	China	Experimental	Dust effect on PV module	As dust deposition density increases from 0 to 22 g, m-2 caused a degradation of efficiency from 0 to 26%
Mekhilef et al.	2012	Malaysia	Review	Effect of dust, humidity, and air velocity on PV systems	Each factor is investigated separately, then lumped together. The authors recommend studying these factors all at once
Chandra et al.	2018	India	Experimental and theoretical study	Wind and temperature on a stand-alone polycrystalline silicon PV	The cooled PV module performance ratio is improved by 7.14% with net energy gain of 7.69% in considered time
Dubey et al.	2012	Singapore	Review	Effect of temperature on PV efficiency	PR drops with latitude as a result of high temperature.
Shiroudi et al.	2012	Taleghan, Iran	Numerical Simulation, using HOMER	Sizing and optimization of hybrid PV-Wind battery system	The initial capital, net present, and energy costs are 22,998 US$, 24,623 US$, and 1.655 US$/kWh, respectively
Baghaee et al.	2016	Iran	Numerical simulation: novel method, using MATLAB	Hybrid PV-Wind	Authors conclude that system cost is directly dependent on the reliability of its components

(continued)

Table 1.3 (continued)

Author(s)	Year	Location	Research approach	Research area	Findings
Prévost	2018	Indonesia	Numerical simulation, using Alkimos software	Hybrid PV-Biomass Power Plant	Optimum system is composed of 1600 kWp PV, 450 kW gasification, and 1275 kWh storage capacity. The levelized cost of energy is around US$ 141/MWh
Bright et al.	2018	Australia	Intuitive and mathematical	PV application – satellite-derived PV power nowcasting	The fourth case improves the performance of PV power estimates by 26%, 25%, 14%, and 14% for MBE, rMBE, RMSE, and rRMSE, respectively
Alwaeli et al.	2017	Sohar – Oman	Numerical and intuitive, using HOMER and MATLAB	PV application – PV water pumping system	The CoE of the proposed system is around US$ 0.309/ kWh, while diesel generator system has CoE of US$ 0.79/ kWh
COLAK et al.	2015	Turkey	Numerical design and experimental measurement	PV applications – solar charging station	Authors recommend using worst case for solar irradiation and temperature to model the output of PV
Tiwari et al.	2018	India	Experimental	PV applications – PV water pumping system	Best efficiency is achieved when total head is 10-bar with a helical rotor submersible pump
Biyik et al.	2017	–	Review	Techno-economic and sizing of a BIPV system	The authors observed more study aim to study energetic as appose to exergetic

to NREL [53]. Many combinations of solar thermal systems are possible, and hence it is necessary to classify this technology to be able to tackle each aspect independently. Different classifications are made depending on the basis to the classifications. Some classify thermal systems based on operation to "active vs. passive," some classify them based on concentration to "concentrating vs. non-concentrating," while others classify them based on achievable temperature [54]. Figure 1.26a, b, and c shows classification of solar thermal systems based on operation, achievable temperature, and type of concentration, respectively. The opera-

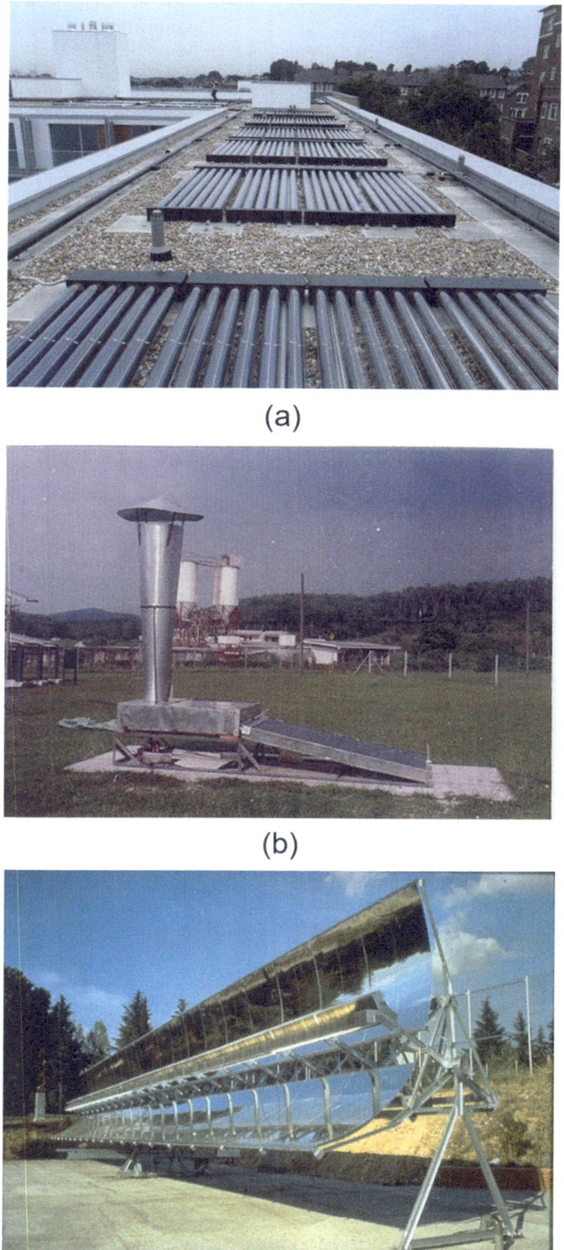

Fig. 1.25 (**a**) Solar water heater, (**b**) solar air heater, (**c**) solar parabolic troughs

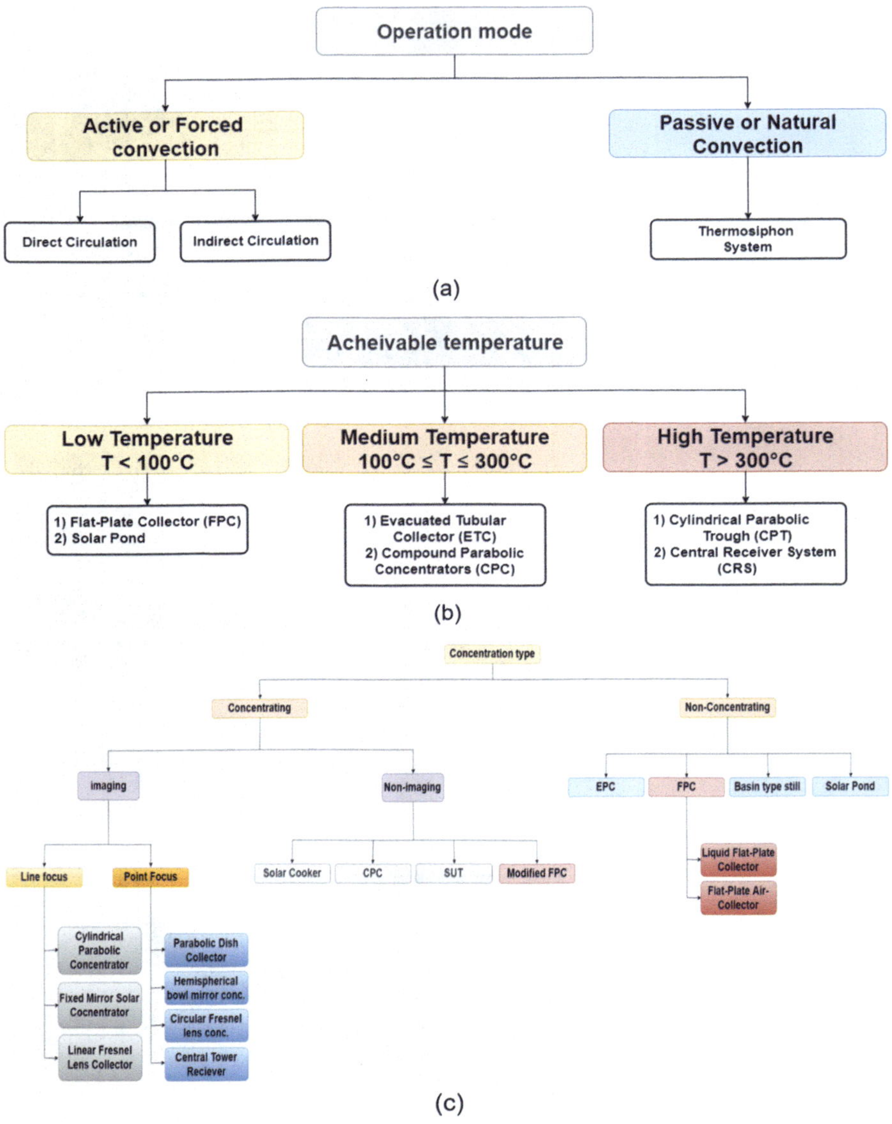

Fig. 1.26 Classification of solar thermal systems based on (**a**) operation mode, (**b**) achievable temperature, (**c**) concentration type

tion mode is subdivided into active vs. passive or forced convection vs. natural convection. In the active mode, the system utilizes pumps and/or fans and other active equipment, in order to carry out the circulation of the fluid which can either be direct or indirect. Direct systems carry the produced fluid directly into the application, where hot water is needed, while indirect systems usually employ water and glycol in a closed system which connects the collector, on the rooftop, and a heat exchanger. Indirect solar water heaters are more successful mainly due to their resistance to cold and heat losses as the solar collector will circulate the hot fluid into the storage tank in, usually, a basement or utility area. Passive systems are quite different in that it mainly utilizes gravity and change of pressure and temperature. These systems are cheaper in cost but less efficient. Some integrate such systems with collector/storage system or thermosyphon systems. These integrated collector/storage systems are commonly used to preheat the water. The tanks of thermosyphon systems must be located higher than their collectors. This is because after water is heated in the panel it will flow upward to the tank, and the cooler water returns to the collector for heating [55].

The achievable temperatures from solar thermal collectors can be divided into three categories, as illustrated in Fig. 1.26b, namely, low, medium, and high temperatures, which range from 0–100 ° C, 10–300 °C, to above 300 °C, respectively. Flat-plate collectors mostly fall in the low temperature range, and hence it is more suitable for residential applications and consumption. They also require small installation areas. On the other side of the spectrum, central receiver towers are the highest in terms of temperature production. This form of solar collectors is not optimal for residential or commercial uses, given that it required large space and high solar irradiance. According to Figure 1.26c, this type of collector is considered an imaging point-focused type, while flat-plate collectors fall under the non-concentrating collector classification, although they can be modified to the concentrating non-imaging thermal design. Another type that is competing with FPC is the evacuated tube collector (EPC). Although FPC exhibits a higher thermal efficiency, peaking at around 70–74%, EPC exhibits lower peak efficiency of 40–50%, yet it can maintain its overall efficiency curve across increase of temperature difference, explained in next section, while FPC efficiency drops below with increase of temperature difference [56].

It is noteworthy to mention that this section will focus on direct non-concentrating type flat-plate solar collectors (FPC) with low-to-medium achievable temperature. This type of collector is chosen as its generally used to form PV/T collectors (which are illustrated in Chap. 2). This section will illustrate the concept, mathematical model, and performance parameters of FPC, along with recent literature review on this technology.

1.4.1 Flat-Plate Collector (FPC)

As explained in the previous section, flat-plate collectors are non-concentrating type thermal systems with low-to-medium achievable temperature. They are mostly used for residential purposes, and their peak thermal efficiency is around 70–87% [57]. Figure 1.27 shows a schematic drawing of a typical FPC with illustration of its components.

This type of thermal collectors is composed of five main components:

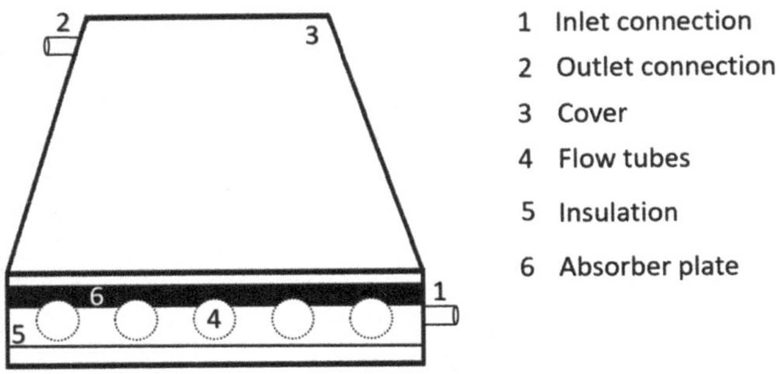

Fig. 1.27 Flat-plate collector (FPC)

1. Cover plate, which is commonly made of glass or plastic. It is used to reduce convective and radiative heat losses from the absorber and allows solar irradiance to pass through (number 3 in Fig. 1.27).
2. Absorber plate, which is made of metal such as copper, steel, or plastic with its surface covered with flat black material to increase the absorptance (number 6 in Fig. 1.27).
3. Flow passages are used to convey the working fluid which is heated through heat transfer from absorber plate to the passage. Type of working fluid affects the design, where air is most suitable below the absorber to minimize heat, while liquids are in a tube which could be attached to the, or part of the, absorber. Different designs for flow passages are made to enhance the heat transfer process (numbers 1, 4, and 2 in Fig. 1.27).
4. Insulation is placed at the back and sides of the absorber to minimize heat losses. Fiberglass material is a common type of insulation material (number 5 in Fig. 1.27).
5. Enclosures carry all collector components within and hold them together. It also provides an appropriate frame. Table 1.4 shows different types of components used for different flat-plate collectors.

1.4.2 Energy Balance of Flat-Place Collector (FPC)

To analyze the performance of FPC, it is necessary to understand its energy balance equations to be able to calculate important parameters such as thermal efficiency, heat gain, thermal losses, absorbed solar radiation, heat removal factor, etc. These factors help researchers and manufacturers optimize these collectors by targeting the elements needed to enhance (e.g., absorbed solar energy) or eliminate (e.g., thermal losses). The steady state equations for FPC are provided below [16]:

$$\text{Useful energy}\left(Q_{\mathrm{u}}\right) = \text{Absorbed solar energy} - \text{thermal losses} \quad (1.14)$$

Table 1.4 Different components of FPC

Cover plate						
Material	Crystal glass	Window glass	Acrylate, plexiglass	Polycarbonate	Polyester	Polyamide
Transmittance	0.91	0.85	0.84	0.84	0.84	0.80
Absorber plate						
Material	Copper/aluminum/steel					
Absorber color	White	Fresh snow	White enamel	Green paint	Red brick	Flat Black
Absorbance	0.07	0.13	0.35	0.5	0.55	0.98
Flow passages						
Configuration	Direct flow	Parallel flow	Serpentine flow – internal headers	Serpentine flow – external headers	Tube formed in metal sheet	Trickle type corrugated sheet
Pipe shape	Circular, semi-circular, square, rectangular, triangular, elliptical, etc.					
Insulation						
Material	Polystyrene	Fiberglass	Mineral Wool	Cellulose	Polyurethane foam	Aerogel
Advantage	Diverse insulation material	Cheap	Effective	Fire resistant, eco-friendly	Good overall insulation	Best type of insulation
Fluid						
Material	Water	Air	Ethylene glycol	Nanofluids	Pentane	Butane

From this equation, all design steps begin, to either increase absorbed solar energy or reduce thermal losses. Thermal efficiency of the collector is a very important evaluation parameter and is provided in Eq. 1.15.

$$\eta = \frac{Q_u}{I_T \cdot A_C} \tag{1.15}$$

where Q_u, A_C, and I_T represent the heat gain, collector area, and solar irradiance, respectively. The denominator represents the total input energy. The numerator of this equation represents the useful energy as output from the collector. This energy, referred to as heat gain, is illustrated bellow:

$$Q_u = A_c\left[S - U_L\left(T_{pm} - T_a\right)\right] \tag{1.16}$$

where S and U_L represent the absorbed solar radiation and heat transfer coefficient, respectively. T_{pm} and T_a represent the plate and ambient temperatures, respectively. The mean plate temperature, in Eq. 1.16, can be used to calculate the useful heat gain. The absorbed solar radiation (S) is calculated using Eq. (1.17).

$$S = (\tau.\alpha)_{av}.I_T \quad (1.17)$$

where $(\tau.\alpha)_{av}$ is the product of transmittance of the collector cover and absorptance of the plate averaged over different types of radiation, where incident solar radiation is denoted with I_T. Figure 1.28 shows the efficiency of FPC across.

The absorbed and lost energies are illustrated in Eqs. (1.18) and (1.19), respectively.

$$\text{Absorbed energy} = A_C.F_R.S \quad (1.18)$$

where F_R is the heat removal factor, which is calculated in Eq. (1.24).

$$\text{Lost energy} = A_C.F_R.U_L(T_i - T_a) \quad (1.19)$$

where U_L is the heat transfer loss coefficient, T_i and T_a are the inlet fluid temperature and ambient temperature, respectively.

Substituting the Eqs. (1.18) and (1.19) in Eq. (1.14) to get:

$$Q_u = A_C.F_R.S - A_C F_R U_L(T_i - T_a) \quad (1.20)$$

which can be simplified to

$$Q_u = A_C.F_R.[S - U_L(T_i - T_a)] \quad (1.21)$$

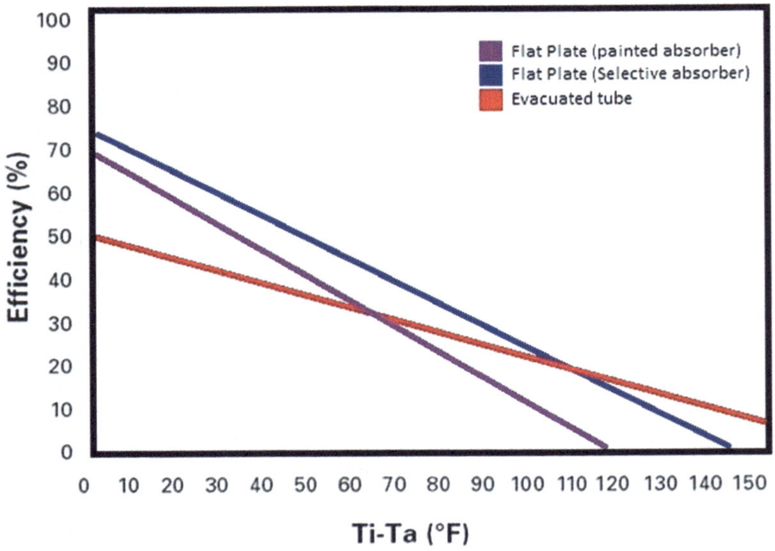

Fig. 1.28 Efficiency of solar thermal collectors

To assess the collector performance using experimental data, Eq. 1.22 is useful. This is done by running the base fluid through the collector and measuring both inlet and outlet temperatures along with mass flow rate.

$$Q_u = \dot{m}.C_P.(T_o - T_i) \tag{1.22}$$

where $\dot{m}$ and C_P are the mass flow rate and specific heat of the fluid, respectively, while T_o and T_i are the outlet and inlet temperatures of the fluid, respectively. Substituting this equation in Eq. (1.15) Results in the instantaneous efficiency of the collector, provided in Eq. 1.23.

$$\eta_i = \frac{\dot{m}.C_P.(T_o - T_i)}{A_C.\ G_T} \tag{1.23}$$

As shown in Eq. 1.16, the highest achievable useful energy gain is when the collector has the same temperature as the inlet fluid. This leads to minimization of heat losses. However, it is hard to achieve such condition. The heat removal factor (FR) is used to describe effective, actual, and useful energy gain from FPC. This factor is calculated in Eq. 1.24.

$$F_R = \frac{\dot{m}.C_P.(T_o - T_i)}{A\left[I(\alpha\tau) - U_L(T_i - T_a)\right]} \tag{1.24}$$

Heat removal factor is also the product of collector efficiency factor (F') and flow factor (F'').

$$F_R = F' \times F'' \tag{1.25}$$

The collector efficiency factor can be defined as the ratio of the actual useful energy gain to the useful gain if the collector absorbing surface had been at the local fluid temperature.

$$F' = \frac{1/U_L}{W\left[\frac{1}{U_L\left[D + (W - D)F\right]} + \frac{1}{C_b} + \frac{1}{\pi.\ D_i.h_{fi}}\right]} \tag{1.26}$$

where W, D, C_b, D_i, and h_{fi} represent the distance between parallel tubes, tube outer diameter, the bond conductance, inner tube diameter, and heat transfer coefficient between the fluid and the tube wall, respectively. Equations 1.27 and 1.28 describe the heat transfer coefficient (H_{fi}) and bond conductance (C_b).

$$h_{fi} = \frac{N_u . k}{D_h} \tag{1.27}$$

where N_u, k, and D_h represent the Nusselt number, thermal conductivity, and hydraulic diameter of pipe.

$$C_b = \frac{k_b \, b}{\gamma} \tag{1.28}$$

where K_b, b, and γ are the average bond thickness, bond width, and thermal conductivity of bond, respectively.

Finally, Eq. 1.29 describes the function of collector flow factor.

$$F'' = \frac{\dot{m}. \, C_p}{A_C. \, U_L. \, F'} \times \left[1 - \exp\left(\frac{-A_C. \, U_L. \, F'}{\dot{m}. \, C_p} \right) \right] \tag{1.29}$$

The process could be summarized by observing the solar irradiance, which will strike the solar thermal collector. Some of this irradiance will be reflected, radiated, and absorbed by the cover, while the remaining will pass through to the absorber (commonly painted or made with selective material), which in turns absorb the heat. A tube configuration is attached at the back of the absorber. Heat will transfer from the absorber to the tube by conduction. A fluid will flow within the tube, or flow passage, and gain that heat, by heat transfer. Essentially, fluid extracts this heat and is either taken directly to the load or passed through a heat exchanger. The amount of heat that remains after the thermal losses is referred to as useful energy gain. The thermal efficiency represents the measure of the collector's performance and is defined as the ratio between useful energy gain to incident solar energy over a particular time period.

Figure 1.28 shows higher optical efficiency associated with flat-plate collectors, with highest for collectors with a painted absorber. The optical efficiency is the maximum achievable efficiency by the solar collector. It is defined as "the rate of optical (short wavelength) energy reaching the absorber, divided by the energy coming from the solar resource." This occurs when the fluid inlet temperature equals ambient temperature ($T_i - T_a = 0$ °F). The point where the efficiency reaches zero is when no useful heat gain can be produced by the collector, which occurs if fluid flow through the collector stops (power failure). This point is defined as the stagnation temperature.

The theory, analysis, and mathematical modeling of solar thermal collectors are rich area of study and require different methods and techniques. The aforementioned equations are merely an introduction to the basics of this technology. However, Eqs. 1.14 to 1.29 [16] are the foundation of energy balance of flat-plate collectors.

1.4.3 *Flat-Plate Collector (FPC) Literature Review*

The literature in flat-plate collectors is extensive and comprehensive. The research discussed in this section is merely a sample of the research made in this field. The section classifies research discussed to three classifications: research on collector designs, research on modeling of FPC's, and recent advances and designs of FPC's.

Design of FPC

The type of the absorber design, working fluid, and auxiliary system is important and causes lots of changes in the performance of the collector. This opens up an entire field of studies targeting the search for optimal collector design and performance.

Fan et al. [58] presented a study of a new design for liquid-based flat-plate solar collector by employing a V-corrugated absorber with multi-channels (VFPC). The absorber is made of aluminum, and channels are triangular shaped. The purpose of this absorber is to improve the collector's performance. The authors used both numerical model and experiments to test the performance, under different operational conditions, of the proposed design against a tube-and-sheet FPC (TFPC) and a tube-and-sheet with VFPC (VTFPC). The heat transfer model is done with the aid of ANSYS Fluent software.

The experimental setup is situated outdoors and is composed of main thermal system components such as collector, heat exchanger, circulating pump, and a thermostatic water tank. The setup is also composed of necessary sensing equipment such as pyranometer, mobile microclimatic station, platinum resistance sensors, and a hygrometer. The study uses data from previously made indoor tests, along with the tests from outdoors experiments. The main parameters to be used by the module are inlet and outlet liquid temperature, mass flow rate, solar irradiance, and wind speed. The evaluation parameters in the study include thermal/optical efficiency, exergy efficiency, pressure drop, and pump power requirements. The main findings of the study suggest that proposed system exhibits superior daily average optical, thermal, and exergy efficiencies to the conventional counterpart. Those efficiencies are 84.9%, 69.4%, and 3.8%, respectively. The conventional collector is 4.1 times higher than the proposed collector in terms of pressure drop and pump power consumption. Increase of mass flow rate from 10 to 90 g/s led to decrease in pump power consumption from around 0.16 to 0.06 Wp. Moreover, the study shows a 10.1%/15.8% difference in daily average optical efficiency between VFPC/VFPCT and TFPC, which highlights the utility of V-corrugated absorber in terms of increasing solar radiation absorption. Although V-corrugated absorber shows a better exegetic efficiency, it is not significantly higher than conventional collector.

Karim and Hawlader [59] conducted an experimental and theoretical investigation on three air heating collectors of flat-plate, finned, and V-corrugated and in single and double-pass modes for performance enhancement purposes. The experiments examine different operating and design parameters. The experiments were conducted according to ASHRAE guidelines and standards. The test rig is composed of heater, damper, fan, support frame, collectors, and sensing equipment such as pressure transducer, anemometer, thermocouple, and axial flow. The numerical simulation employed meteorological data, collector geometry, and material property. The results obtained indicate increase of flat-plate thermal efficiency from 26% to 69% as mass flow rate is raised from 0.01 to 0.06 kg/m^2 s. For a mass flow rate of 0.056 kg/m^2 s, the flat-plate collector with double-pass operation has an optical efficiency of around 78%. The experimental results closely resemble the numerical, with some variations which are normal and are attributed to some, unaccounted, parameters and losses that occur during experiments. The authors put good effort in defining the relative uncertainty and error analysis. The study concludes that double-pass mode leads to higher efficiency for all tested collectors, especially for flat-plate collector (as appose to v-groove collector). Overall, the most efficient collector was the V-corrugated while the least was the flat-plate with difference of efficiency from 10 to 15% for single-pass mode.

Alvarez et al. [60] performed an experimental and numerical analysis of a flat-plate collector with corrugated channels and a serpentine flow path. The experimental rig employed the proposed collector, which was pointed southward at an angle of 35 °C, connected to a storage tank, adjustable speed pump, and proper sensing equipment with measurements of fluid temperature taken every 5 minutes. A hydrodynamic model of the collector was done and validated with a fin-and-tube type collector. Experimental heating curves were made to analyze thermodynamic efficiency of the collector. The findings show that experiments fit the predictions of the simulation. The optical efficiency is around 84%, and efficiency during the day is 65%.

Rojas et al. [61] conducted experimental work, steady-state tests, on single-glazed selective surface flat-plate collector using both EN12975–2 and ASHRAE 93 standards, to compare between the two. In addition to steady-state tests, the authors conducted quasi-dynamic tests. In addition, ISO9806-1 standard was used as well. The setup is composed of three electric heaters for control of inlet fluid temperature, a weather station for wind speed, ambient temperature, and relative humidity. Incident angle modifier coefficient, ASHRAE method, was found to be more accurate than transient method, EN12975–2 standard. However, transient model of the collector is more refined and includes the terms like wind speed dependence and the collector thermal capacitance. The numerical work employed TRNSYS software for simulation of long-term performance of the proposed system. The study concludes that use of ASHRAE 93 standard is feasible for solar thermal collector tests. The number of suitable days for testing is strongly dependent on climate conditions of the test site. For both standards, the solar fractions obtained by simulation are within 7%.

Modeling of FPC

Modeling FPC is very important to understand and simulate their behavior for different conditions and cases. Modeling is an effective method to save costs and time on extensive experiments. Throughout the literature on this topic, researchers use different models depending on the method and targeted analysis. Further praise is directed to research with numerical modeling and experimental validation.

Aleksiejuk et al. [62] proposed an analog dynamic model, using equivalent thermal network method, of a flat-plat solar collector. This model is compared to a digital model determined based on an experiment. Important parameters include solar intensity, inlet temperature of medium, and ambient temperature. The output signal is the temperature on the collector outlet. From this study, it is safe to conclude that construction and operating parameters are highly important for the dynamic modeling of flat-plate collectors, both for analog and digital models.

Ampuño et al. [63] analyzed the effect of transport delay in FPC and how to model it dynamically for control, using a simplified model, which was found to be reliable when comparing the simulation results with real data from experiments and modeling errors which were found to be quite low when changing values of flow rate, inlet loop temperature, and solar irradiance.

Tian et al. [64] used the flat-plate collector in series with a parabolic trough collector, as solar heating plants for district heating networks. The study utilizes TRNSYS-GenOpt model for parameter optimization. Among plant design considerations are areas of both collector types, storage size, orientation of the parabolic trough collectors, etc. The model includes flat-plate collectors, parabolic trough collectors, the storage tanks, natural gas boilers, and so on. The flat-plate collector is studied in two cases: with and without FEP foils. The study uses net levelized cost of heat to evaluate the system performance, as well as employing it to introduce a generic method to optimize the system. The annual energy output of the flat-plate collector field is around 449 kWh/m2 for a zero-heat demand. The findings show 5–9% reduction in system levelized cost of heat when using solar collectors, compared to not using them. Moreover, the LCOH of the proposed system could reach 0.36 DKK/kWh.

Farahat et al. [65] presented a technical note of a numerical simulation for exergetic optimization of flat-plate solar collector for optimal performance and design parameters. The work employs energy and exergy analysis for evaluation of thermal and optical performance as well as losses associated with flat-plate collectors for specified operating conditions. Among parameters used for analysis are absorber plate area and collector dimensions, mass flow rate, temperature of inlet and outlet fluid, etc. The authors conclude that increase of solar irradiance absorbed by the plate leads to increase of its exergy efficiency, while increase of ambient temperature and wind speed causes the opposite effect, decrease.

Jafarkazemi and Ahmadifard [66] presented a theoretical model for energy and exergy analysis of flat-plate collectors to examine the effect of the design and operation parameters, such as thickness of back insulation and flow rate, on energy and exergy efficiencies. The models were confirmed with experimental data. The tests were done on an open-loop system with an electric heater, circulation pump, and proposed collector with water as working fluid. The experiments utilized meteorological equipment such as pyranometers, ambient temperature sensors, and anemometer. The experiments were conducted for mass flow rates of 0.03, 0.04, and 0.05 kg/s. The same was done for the theoretical model. The findings show that having inlet water temperature of 40 °C more than ambient temperature and lowering flow rate leads to enhanced overall performance. Furthermore, the efficiency of the collector increased with decrease of overall heat loss coefficient as a consequence of decreasing the absorber plate's temperature, by increasing flow rate to around 0.01 kg/s. Moreover, the thermal efficiency is found to drop with increase of inlet water temperature. The authors explain this is because it, inlet water, causes increase in temperature of absorber plate and hence increase in temperature gradient between the absorber plate and surrounding environment which ultimately leads to increase of overall heat loss coefficient. However, it does lead to increase of overall exergy efficiency, for a better overall performance. The study shows the root mean square error (RMSE) used for estimation of errors in the research. Energy efficiency errors are around 9.51% for mass flow rate 0.03 kg/s. For the same flow rate, the exergy efficiency error is around 1.12%. RMSE at other mass flow rates for energy and exergy efficiencies is calculated as well.

Villar et al. [67] developed a transient 3D mathematical model of a flat-plate collector by setting mass and energy balance on finite volumes. The model can compare between different configurations such as parallel tubes (PTC), serpentine tube (STC), two-parallel plate (TPPC), and other types. Furthermore, the author studies the collector performance when flow rate in the risers is nonuniform. Experimental validation was done using a commercial PTC. The main advantage of this model is its flexibility which allows it to deal with different parameters (thermal properties, initial temperature, inlet and outlet temperature, and climate conditions), types (collector dimension and tilt), and configurations of flat-plate collectors. Furthermore, the model tests performance of collector when flow rate in the riser is nonuniform. The optical efficiency of the system is found somewhere around 77%. The results also indicate a correlation between nonuniformity of fluid flow and efficiency deterioration; higher nonuniformity leads to higher deterioration. At a flow rate of 80 kg/h and nonuniformity ($\varnothing$) of 0.0612, the efficiency is decreased by 2%. The authors used Eq. (1.30) from Chiou [68] to calculate the nonuniformity factor.

$$\varnothing = \sqrt{\frac{\sum_{i=10}^{10} (\beta_i - 1)^2}{10}} \tag{1.30}$$

where (β_i) is a parameter defined, by Eq. (1.31), as ratio between flow through ith tube (Q_i) and total flow (Q_o):

$$\beta_i = \frac{Q_i}{Q_o} \tag{1.31}$$

This paper illustrates the importance of considering all design parameters for optimum performance evaluation.

Advanced Concepts and Designs of FPC

Throughout the year, advanced concepts and designs have been proposed by researchers. This section shows the use of nanofluids and phase change material in different flat-plate collectors. Nanofluids are mixtures of nano-sized particles and a base fluid such as water. These mixtures have better thermophysical properties and can be used as working fluids. Research in this area studies the effect of employing this technology on heat transfer enhancement. Moreover, study of appropriate mass concentration (%) of the nanoparticles out of the base fluid is very crucial for selection optimum nanofluid. Phase change material (PCM) is also important due to their ability of latent heat and storing and releasing large amount of heat. A prime example is the organic PCM "paraffin wax." Further exploration on this topic is made in Chap. 3.

Verma et al. [69] performed an experimental investigation to test the effect of replacing water with nanofluids, of different types (e.g., CuO, MgO, MWCNTs, hybrid CuO with MWCNTs, and MgO with MWCNTs) for different concentrations, varying from 0.25% to 2% and different mass flow rates, from 0.5 lpm to 2.0 lpm. The evaluation parameters contain both quantitative and qualitative observations of the FPC behavior. The experimental setup contained the collector, pumps, and proper sensing equipment. Testing lasted for 1 year, with 3 hours of continuous running of two 500 W pumps with capacity of 5 liters per minute. The findings in general indicate that at constant mass flow rate and solar irradiance, the FPC efficiency will increase with the increase of nanofluid particle concentration from 0.75% to 1%. The highest enhancement in collector efficiency is when using a MWCNT type nanofluid as base fluid. At 0.75% particle concentration of nanofluid, varying mass flow rate shows increase in collector efficiency and peaking at 0.025–0.03 kg/s. The exegetic efficiency of FPC with MgO and CuO nanofluids is around 71.54% and 70.63%, respectively. The thermal efficiency for both is around 70.55% and 69.11%, respectively. The results also show a proportional relationship between solar intensity and collector efficiency. Among evaluation parameters is the increase of Bejan number, which is defined by the authors as "increase of Bejan number is an indication of system's quality credit. It enhances productive entropy due to transfer of heat caused by temperature difference and suppresses production of entropy, arises by systems irreversibility." The increase of mass flow rate is found to cause reduction in exergy efficiency and Bejan number. Highest rise in Bejan number is around 0.97 and achieved when using a MWCNTs/water nanofluids. Finally, the pumping power loss ratio is higher for CuO/water nanofluid than MWCNTs/water nanofluid. This is due to CuO's higher density and poorer rheological property. The authors conclude that hybrid MgO, with MWCNTs, leads to the best overall performance of FPC, as appose to CuO hybrid. Performing parameters of the hybrid MgO is closer to MWCNT/water fluid.

Baharin and Mohammad [70] experimentally investigated the performance of a single slope solar still coupled with a flat-plate collector under the climate conditions of Malaysia. This is to compare between passive and active solar still. The system is intended for use of desalination. The setup is composed of a solar still and an FPC. The solar still consist of two sections: evaporator and collector basins. In addition, the structure has a condenser plate and a separator wall. The flat-plate solar collector is located on both sides of the solar still. Four cases were investigated for the experiments: three solar stills with solar collectors having different copper rod diameters (3, 4, and 5 mm) and one solar still without a solar collector. The experiment period is 24 hours with measurement of the stills output at 2-hour intervals. Measurements included ambient temperature, relative humidity, and water salinity. The results indicate better performance of stills coupled with solar collectors than without. The highest volume of fresh water is yielded by the still with a 5 mm copper rod FPC with around 125 ml which is twice the amount produced by a single slop solar still which is 62.5 ml.

Sakhaei and Valipour [71] reviewed, comprehensively, studies of different flat-plate collectors (FPC) in terms of thermal performance enhancement. The review encompassed performance models, numerical simulations, and experimental works that aim to improve the collector's thermal efficiency. The authors discussed thermal models such as heat transfer models for convection, conduction, and radiation, as well as one- and two-dimensional thermal models. The authors argue that to find the irreversibility sources to approach an optimal design of the FPC and thermal performance enhancement, it is important to develop the exergy model. Dynamic models such as lumped capacitance models and discretized models are also discussed in the review. The need for these dynamic models is a must due to variation of external conditions during tests such as solar irradiation. The FPC design parameters, for components like glass cover, absorber plate, air gap, riser pipe, and insulation material, and their effect are discussed in detail. Advanced research of experimental and numerical investigations focuses on the use of turbulators, porous material, phase change material, and nanofluids.

Raj and Subudhi [72] reviewed and investigated performance enhancement in flat-plate collectors and direct absorption solar collectors (DASC) when using nanofluids as base fluids. Various studies and concepts associated with nanofluids' enhancement of solar collectors in terms of thermal performance are present in the literature. The authors classify parameters that affect FPC to three classifications: design parameters (such as glass cover, absorber plate material, and absorber plate design), operational parameters (such as fluid inlet temperature, type of absorber fluid, and collector tilt), and meteorological and environmental parameter (such as ambient temperature, wind speed, and solar flux). Furthermore, the review provides a brief introduction on nanofluids the two-step method to prepare them. The authors put forth effective arguments to justify use of nanofluids from a technical prospective. Among the study conclusions, the authors state that increase of addition of surfactant to nanofluids is positively correlated to thermal efficiency of collector. Also, proper dispersion of particles leads to long-term stability of the fluids.

A.M. Genc et al. [73] conducted a transient heat transfer study for a nanofluid-based flat-plate solar thermal collector to determine the thermal inertia of its components. The authors compared between four cases where three different volumetric concentrations of Al2O3 nanofluids and water as base fluids, which is tested for mass flow rates ranging 0.004 to 0.06 kg/s. The daily average outlet temperature of the collector for the four cases is illustrated for months of January, April, July, and October. The 3% volume faction-nanofluid was found to achieve highest enhancement in the collector's exergy efficiency. This case, and at 0.004 kg/s, led to increase of outlet temperature by 7.2%, in July. Employing a 1% (vol.) nanofluid-based fluid and at 0.06 kg/s mass flow rate led to highest FPC thermal efficiency at 83.90%. For all cases, increase of mass flow rate shows to increase thermal efficiency. However, increase of mass flow rate led to decreased exegetic efficiency. The highest outlet temperatures are recorded in period from 11:30 to 12:00 o'clock for different mass flow rates. The highest daily average outlet temperature is recorded for the month of October. This study shows the importance of the base fluid to a flat-plate collector, as well as its mass flow rate. These two topics are well studied throughout the literature.

Carmona and Palacio [74] performed both experimental and numerical study of latent heat storage of a flat-plate collector integrated with PCM. The numerical aspect is developed to evaluate the performance of the collector by creating a thermal dynamic model based on simplified semiempirical correlations and heat transfer for each part of the collector. The outdoor experiments are used to validate the model. Experimental setup was composed of the proposed collector, water reservoir, pump, rotameter, weather station, reception tank, and a data acquisition system. The thermal insulation of the collector is made of poliurethane. After validation, the model proved its ability of estimating global thermal performance of the proposed device with good accuracy. The comparison between model and experiments data for 20 days was presented encompassing days of high and low incident radiation. The maximum error for the predicted collector temperatures is around 4.62%. Two different types of PCM, different melting points, were analyzed and compared to collector without PCM in terms of temperature values and efficiency.

Khan et al. [75] conducted a review to investigate recent works that integrates phase change material (PCM) in solar collectors of different types such as flat-plate collector (FPC), evacuated tube collectors (ETC), compound parabolic collectors (CPC), and photovoltaic/thermal (PV/T) solar collectors. The authors claim that PCM is used for better stability and performance, stating that "stabilizes the intermittent temperature fluctuations as well as extends the operating hours." The authors argue that use of PCM in flat-plate collectors is rather an old technique and provides research study that dates to 1987. In addition, the study presents schematic diagrams of different solar water heating systems with integrated phase change material. The same was done for solar air heaters and PV/T systems. Overall, the review argues in favor of incorporating PCM in solar thermal collectors. For FPC's the authors emphasize the importance of surface contact between PCM and absorber plate, use of V-corrugated surface which exhibits higher exposure in the same aperture area, and fins/metal foams which add possibility of thickening the PCM layer.

Summary of FPC Literature

A detailed summary of the literature is provided in Table 1.5. illustrating year of publication, study area, research approach, working fluid, design of collector, and some findings.

1.5 Chapter Summary and Conclusions

In this chapter, the chronology, theory, concepts, and literature associated with photovoltaic (PV) and solar thermal (ST) technologies have been discussed. The first section provides a broach introduction to the topic, while the other two are deeper in nature discussing physical, electrical, and thermodynamic elements of these technologies. This section will revisit some of the main topics, of this chapter, as well as present the conclusions and recommendations.

1.5.1 PV Summary, Conclusions, and Recommendations

The solar cell is a P-N junction, semi-conductor device that converts the solar irradiance into direct current (DC) electricity. Solar (photovoltaic – PV) cells are connected to produce a module, connected modules are viewed as a panel, and panels are connected to make up arrays. Series-connected modules are known as strings, and parallel connection of this entire circuit leads to forming the array. Series connection of cells/modules/panels leads to increase of voltage, while parallel connection leads to increase of current. The power conversion efficiency of PV is the ratio of maximum produce power to the incident solar energy over the area of collector. PV systems are usually divided into two types: stand-alone and grid-connected. Stand-alone systems are more autonomous and usually employ charge controllers and batteries for storage. Grid-connected systems feed the excess generated energy back to the grid. PV systems can also be hybrid with other renewable technologies, such as wind turbines and solar thermal systems, and even with fossil fuel-based technologies like diesel generators. PV systems are affected by the efficiency of their sub-components as well as surrounding environment. These effects are as follows:

1. Increase of solar irradiance leads to increase in overall efficiency of the PV panel.
2. Increase of cell temperature leads to decrease in overall efficiency of the PV panel.
3. Increase of cell temperature leads to increase of short-circuit current.
4. Increase of cell temperature leads to decrease of open-circuit voltage.

Table 1.5 Summary of research conducted in the literature

Author(s)	Year	Location	Research approach	Working fluid	Design of collector	Findings
Fan et al.	2019	China	Numerical and experimental	Water	V-corrugated absorber with multi-channels (triangular) flat-plate collector	10.1%/15.8% difference in daily average optical efficiency between VFPC/ VFPCT and TFPC
Khan et al.	2018	India	Review	Nanofluid	FPC, ETC, CPC, and PV/T	Study emphasizes the importance of surface contact between PCM and absorber plate
Ampuño et al.	2019	Spain	Numerical model and experimental tests	Water	Conventional FPC	For large solar fields, it is mandatory to define an irradiance-output transport delay as a function of the flow rate
Aleksiejuk et al.	2018	Poland	Numerical analog dynamic model	Water	flat-plate collectors, vacuum collectors	Step response characteristics for transfer function G1(s) indicate that the solar collector responds to a step function like a typical thermal object
Carmona and Palacio	2019	Colombia	Experimental and numerical (MATLAB/ Simulink)	Water	FPC with PCM	After validation, the model proved its ability of estimating global thermal performance of the proposed device with good accuracy

(continued)

Table 1.5 (continued)

Author(s)	Year	Location	Research approach	Working fluid	Design of collector	Findings
Tian et al.	2018	Denmark	Numerical (TRNSYS – GenOpt)	35% propylene glycol/ water	FPC and PTC in series	The annual energy output of the flat-plate collector field is around 449 kWh/ m2 for a zero-heat demand
A.M. Genc et al.	2018	Turkey	Numerical (MATLAB)	Nanofluids	FPC	Employing a 1% (vol.) nanofluid-based fluid and at 0.06 kg/s mass flow rate led to highest FPC thermal efficiency at 83.90%
Raj and Subudhi	2018	India	Review	Nanofluids	DASC and FPC	Increase of addition of surfactant to nanofluids is positively correlated to thermal efficiency of collector
Sakhaei and Valipour	2019	Iran	Review	Nanofluid	Dynamic and exergy models, nanofluid, PCM, and porous media-based FPC	The need for these dynamic models is a must due to variation of external conditions during tests
Baharin and Mohammad	2018	Malaysia	Experimental (outdoor)	Water	Single slope solar still coupled with a flat-plate collector	5 mm copper rod FPC produce around 125 ml volume of fresh water
Verma et al.	2018	India	Experimental	Nanofluids (CuO, MgO, and MWCNT)	FPC	MWCNT led to highest enhancement of exegetic efficiency. All types are better than using water

(continued)

Table 1.5 (continued)

Author(s)	Year	Location	Research approach	Working fluid	Design of collector	Findings
Alvarez et al.	2010	Spain	Experimental and numerical	30% water–ethylene glycol mixture	V-corrugated flat-plate collector with serpentine flow-path	The optical efficiency is around 84%, and efficiency during the day is 65%
Farahat et al.	2009	Iran	Technical note, numerical simulation	Water	FPC	Increase of solar irradiance absorbed by the plate leads to increase of its exergy efficiency
Jafarkazemi and Ahmadifard	2013	Iran	Theoretical model and experimental	Water, ethylene glycol, and mixture of water – propylene glycol	FPC	Having inlet water temperature of 40 °C more than ambient temperature and lowering flow rate leads to enhanced overall performance
Karim and Hawlader	2006	Singapore	Experimental and theoretical	Air	FPC, FPC Finned, FPC V-corrugated	The most efficient collector was the V-corrugated, while the least was conventional the flat-plate
Rojas et al.	2008	USA	Experimental and numerical (TRNSYS)	Water	FPC	ASHRAE 93 standard is feasible for solar thermal collector tests
Villar et al.	2009	Spain	Numerical with experiments	NM	FPC, PTC, STC, TPPC	The results also indicate a correlation between nonuniformity of fluid flow and efficiency deterioration

5. High ambient temperature, relative humidity, and low wind speed are associated with reduction of overall PV efficiency.
6. High wind speed is associated with increase of overall PV efficiency. However, caution must be exercised in the installation process and design of support structure for windy locations, which should also be a point of assessment cost-wise.

7. Dust accumulation and shading cause massive reduction in power conversion efficiency and the lifetime of the device.
8. One of the main issues associated with PV systems is storage, and research into enhancement of storage (batteries) technology is critical for the growth of this field.
9. Optimum design of PV systems is achievable through numerical approach.
10. Main cost evaluation parameters are life cycle costs (LCC), cost of energy (COE), and payback period (PBP). The best systems are those with lowest possible LCC, PBP, and COE.
11. Feed-in tariff (FiT) program is highly useful for encouragement of investment in grid-connected PV systems.

Recommendations for PV systems:

1. Optimum design of PV systems for the set load is recommended using numerical methods and software, employing techniques such as loss of load probability (LLP) and software such as HOMER [36, 45].
2. Design of customized numerical software for the system location, such as REPS. OM [49] and spreadsheet software [31], is better than employing a global tool, such as HOMER software.
3. Determination of optimum tilt angle for the installation location of the PV system is very useful and can lead to increase of gain of collected energy by 24.6% [27]. Furthermore, it is more useful to employ two-axis solar tracking, which is found to increase the daily average gain in solar irradiation and generated power by around 29.3% and 34.6%, respectively.
4. Studies for the performance of PV systems' applications, such as PV water pumping systems, are recommendation for site-specific studies and comparison studies with fossil fuel energy sources, such as diesel generator pumping system.
5. Longevity and reliability of the PV system components are key to keeping the LCC, PBP, and CoE low. This could be achieved by focusing the efforts into enhancing and specializing the different components such as batteries, inverters, charge controllers, solar tracking systems, etc.
6. Cooling of PV modules is very important for enhancing its performance in outdoor conditions. It is recommended to conduct studies on hybrid designs that add cooling elements such as employment of PCM containers, hybrid PV/T systems, and floating PVs.
7. The use of artificial intelligence (AI), artificial neural networks (ANN), deep learning (DL), and Internet of things (IoT) is highly recommended to transform research in PV systems to reach goals of long-term performance prediction and data mining.

1.5.2 FPC Summary, Conclusions, and Recommendations

Solar thermal systems can be classified depending on their concept of operation, construction, achievable temperature, and working fluid. Flat-plate collectors (FPC) are glazed, non-concentration, low temperature ranged-collectors. FPCs are composed of a cover, absorber, flow passage, insulation, and enclosure to reduce thermal losses, absorb heat, carry the working fluid, minimize heat losses, and carry all collector components, respectively. Performance of flat-plate collectors is evaluated based on parameters such as optical and thermal efficiencies, thermal exergy efficiency, overall heat gain, and levelized cost of heat (LCOH). The parameters investigated in research on FPCs are divided into three areas: design parameters, operational parameters, and metrological parameters.

Design parameters:

1. FPC components shape
2. FPC components material
3. FPC tube configuration
4. FPC combined system

Operational parameters:

1. Fluid inlet temperature
2. Type of working fluid
3. Collector tilt angle

Metrological parameters:

1. Solar irradiance
2. Ambient temperature
3. Wind speed

Among discussed types of possible insulation material were polystyrene, fiberglass, mineral wool, cellulose, polyurethane foam, and aerogel. Polyurethane performs very well due to high ratio of strength to weight at low temperatures.

Critical parameters affecting FPC performance (heat removal factor, efficiency, etc.) include:

1. Thickness and coating of the glass cover
2. Material and thickness of absorber plate
3. Air gap between absorber plate and glass cover
4. Diameter, thickness, and material of pipes (or flow passage) and their configuration
5. The insulation material

Testing of FPC's must account for the variations which are a part of the external conditions such as solar irradiance, ambient temperature, and wind speed. All of which are major factors for the dynamic behavior of solar collectors. Optimum design of FPC is achievable through dynamic, modeling, analysis of the proposed collector. These models include lumped capacitance and discretized models.

From the findings and supported theory, the following conclusions are made:

1. Increase of solar irradiance leads to increase in thermal efficiency.
2. Increase of mass flow rate leads to increase in electric efficiency.
3. Optical efficiency is achieved when the fluid inlet temperature equals the ambient temperature.
4. Increase of ambient temperature and wind speed leads to rapid decrease of exergy efficiency.
5. The exergy efficiency of solar collectors is highly dependent on solar irradiance, while thermal efficiency is highly dependent on mass flow rate.

Recommendations

1. It is recommended to use aluminum as material for the absorber plate, unless corrosion is a consideration, in which copper may be a better option.
2. It is recommended to employ phase change material (PCM) with high latent heat and surface area to achieving a better solar thermal heater. Moreover, investigations into different types, in-organic, of PCM material and nano-PCM are recommended.
3. It is recommended to combine passive methods and nanofluids to increase thermal performance of FPCs and PTCs. Research studies should be carried out to investigation different types, mixtures, and concentration ratios, of nanofluids and the effect they have on thermal performance of FPCs. Moreover, studies to determine the optimum mass flow rate and consequential pump power consumption are necessary for this topic.
4. It is recommended to consider as much related parameters as possible for a more accurate analysis using the multi-objective genetic algorithm, employing deep learning (DL) and artificial neural network (ANN) techniques for long-term performance prediction.
5. Theoretical evaluation of flat-plate collector performance is recommended using dynamic models, along with validation using experimental procedures.
6. ASHRAE standards are highly recommended for testing and evaluations of FPCs.
7. Exergy analysis is highly recommended for future designs of FPCs.
8. Use of numerical software TRNSYS is highly recommended for theoretical studies of FPC performance.
9. More studies discussing the levelized cost of heat (LCOH), life cycle assessment (LCA), and carbon-footprint of this technology are recommended.
10. Use of nanofluid-based double-pass, V-grooved/V-corrugated FPCs is recommended for future theoretical and experimental investigations.

References

1. C.D. Keeling, Industrial production of carbon dioxide from fossil fuels and limestone. Tellus **25**(2), 174–198 (1973)
2. M.M. Halmann, M. Steinberg, *Greenhouse Gas Carbon Dioxide Mitigation: Science and Technology* (CRC Press, Boca Raton, 1998)
3. D.A. Lashof, D.R. Ahuja, Relative contributions of greenhouse gas emissions to global warming. Nature **344**(6266), 529 (1990)
4. Renewables. Global Status Report (REN21). A comprehensive annual overview of the state of renewable energy (2018), Retrieved on January 5th 2019 – from Ren21.net
5. Renewables 2018 (IEA). Market analysis and forecast from 2018 to 2023 (2018), Retrieved on January 5th 2019 – from IEA.org
6. Renewables 2018 (IEA). Solar energy. Retrieved on January 5th 2019 – from IEA.org
7. L. Lakatos, G. Hevessy, J. Kovacs, Advantages and disadvantages of solar energy and windpower utilization. World Futures **67**(6), 395–408 (2011)
8. G.B. Dalrymple, The age of the earth in the twentieth century: A problem (mostly) solved. Geol. Soc. Lond., Spec. Publ. **190**(1), 205–221 (2001)
9. M. Pidwirny, *Surface Area of Our Planet Covered by Oceans and Continents* (University of British Columbia, Okanagan, 2006), Retrieved 2019-11-2
10. J.W. Morgan, E. Anders, Chemical composition of earth, Venus, and mercury. Proc. Natl. Acad. Sci. **77**(12), 6973–6977 (1980)
11. M. Emilio, S. Couvidat, R.I. Bush, J.R. Kuhn, I.F. Scholl, Measuring the solar radius from space during the 2012 Venus transit. Astrophys. J. **798**(1), 48 (2014)
12. M. Asplund, N. Grevesse, A.J. Sauval, P. Scott, The chemical composition of the Sun. Annu. Rev. Astron. Astrophys. **47**, 481 (2009)
13. D.B. Guenther, Age of the sun. Astrophys. J. **339**, 1156–1159 (1989)
14. J.E. Braun, J.C. Mitchell, Solar geometry for fixed and tracking surfaces. Sol. Energy **31**(5), 439–444 (1983)
15. F. Vignola, J. Michalsky, T. Stoffel, *Solar and Infrared Radiation Measurements* (CRC Press, Boca Raton, 2016)
16. J.A. Duffie, W.A. Beckman, *Solar Engineering of Thermal Processes* (Wiley, Hoboken, 2013)
17. J.W. Spencer, Fourier series representation of the position of the sun. Search **2**(5), 172–172 (1971)
18. Global Solar Atlas (2018), Retrieved on January 9th 2019 – from globalsolaratlas.info
19. K.H. Solangi, M.R. Islam, R. Saidur, N.A. Rahim, H. Fayaz, A review on global solar energy policy. Renew. Sust. Energ. Rev. **15**(4), 2149–2163 (2011)
20. Report IEA PVPS T1–33:2018. Snapchat of Global Photovoltaic Markets 2018, Retrieved on 10th January 2019 from iea-pvps.org
21. A. Goetzberger, J. Knobloch, B. Voss, *Crystalline Silicon Solar Cells* (Editorial Wiley, Chichester, 1998), p. 1
22. Tonio Buonassisi, *2.627 Fundamentals of Photovoltaics*. (Massachusetts Institute of Technology: MIT OpenCourseWare, Fall 2013), https://ocw.mit.edu. License: Creative Commons BY-NC-SA
23. A. Luque, S. Hegedus (eds.), *Handbook of Photovoltaic Science and Engineering* (Wiley, Chichester, 2011)
24. E. Lorenzo, *Solar Electricity: Engineering of Photovoltaic Systems* (Earthscan/James & James, 1994)
25. D.P. Kaundinya, P. Balachandra, N.H. Ravindranath, Grid-connected versus stand-alone energy systems for decentralized power—A review of literature. Renew. Sust. Energ. Rev. **13**(8), 2041–2050 (2009)

26. B. Marion, J. Adelstein, K.E. Boyle, H. Hayden, B. Hammond, T. Fletcher, B. Canada, D. Narang, A. Kimber, L. Mitchell, G. Rich, Performance parameters for grid-connected PV systems, in *Conference Record of the Thirty-first IEEE Photovoltaic Specialists Conference, 2005*, (IEEE, Lake Buena Vista, 2005, January), pp. 1601–1606
27. H.A. Kazem, T. Khatib, A.A. Alwaeli, Optimization of photovoltaic modules tilt angle for Oman, in *Power Engineering and Optimization Conference (PEOCO), 2013 IEEE 7th International*, (IEEE, Langkawi, 2013, June), pp. 703–707
28. F. Belhachat, C. Larbes, A review of global maximum power point tracking techniques of photovoltaic system under partial shading conditions. Renew. Sust. Energ. Rev. **92**, 513–553 (2018)
29. M. Kacira, M. Simsek, Y. Babur, S. Demirkol, Determining optimum tilt angles and orientations of photovoltaic panels in Sanliurfa, Turkey. Renew. Energy **29**(8), 1265–1275 (2004)
30. A. Mellit, Sizing of photovoltaic systems: A review. Rev. Energ. Renouv. **10**(4), 463–472 (2007)
31. B. Herteleer, J. Cappelle, J. Driesen, An autonomous photovoltaic system sizing program for office applications in Africa. The Renew. Energ. & Power Quality Journal **1**, 728–733 (2012)
32. I. Zanesco, A. Moehlecke, G.S. Medeiros, T.C. Severo, D. Eberhardt, S.S. Junior, E.A. Zenzen, Experimental evaluation of an analytic method for sizing stand-alone PV systems, in *X Congresso Brasileiro de Energia, Rio de* (WIP Renewable Energies, München, 2004), p. 530
33. B. Burger, R. Rüther, Inverter sizing of grid-connected photovoltaic systems in the light of local solar resource distribution characteristics and temperature. Sol. Energy **80**(1), 32–45 (2006)
34. M.S. Adaramola, Techno-economic analysis of a 2.1 kW rooftop photovoltaic-grid-tied system based on actual performance. Energy Convers. Manag. **101**, 85–93 (2015)
35. K. Jeong, T. Hong, C. Ban, C. Koo, H.S. Park, Life cycle economic and environmental assessment for establishing the optimal implementation strategy of rooftop photovoltaic system in military facility. J. Clean. Prod. **104**, 315–327 (2015)
36. M.A. Ramli, A. Hiendro, S. Twaha, Economic analysis of PV/diesel hybrid system with flywheel energy storage. Renew. Energy **78**, 398–405 (2015)
37. S. Saib, A. Gherbi, A. Kaabeche, R. Bayindir, Techno-economic optimization of a grid-connected hybrid energy system considering voltage fluctuation. J. Electr. Eng. Technol. **13**(2), 659–668 (2018)
38. P. Rajput, Y.K. Singh, G.N. Tiwari, O.S. Sastry, S. Dubey, K. Pandey, Life cycle assessment of the 3.2 kW cadmium telluride (CdTe) photovoltaic system in composite climate of India. Sol. Energy **159**, 415–422 (2018)
39. V.J. Fesharaki, M. Dehghani, J.J. Fesharaki, H. Tavasoli, The effect of temperature on photovoltaic cell efficiency, in *Proceedings of the 1st International Conference on Emerging Trends in Energy Conservation–ETEC, Tehran, Iran*, (Civilica, Tehran, 2011, November), pp. 20–21
40. Y. Jiang, J.A.A. Qahouq, M. Orabi, Matlab/Pspice hybrid simulation modeling of solar PV cell/module, in *Applied Power Electronics Conference and Exposition (APEC), 2011 Twenty-Sixth Annual IEEE*, (IEEE, Fort Worth, 2011, March), pp. 1244–1250
41. H. Jiang, L. Lu, K. Sun, Experimental investigation of the impact of airborne dust deposition on the performance of solar photovoltaic (PV) modules. Atmos. Environ. **45**(25), 4299–4304 (2011)
42. S. Mekhilef, R. Saidur, M. Kamalisarvestani, Effect of dust, humidity and air velocity on efficiency of photovoltaic cells. Renew. Sust. Energ. Rev. **16**(5), 2920–2925 (2012)
43. S. Chandra, S. Agrawal, D.S. Chauhan, Effect of ambient temperature and wind speed on performance ratio of polycrystalline Solar photovoltaic module: An experimental analysis. International Energy Journal **18**(2) (2018)
44. S. Dubey, J.N. Sarvaiya, B. Seshadri, Temperature dependent photovoltaic (PV) efficiency and its effect on PV production in the world–a review. Energy Procedia **33**, 311–321 (2013)

45. A. Shiroudi, R. Rashidi, G.B. Gharehpetian, S.A. Mousavifar, A. Akbari Foroud, Case study: Simulation and optimization of photovoltaic-wind-battery hybrid energy system in Taleghan-Iran using homer software. Journal of Renewable and Sustainable Energy **4**(5), 053111 (2012)
46. H.R. Baghaee, M. Mirsalim, G.B. Gharehpetian, H.A. Talebi, Reliability/cost-based multi-objective Pareto optimal design of stand-alone wind/PV/FC generation microgrid system. Energy **115**, 1022–1041 (2016)
47. Prévost, C., Hybrid PV-Biomass Power Plant Design for an Indonesian Village (KTH School of Industrial Engineering and Management, Stockholm, 2018)
48. J.M. Bright, S. Killinger, D. Lingfors, N.A. Engerer, Improved satellite-derived PV power nowcasting using real-time power data from reference PV systems. Sol. Energy **168**, 118–139 (2018)
49. A.H. Al-Waeli, A.H. Al-Kabi, A. Al-Mamari, H.A. Kazem, M.T. Chaichan, Evaluation of the economic and environmental aspects of using photovoltaic water pumping system, in *9th International Conference on Robotic, Vision, Signal Processing and Power Applications*, (Springer, Singapore, 2017), pp. 715–723
50. I. Colak, R. Bayindir, A. Aksoz, E. Hossain, S. Sayilgan, Designing a competitive electric vehicle charging station with solar PV and storage. 2015 IEEE International Telecommunications Energy, 2015, pp. 1–6
51. A.K. Tiwari, V.R. Kalamkar, Effects of total head and solar radiation on the performance of solar water pumping system. Renew. Energy **118**, 919–927 (2018)
52. E. Biyik, M. Araz, A. Hepbasli, M. Shahrestani, R. Yao, L. Shao, E. Essah, A.C. Oliveira, T. del Caño, E. Rico, J.L. Lechón, A key review of building integrated photovoltaic (BIPV) systems. Eng. Sci. Technol. Int. J. **20**(3), 833–858 (2017)
53. solarpowerworldonline. NREL: This parabolic trough 73% efficient By Solar Power Engineering (2010, Sep 3rd), https://www.solarpowerworldonline.com. Accessed 2nd Feb 2019
54. Selvakumar, N., Barshilia, H.C. and Rajam, K.S., Review of Sputter Deposited Mid-to High-Temperature Solar Selective Coatings for Flat Plate/Evacuated Tube Collectors and Solar Thermal Power Generation Applications, (National aerospace laboratories, Bangalore, 2010)
55. S. Jaisankar, J. Ananth, S. Thulasi, S.T. Jayasuthakar, K.N. Sheeba, A comprehensive review on solar water heaters. Renew. Sust. Energ. Rev. **15**(6), 3045–3050 (2011)
56. E. Zambolin, D. Del Col, Experimental analysis of thermal performance of flat plate and evacuated tube solar collectors in stationary standard and daily conditions. Sol. Energy **84**(8), 1382–1396 (2010)
57. VREC Solar. What is the best type of solar thermal panel in terms of performance? http://www.vrec.ca, Accessed 23rd Feb 2019
58. M. Fan, S. You, X. Gao, H. Zhang, B. Li, W. Zheng, L. Sun, T. Zhou, A comparative study on the performance of liquid flat-plate solar collector with a new V-corrugated absorber. Energy Convers. Manag. **184**, 235–248 (2019)
59. M.A. Karim, M.N.A. Hawlader, Performance investigation of flat plate, v-corrugated and finned air collectors. Energy **31**(4), 452–470 (2006)
60. A. Alvarez, O. Cabeza, M.C. Muñiz, L.M. Varela, Experimental and numerical investigation of a flat-plate solar collector. Energy **35**(9), 3707–3716 (2010)
61. D. Rojas, J. Beermann, S.A. Klein, D.T. Reindl, Thermal performance testing of flat-plate collectors. Sol. Energy **82**(8), 746–757 (2008)
62. J. Aleksiejuk, A. Chochowski, V. Reshetiuk, Analog model of dynamics of a flat-plate solar collector. Sol. Energy **160**, 103–116 (2018)
63. G. Ampuño, L. Roca, J.D. Gil, M. Berenguel, J.E. Normey-Rico, Apparent delay analysis for a flat-plate solar field model designed for control purposes. Sol. Energy **177**, 241–254 (2019)
64. Z. Tian, B. Perers, S. Furbo, J. Fan, Thermo-economic optimization of a hybrid solar district heating plant with flat plate collectors and parabolic trough collectors in series. Energy Convers. Manag. **165**, 92–101 (2018)

65. S. Farahat, F. Sarhaddi, H. Ajam, Exergetic optimization of flat plate solar collectors. Renew. Energy **34**(4), 1169–1174 (2009)
66. F. Jafarkazemi, E. Ahmadifard, Energetic and exergetic evaluation of flat plate solar collectors. Renew. Energy **56**, 55–63 (2013)
67. N.M. Villar, J.C. López, F.D. Muñoz, E.R. García, A.C. Andrés, Numerical 3-D heat flux simulations on flat plate solar collectors. Sol. Energy **83**(7), 1086–1092 (2009)
68. J.P. Chiou, The effect of nonuniform fluid flow distribution on the thermal performance of solar collector. Sol. Energy **29**(6), 487–502 (1982)
69. S.K. Verma, A.K. Tiwari, S. Tiwari, D.S. Chauhan, Performance analysis of hybrid nanofluids in flat plate solar collector as an advanced working fluid. Sol. Energy **167**, 231–241 (2018)
70. Z.A.K. Baharin, M.H. Mohammad, Experimental investigations on the performance of a single slope solar still coupled with flat plate solar collector under Malaysian conditions. J. Mech. Eng. **5**, 16–24 (2018)
71. S.A. Sakhaei, M.S. Valipour, Performance enhancement analysis of the flat plate collectors: A comprehensive review. Renew. Sust. Energ. Rev. **102**, 186–204 (2019)
72. P. Raj, S. Subudhi, A review of studies using nanofluids in flat-plate and direct absorption solar collectors. Renew. Sust. Energ. Rev. **84**, 54–74 (2018)
73. A.M. Genc, M.A. Ezan, A. Turgut, Thermal performance of a nanofluid-based flat plate solar collector: A transient numerical study. Appl. Therm. Eng. **130**, 395–407 (2018)
74. M. Carmona, M. Palacio, Thermal modelling of a flat plate solar collector with latent heat storage validated with experimental data in outdoor conditions. Sol. Energy **177**, 620–633 (2019)
75. M.M.A. Khan, N.I. Ibrahim, I.M. Mahbubul, H.M. Ali, R. Saidur, F.A. Al-Sulaiman, Evaluation of solar collector designs with integrated latent heat thermal energy storage: A review. Sol. Energy **166**, 334–350 (2018)

Chapter 2
PV/T Principles and Design

2.1 Background

In this chapter, a general overview of photovoltaic thermal (PV/T) collectors is presented by first introducing the technology and the rationale behind it and then providing a brief chronology and classification of PV/T collectors. In addition, more detailed descriptions of the design aspects and performance evaluation criteria are provided. The bulk of this chapter is made over two parts which are (i) detailed analysis of electrical and thermal performance of PV/T, which is carried out to highlight the critical parameters affecting systems' behavior, and (ii) a comprehensive literature review of PV/T systems from different classification parameters such as different base fluids, absorber designs, etc. The theory and literature review of this chapter are provided to gain proper understanding of the dynamic aspects of PV/T collectors.

2.2 Introduction

As briefly mentioned in Chap. 1, hybrid PV/T collectors (also abbreviated PV-T or PVT) can serve in maintaining the electrical efficiency of photovoltaic modules during daytime by cooling them [1]. The literature review on effect of rise in cell temperature on PV conversion efficiency is provided in Chap. 1 as well.

According to many studies in the field [2–5], increase in cell temperature has led to decrease in the voltage of the PV module, which has caused reduction in overall electrical efficiency. To counteract these phenomena, cooling methods are exercised to maintain the voltage and hence electrical efficiency. The use of PV/T collectors for cooling is not only helpful to maintain electrical yield of PV modules but also offers thermal yield, as it employs a solar thermal absorber. Electrical power and heating of fluids simultaneously, if cost-effective, can be very beneficial to consumers.

A. H. A. Al-Waeli et al., *Photovoltaic/Thermal (PV/T) Systems*,
https://doi.org/10.1007/978-3-030-27824-3_2

This chapter aims to present the theory and fundamentals of PV/T collectors, as well as review most significant developments in this research field. The focus of literature review will be directed to type of fluid and absorber design in PV/T systems. Thermal and electrical analyses are presented by observing the behavior of their parameters for different designs and operational conditions. In addition, conclusions and recommendations are made in the chapter summary section.

2.3 PV/T Concept

PV/T is a hybrid collector which combines the mechanism of solar thermal absorbers and photovoltaic modules. The collector aims to cool down the photovoltaic module through means of heat transfer while simultaneously using its heat to produce thermal energy. Figure 2.1 shows the components of a typical PV/T collector with a flat-plate absorber (numbered).

The components in Fig. 2.1 are typically used in every PV/T collector with some exceptions in the literature, for example, to the use of enclosure. Figure 2.2 shows another drawing of a typical PV/T collector.

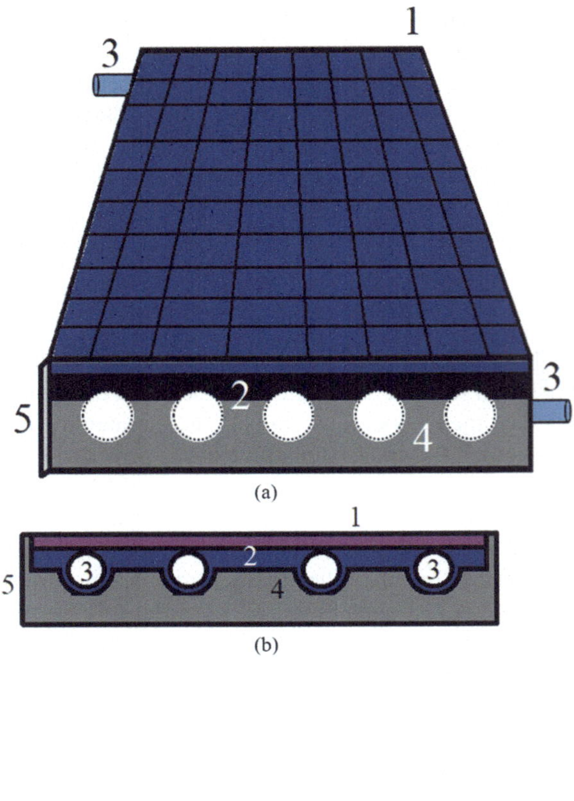

Fig. 2.1 Components of a typical PV/T collector (**a**) front 3D view drawing (**b**) cross-sectional drawing

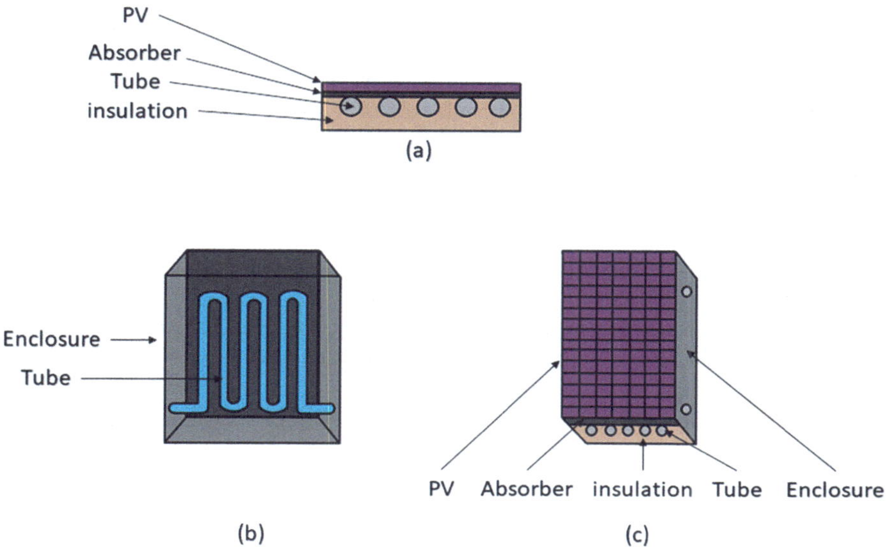

Fig. 2.2 Drawing of typical PV/T (**a**) cross-sectional side view, (**b**) inside the collector, and (**c**) 3D top view

1. Photovoltaic module: any type of photovoltaic modules (polycrystalline, monocrystalline, amorphous silicon, etc.).
2. Thermal absorber: attached to the back of the photovoltaic module, generally made of a metal with high thermal conductivity such as copper, aluminum, galvanized steel, etc. The absorber may refer to flow tubes or absorber plate and flow tubes. Different absorber designs both of tube geometry [6] and flow passage configuration [7] are used throughout the literature.
3. Inlet and outlet connections: pipes or hoses, either hard or flexible, are connected to the inlet and outlet connections which represent the two ends of the flow tubes in which fluid is conveyed.
4. Insulation: placed at the bottom and sides of the collector to minimize heat losses via conduction and convection. This is very important to maximize thermal yield, efficiency, and overall PV/T efficiency.
5. Enclosure: carry all collector components for mainly protection purposes and to ensure correct placement and hence operation of the PV/T collector.

The incident solar irradiance is partially absorbed by the photovoltaic module. Losses occur due to cover glass reflection and inner PV module losses. The PV module absorbs and converts the solar irradiance into electrical power, while its cell temperature begins to rise due to losses and heat of surrounding atmosphere. Some of this temperature is transferred back onto the surrounding atmosphere via convection, while the remaining is transferred to the thermal absorber via conduction. The base fluid will enter from a fluid source into the PV/T collector and flow within the flow passage or absorber tubes. This fluid will absorb the heat from the walls of the tubes and carry them into the outlet connection. Different components are used in a typical PV/T collector [8, 9]. The elements of the PV/T system are further illustrated in Fig. 2.2. Photograph of the back side of a PV/T collector is provided in Fig. 2.3 [10]. The heated working fluid can be used directly as thermal energy. For example, if the working fluid is water, then the produced hot water could be used directly for dish washing, hot shower, etc.

If a special type of working fluid is being used, such as ethylene glycol or nanofluids, it is necessary to use a heat exchanger to keep it within a closed-loop and separate it from clean water.

PV/T systems can be designed to be passive or active, also just like solar thermal systems, and could be made direct or indirect (closed-loop). Figure 2.4 shows different designs of PV/T systems.

Figure 2.4a shows a direct passive PV/T system which is dependent by natural convection to circulate the water. Active systems employ pumps and electrical components to exert circulation and can be either direct or indirect. Figure 2.4b shows a direct active PV/T system where pump circulates the water. The systems portrayed in Fig. 2.4b, c are very simplified; commercially more complex components are integrated in such systems such as thermostat control. Indirect systems are more suitable for freezing climates where a different type of fluid is used to avoid it freezing inside the pipes. This fluid gains heat and transfers it to water via heat exchanger.

Aside for the electrical efficiency of PV module described in Eq. (1.12) [11], the efficiency can be described using Eq. (2.1):

$$\eta_{el} = \frac{P_{max}}{G_s\, A_C} \tag{2.1}$$

where η_{el} is the electrical efficiency of the PV module, while P_{max}, G_S, and A_C represent the maximum power (W), solar irradiance (W/m^2), and collector area (m^2), respectively. The thermal efficiency of solar thermal collector is provided in Eq. 1.15.

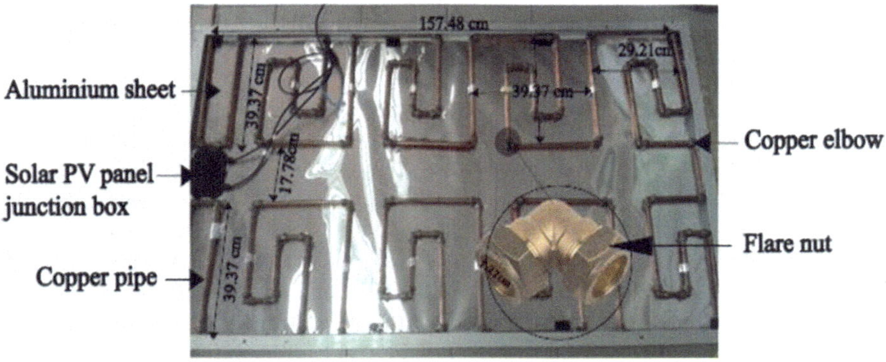

Fig. 2.3 Photograph of PV/T collector's back side and layout [10]

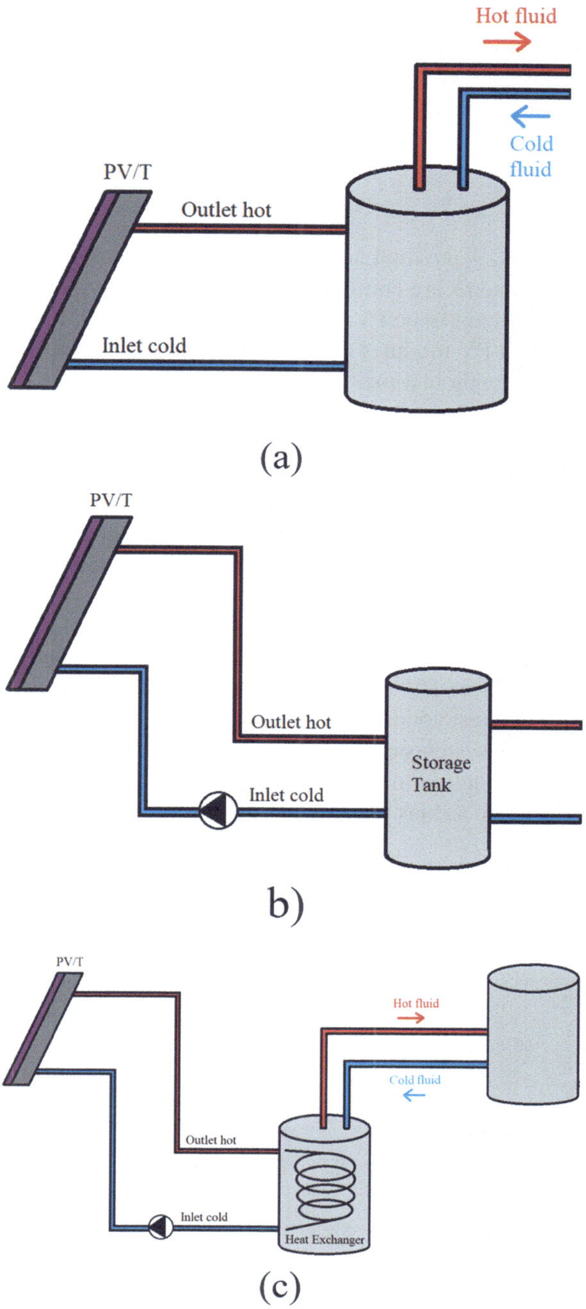

Fig. 2.4 Different types of PV/T systems (**a**) direct passive, (**b**) direct active, (**c**) indirect active

Given that both equations have the same denominator, the addition of both fractions can be done using algebraic sum of the nominator. Hence, the total efficiency of PV/T is the algebraic sum of the electrical efficiency (η_{el}) and thermal efficiency (η_{th}). Equation 2.2 shows the total PV/T efficiency [12]:

$$\eta_{PV/T} = \eta_{thermal} + \eta_{electrical} = \frac{Q_u + P_{max}}{G_S \times A_T} \tag{2.2}$$

where A_T is the total area of the collector. In this case it is the same area, given that both systems are combined.

The main issue PV/T collectors are intended to solve is the rise of cell temperature that PV modules experience which leads to performance degradation instantly and throughout time. Chow et al. [13] claimed a 10 °C rise in cell temperature causes a drop in electrical power of around 5% in crystalline silicon PV cells, which is inventible in regions with high ambient temperatures and solar irradiance levels.

Radziemska [14] claims a 0.65% decrease in output power per increase of 1 kelvin in temperature. In addition, conversion efficiency drops by 0.08% per kelvin. The size of this issue is highlighted well in Sect. 1.3.8. in Chap. 1. Figure 2.5 shows degradation of electrical power throughout time for different PV modules, from different manufacturers.

Across all produced types of PV modules, the degradation in electricity generation is quite significant. Hence, the role of PV/T is crucial to restore the electrical efficiency of PV modules to reasonable levels that approximate the efficiency at standard test conditions (STC).

PV/T systems can be designed to significantly improve electrical efficiency with some reduction in thermal and optical efficiencies of the thermal collector. This is referred to as bias towards electrical generation. Another mode is to design the PV/T

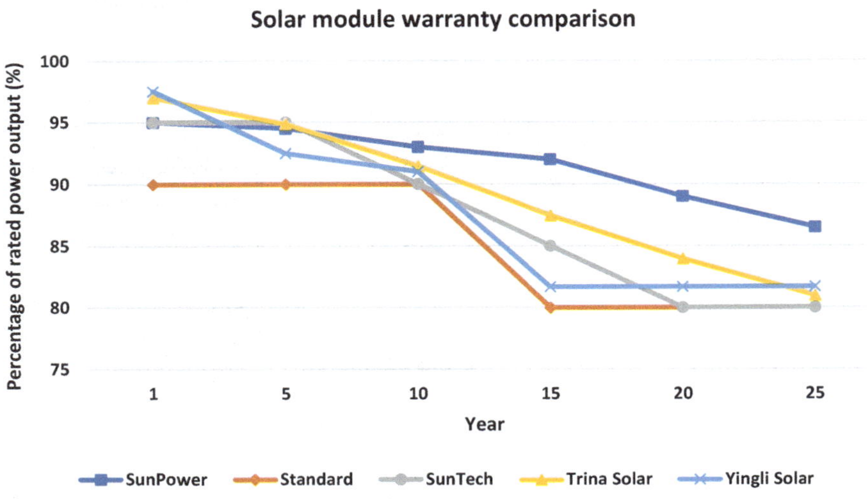

Fig. 2.5 Degradation in PV output power throughout its lifetime

to significantly improve thermal efficiency with some reduction in electrical efficiency of the PV module. This is referred to as bias towards thermal generation. The decision-making in this process is highly dependent on the nature of load (electric/thermal) and of solar potential for each type of collectors.

2.3.1 Rationale A: Enhanced Combined Efficiency per Area

One of the important parameters to consider is the installation area which if increased will lead to massive increase in costs. Due to this area being limited for residential installation, for example, rooftops, it is necessary to view the system with respect to combined efficiency generation.

Rationale (A) aims to compare between the combined efficiency of PV/T collector vs. separate PV and solar thermal collectors. Figure 2.6 shows a drawing illustration of the two scenarios.

From Fig. 2.6, a comparison between the two scenarios can be made mathematically. The rooftop area of the separate systems is described in Eq. 2.3, while that of PV/T is described in Eq. 2.4.

$$\text{Total collector area} = \text{Area of PV} + \text{Area of solar thermal} \tag{2.3}$$

$$\text{Area of PV / T collector} = \text{Area of PV} = \text{Area of solar thermal} \tag{2.4}$$

Hence, the total efficiency of the PV/T collector is the addition of the thermal and electrical efficiencies over the entire area, as shown in Eq. 2.2, while the total efficiency of separate PV and solar thermal collectors for the same area is divided by 2, given that the area is split in half, as shown in Eq. 2.5.

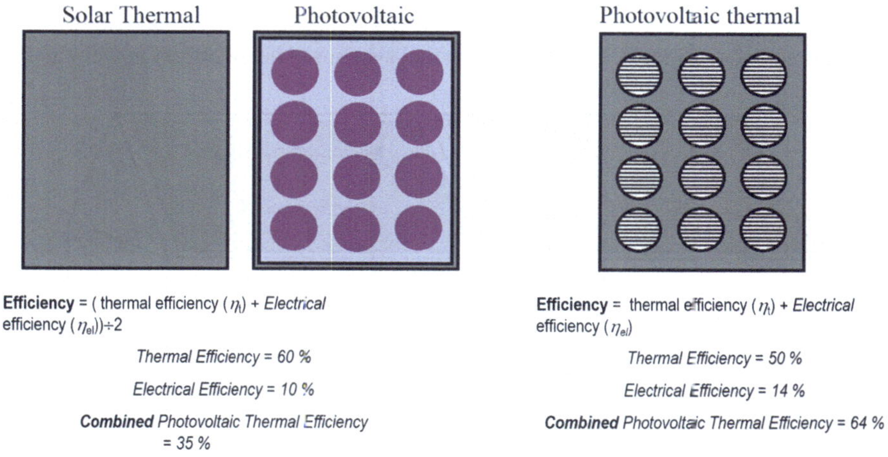

Fig. 2.6 Solar thermal and PV separate collectors vs. PV/T collector

$$\text{Total efficiency} = \frac{(\text{PV efficiency} + \text{solar thermal efficiency})}{2} \tag{2.5}$$

Hence, a hybrid PV/T collector is a better solution for overall conversion efficiency than separate PV and solar thermal systems which is a great solution for rooftop (residential) installation. However, if thermal energy generation is required, favorability of this system is debatable.

The use of thermoelectric generators could be studied as well, and final electrical power generation could be examined to check for the systems' cost-effectiveness.

The point of utilizing entire roof area for both electrical and thermal generation is further elaborated by Lämmle et al. [15]. In the study the authors used an empirical performance model of PV/T with four types of solar thermal systems in four European locations. The empirical model was validated and implemented into TRNSYS software. The simulation uses a flat-plate collector, reference PV module, PV/T with glazing, PV/T without it, and a glazed PV/T with low emissivity. It is found from the assessment that either high electrical or thermal yields are achieved by the systems. The highest overall yield is done by the glazed PV/T with low emissivity. The authors claim these collectors achieve three times the electric output of FPC and PV modules with equal thermal output. Figure 2.7 shows the limited area assessment of system for the two scenarios, where it appears to be consistent with Rationale A.

2.3.2 *Rationale B: Overall Cost-Effectiveness*

Given that PV modules are being cooled and voltage is being maintained when employing a hybrid PV/T collector, the power conversion efficiency of this technology is further protected over the devices' lifetime. The overall increase in efficiency generation means that for the renewable input of sunlight, more output can be generated than conventional PV and solar thermal systems. The electrical power can be

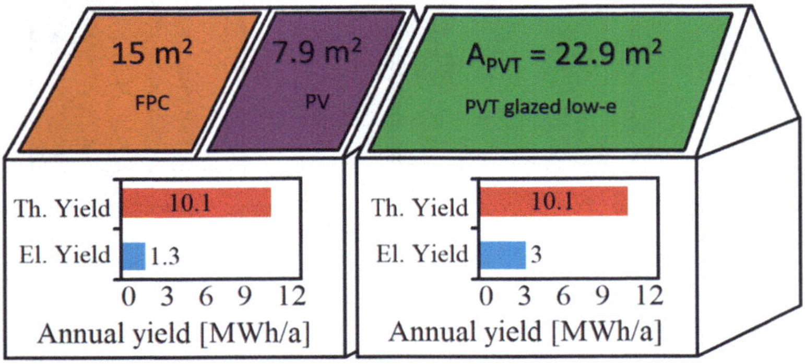

Fig. 2.7 Numerical results of limited area assessment of PV/T vs. PV and ST systems in Ref. [15]

sold back to the grid or considered financial gain by reducing the electricity bill costs. The thermal power can be calculated financially using the levelized cost of heat or consider financial gain by reducing needed energy for boilers, dryers, etc. For the same input, PV/T systems produce higher output which means they have better payback period (PBP), even if they are costlier than separately installed PV and FPC systems. Economic analysis and design are provided in Chap. 4.

PV/T appears to have better return on investment (ROI) than conventional PV or conventional FPC, separately. Hence, this technology under optimum conditions and proper design is more cost-effective which can be very useful to attract decision-makers, investors, and consumers to adopting solar and renewable energy. The aspect of cost-effectiveness is very important to tap into the energy market and compete against fossil fuel energy generation devices such as diesel generators.

The major benefits of PV/T systems are summarized below:

1. A broader utilization of the solar spectrum. Photovoltaic modules only capitalize on the visible spectrum, while thermal collectors absorb infrared waves as well. The use of both solar technologies in tandem can lead to more effective utilization of incident irradiance.
2. The structure design will be unified, and hence the price can be reduced for the hybrid unit installation, while total cost compared to installing two individual systems.
3. Total needed space of installation is lesser for PV/T than it is for each system, PV and FPC, individually.
4. PV/T systems increase the isolation of construction and surface shading during summer which can lead to reduction in thermal load.
5. Good visual impact is achieved when using a hybrid PV/T collector with a proper enclosure. It may not be the case if using flat-plate collectors and PV panels separately. However, combining the two allows to only display the outer layer (which is the top of PV module) to viewers of the buildings' roof.
6. The lifetime of the PV module is improved compared to a conventional PV module, given that it continuously experiences less cell temperature rise and hence better operating conditions. Throughout long periods the material can maintain its reference efficiency better than conventional PV modules.

The disadvantages of PV/T collectors are also summarized below:

1. The trade-off between electrical efficiency enhancement and optical efficiency reduction, or vice versa, cannot be eliminated. Losses in incident irradiance are inevitable if a PV panel is the cover of a flat-plate collector. The same can be said if the FPC is prioritized, where PV temperature may be reduced to accommodate the thermal collector process.
2. Although in theory it is economically better, given that installers have not adopted this technology, it is still costly in comparison to PV and FPC. The processes of installing PV and FPC or thermosiphon solar water heaters are quite understood and practiced throughout the solar energy industry. The same cannot be said about PV/T for many countries.

3. Although saving space is an advantage when employing hybrid PV/T units, it is noteworthy to mention that the weight of the combined PV/T is higher than that of separate PV and solar thermal systems. Higher weight may present some issues during installation and some limitations for the specific building it is intended to be installed on.
4. Although energy generation is higher for PV/T when comparing it to a conventional PV system, PV systems have the advantage of zero mechanical movement, emissions, and noise. The same cannot be said about PV/T given that in the active mode, the system needs to be driven by a pump.
5. For advanced designs which utilize nanofluids, the price of acquiring and mixing of those fluids is still quite high, and they are not readily available in the market. Hence, majority of studies use the price spent on preparation and testing.

Although the disadvantages may seem too many, it is noteworthy to mention that this technology is still fairly new, and many novel designs are being proposed at the moment (2019). Once commercialized and full-scale economy comes into effect, it is predicted to be more cost-effective and consumer-friendly. Research in PV/T systems has the potential in being the future of the solar market, which is described in Sects. 1.1 and 1.2 of Chap. 1.

Finally, from the literature review which is displayed in Sects. 2.7 and 2.8 of this chapter, there is an overwhelming proof and validation to the utility of hybrid PV/T collectors both theoretically and experimentally. The utility of PV/T is attributed to raising electricity generation and restoring thermal losses to capitalize on them for thermal energy generations.

2.4 Overview of PV/T Systems

Before laying out the methods of PV/T system design and performance analysis, both electrical and thermal, it is important to provide research background to show the chronology of this technology and classifications of PV/T systems. This section is subdivided into chronology and classifications.

2.4.1 Chronology

In this section, a brief chronology of PV/T technology is presented, highlighting some the main and most known published works and developments in PV/T technology up to the year 2002. Further developments and more recent works will be reviewed in Sects. 2.7 and 2.8.

Works on PV/T technology date to the early 1960s with Norman et al. [16] who patented the solar energy conversion device that is comprised of glass cover, tray containing fluid, solar cells, tubes, and a heat reservoir with heat-radiating fins. The inventors propose using water or ethylene glycol for the device. In 1975, Dean et al. [17] patented a photovoltaic with concentrator coupled with a heat sink, aimed to allow for maintaining cell efficiency when employing solar concentrator. The invention is composed of cylindrical type solar cells, refractor lens (concentrator), and heat sink with fins. In 1977, Varadi [18] patented the design of a novel PV panel for solar energy cells. The design consists of solar cells, resinous cushion, and enclosure wall. The concept of his invention is to transfer the heat generated by solar cell due to light impingement and power generation to the resinous cushion and onto the enclosure wall and then use inlet and outlet ducts with conduit to pass fluids such as water or air to remove the heat from the wall and generate hot water. Both Norman and Varadi presented hybrid and, at the time, novel designs that show the earliest work in PV/T technology. In the same year, the US government report authored by Aaron Kirpich et al. [19] described a hybrid system using PV/T collector as the energy source. The electrical aspect of the system utilizes a direct charge configuration where batteries are directly connected to the solar cell circuits. The thermal aspect of the design is a single Lexan pane collector panel as the heat source with a mixture of water and glycol as the working fluid which is forced using a pump. After the mixture gains heat, it is directed to the water thermal energy storage (TES) tank using a collector heat exchanger.

In 1978, Kern and Russell [20] experimentally tested hybrid PV/T collectors according to the ASHRAE standards. The study discusses hybrid heating and cooling. In addition, cost estimates were discussed which were used to derive initial, salvage, and lifetime costs for each component in the system and display system performance for different hybrid PV/T configurations under different climate conditions (different cities). A year after, Florschuetz [21] extended the FPC model known as "Hottel-Whillier" to describe the hybrid PV/T collector. Furthermore, the study presented and discussed examples of the thermal and electrical performance of hybrid PV/T as a function of collector design parameters. In 1981, Raghuraman and Hendrie [22] developed two separate one-dimensional models for analysis of flat-plate PV/T with water and with air. The analysis was compared to experimental tests. In 1986, Lalovic [23] developed and tested a hybrid PV/T with amorphous silicon-type PV module. The system consists of a glass cover, glass panel, amorphous layers, adhesive, aluminum fin, copper tube, and insulation. The system was tested outdoors in spring and summer. Garg and Agarwal [24] used finite difference method to solve equations for forced circulation water-based flat-plate PV/T.

In 1995, Johnson [25] invented a PV/T collector with a self-steering heliostat working as solar concentrator. The invention was made to operate as swimming pool heater and air drying. In the same year, Bergene and Lovvik [26] presented a physical model and quantitative prediction algorithms of a water-based, flat-plate PV/T collector to analyze energy transfer and predict amount of heat drawn and power output, respectively. The findings of the model show a total PV/T efficiency of 60–80%.

In 2002, Coventry [27] conducted a numerical simulation using a component of TRNSYS© software, which is based on a dynamic model, to investigate the performance of a concentrating PV/T collector. The equations of the model were used to describe the temperature-dependent energy flow between the collector and its surroundings. The simulation was validated through experiments conducted on the proposed system with a water-based flat-plate PV/T and a conventional PV, all three installed side by side.

It is clear from the highlighted works and studies in the field of PV/T that this technology is relatively new and that more effort into research and development should be afforded to fully develop PV/T systems and utilize their potential. Earlier works use experimental and mathematical intuitive analysis, while throughout time, more attention was paid to experimental and numerical analysis, using computer software. In earlier designs, PV/T systems were more passive in nature, while current designs are more active. The references mentioned in this section are merely a sample of the research, which has grown rapidly throughout the last decade, in PV/T field. Further developments are presented throughout this chapter. Figure 2.8 below shows a simple drawing depicting chronology of PV/T with major developments and achievements.

2.4.2 *Classifications*

Given that the main process that occurs with this type of collector is heat transfer, it is crucial to increase the amount of heat transfer through design and optimization. The components that significantly affect heat transfer are absorber shape and configuration, working fluid, and insulation. Given that there is not much design to be had for insulation, the research only focuses on the thermal absorber and working fluids. Because absorbers are designed to accommodate the working fluid, many authors classify PV/T systems according to the type of working fluid it uses. Figure 2.9a–c shows different ways to classify PV/T collectors.

The type of fluid is a major factor in classifying the type of the PV/T. In addition, the type of working fluid immensely affects the design of the PV/T collector.

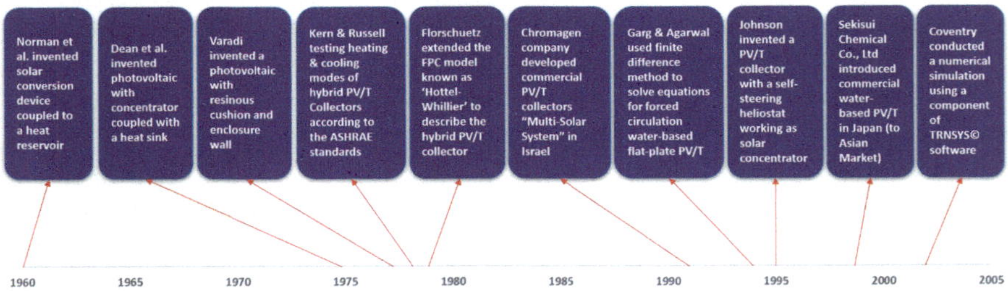

Fig. 2.8 PV/T pivotal moments

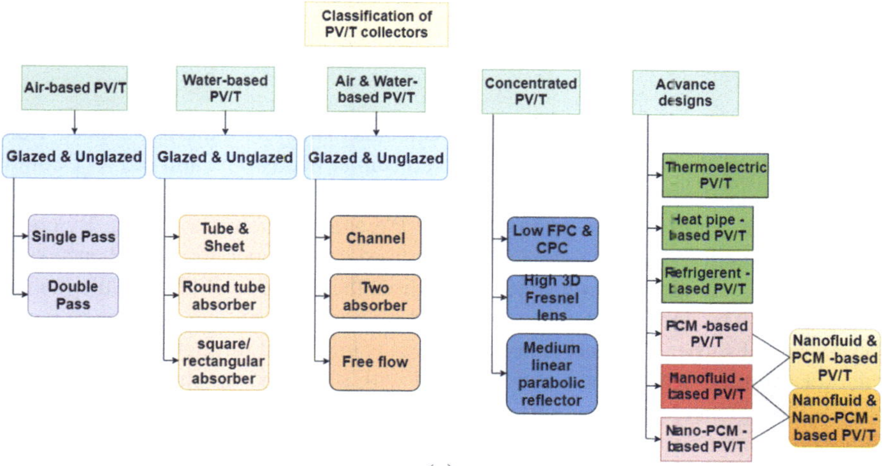

(a)

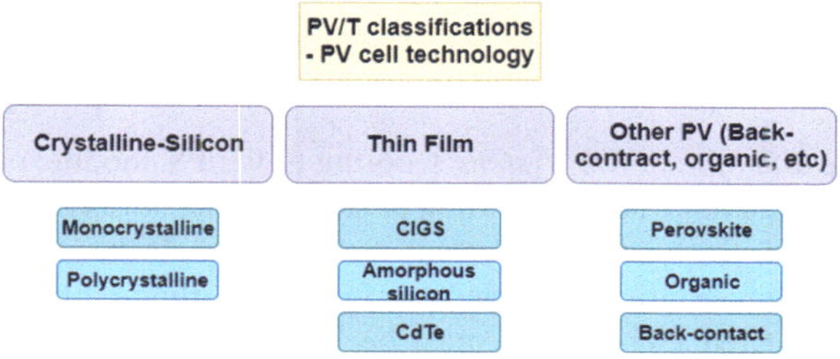

(b)

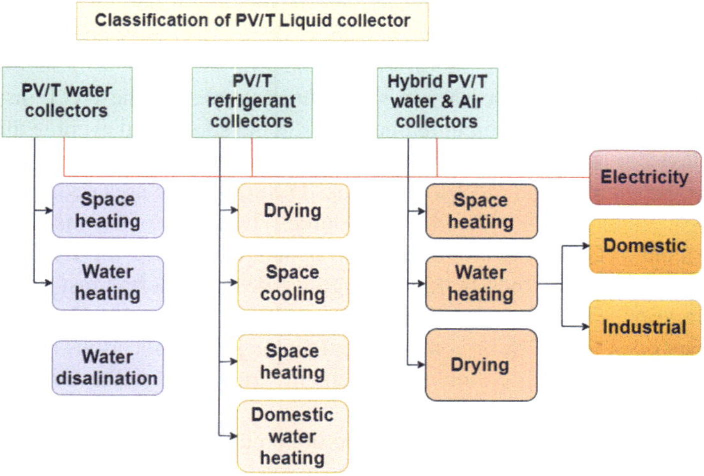

(c)

Fig. 2.9 Classifications of PV/T collectors (**a**) according to design, (**b**) according to PV cell technology, (**c**) according to application

2.5 Design of PV/T Systems

A. *Aims*

The PV/T collector could be designed with bias to the PV or thermal aspect. This is dependent on the load requirements and consumer needs. The aim of PV/T could be to only cool down the PV module using a passive cooling technique. PV/T can also be fabricated to provide both cooling of PV and producing hot water. By cooling, the PV module electrical generation is raised during operation, and hence system is more cost-effective over time. In addition, the life span of PV module device is enhanced. The cooling process is highly dependent on the heat transfer between the PV and absorber. The heat transfer process is improved through different methods, for example, using a heat transfer fluid with better thermophysical properties than water or increasing the surface contact between absorber and PV module. Other considerations include raising the mass flow rate, which is studied in detail in the next sections of this chapter.

B. *Design Considerations*

To design a PV/T system, it is important to consider the following steps:

1. *Deciding the aim of the PV/T system*: Cooling of the PV module or simultaneous cooling of PV and hot water generation of solar thermal absorber. The aim is also a major factor for the type of PV system configuration. If the consumer is aiming to use the system in grid-connected configuration, then equipment associated with grid-connected PV systems must be included such as an inverter, utility meter, AC circuit breaker, etc.
2. *Load profile*: it is important to understand the electrical/thermal demands. Once accurate assessment of the load profile is done, it is easy to size the PV/T system.
3. *Resources*: it is quite necessary to study the resources of the PV/T installation area. Solar irradiance, ambient temperatures, wind speed, and average number of rainy days per month are all highly important to accurately size the PV/T system to meet the thermal and electric demands. The size of the area dedicated for the PV/T system is also considered as a resource. The nature of the installation area is also a major factor. For example, if the system is intended for installation on a rooftop of a building, then building integrated design criteria must be taken into consideration. This would be different than a PV/T system intended for installation in an open field. The area will also affect the pipe connections between storage tank and PV/T collector, which in turn will affect the pumping requirements.
4. *System sizing*: the sizing process can be done either intuitively or numerically. Although no particular standards have been made for sizing of PV/T systems, a simple method could be to size each aspect separately. If limitations occur and compromise must be made to either system, it is preferred to refer back to the

aim and bias of the PV/T system and adjust accordingly. For optimal sizing, cost-effectiveness must be considered as well. This is similar to the concept of numerical analysis for optimal system sizing using HOMER software which is explained in Chap. 1.

5. *PV module*: The type of PV employed is preferred to be of high efficiency and good quality. For fabrication purposes it is important to use frameless PV modules to easily place within enclosure of PV/T collector. Some choose to keep the PV frame and utilize it for the final product (PV/T).
6. *Choosing the type of working fluid and absorber*: these two parameters significantly affect power generation and cost of the PV/T system. Careful consideration of the type of working fluid is a must. For example, using an antifreezing fluid as working fluid is very suitable for cold climates. For rural desert locations, the use of a fluid-based PV/T may be limited. Using passive air-based PV/T for cooling of PV module will be more optimal, although an assessment of wind resources must precede. Furthermore, the material of the absorber will affect the heat transfer process. In addition, the cost of the material should also be considered. Welding of certain metal is very complicated, and hence the possibility for complications in the design must be considered as well. For example, when welding copper or steel into aluminum, brittle intermetallic compounds appear, and hence special techniques must be executed to avoid such outcomes. We recommend for smart designs of enclosure/absorber to be flexible and allow for changing of absorber or even disassemble the collector if needed.
7. *System components*: other components may be included to improve the cooling and control of the system. More control is preferable to ease the operation and control process on the consumer. In addition, check valves, ball valves, and pressure relief valves are very important for safe and uninterrupted operation. It is useful to use drains to drain tanks and heat exchangers from fluid. For research purposes, both in academic and industrial sets, many measurement instruments will be employed to the system such as temperature sensors, flow meter, and power source devices.

C. *Evaluation Criteria*

To evaluate the PV/T collector, it is important for the evaluation to consider both electrical and thermal performance of the collector.

1. *Testing*: the electrical performance is evaluated by investigating parameters such as average values of voltage, current, power, efficiency, exergy, and average daily energy generation of the PV/T system. In case the PV/T is grid-connected, then specific yield and capacity factor must also be included in the evaluation. The thermal performance is evaluated by investigating parameters such as average outlet fluid temperature, exergy, optical and thermal efficiencies, and average daily generated thermal energy. In addition, peak values for both thermal and electrical parameters are important to view the limitation of the technology.

2. *Evaluation*: to assess whether the produced system is successful and cost-effective or not, it is important to do so in view of the aim of the system; as in whether it is biased towards electrical aspect or thermal one. Evaluation of performance alone is not enough; this must be coupled with cost analysis to arrive to the conclusion of its [PV/T] cost-effectiveness. Hence, another step is to analyze the money cash flow both into and out of the PV/T system and predict the life cycle costs, cost of energy, and payback period. This step is further illustrated in Chap. 4. Another step for evaluation is to compare all parameters of the PV/T system to a reference conventional PV system or solar thermal system. It is noteworthy to mention that evaluation can be done prior to system installation and post system installation. The latter is more accurate given that it shows actual values in real conditions. The former [system evaluation pre-installation] can be improved by considering more dynamic aspects and using real metrological conditions such as wind speed, solar irradiance, etc. These data would be recorded prior to installation as well, and hence for more accuracy, we recommend using prediction algorithm using artificial neural networks (ANN) to forecast the long-term metrological conditions.
3. *Assessment*: to view the systems' total efficiency, viability, and potential for the end consumer under optimal conditions, outdoor "real-world" condition, worst-case scenario conditions, and performance over long term. Optimal conditions would be those providing PV output at standard test conditions (STC) and providing flat-plate collector's optical efficiency. Outdoor conditions are those where variation in output of PV/T is due to variations of environmental conditions and other external factors. Worst-case scenario condition is those when the collectors' performance is at its lowest such as during rainy and cloudy days. Another worst-case scenario is when the absorber reaches stagnation temperature. Performance of PV/T over long term can be done either by studying the system for many years or by predicting the output of the system using prediction algorithm such as the prementioned ANN. Although long-term assessment falls more in the category of reliability and risk analysis, assessing the system allows for improving the technology in the future and comparing it to rival technologies.
4. *Development*: the development process is also an area only associated with academic, commercial, and industrial entities. The academic development usually focuses more on developing the concepts, while industrial development usually focuses on scaling the technology, large-scale production, and making it consumer-friendly which is also done by commercial entities, which sell this technology to the end consumer. The research and development are very important to further make this technology viable for commercialization. The industrial developments are illustrated thoroughly in Chap. 4, while the academic development is provided in this Chaps. 2 and 3. The academic research can also cover industrial development by joint research collaborations and numerical investigations using hypothetical data and/or real data from industrial entities. In terms of concept, the academic research aims to analyze the electrical and thermal behavior of the PV/T. A detailed analysis of PV/T is provided in Sect. 2.6.

2.6 Electrical and Thermal Analysis

The electrical and thermal analysis aims to evaluate the electrical and thermal performance, separately, in view of climatic conditions (such as solar irradiance wind speed, ambient temperature, etc.) and design parameters (such as absorber plate design configuration, glazing thickness, etc.), and operating parameters (such as mass flow rate, inlet and outlet fluid temperature, and packing factor). Table 2.1 shows the electrical and thermal analysis for some works in the literature.

2.7 Theory and Literature Review of Passive Cooling PV Systems

Passive cooling methods are used to (i) reduce the PV module temperature, (ii) maintain the temperature uniformity across the PV module, and (iii) increase the power conversion efficiency of the PV module. The idea of thermal regulation without the use of active elements, such as fans, blowers, or pumps, poses interesting advantages. Those advantages include typical simplicity of the design, less power consumption, and maintenance costs associated active components. Among the types of researched passive cooling techniques are (i) immersing the PV module in dielectric mediums; (ii) submerged water cooling of the PV module; (iii) use of buoyancy for airflow cooling; (iv) use of heat sink or dissipater, attaching it to PV module; (v) implementation of phase change material (PCM). and (vi) evaporator cooling technique; etc.

Cuce et al. [31] performed indoor experiments on two polycrystalline PV cells, one with fins (aluminum heat sink) and one without, to investigate the effects passive cooling on the PV electrical performance. Steady-state heat transfer analysis was used to determine the dimensions of the heat sink. Different climate parameters were varied such as solar irradiance, 400 W/m^2–800 W/m^2, and ambient temperature, 15 °C–35 °C. Using fins was found to significantly affect the PV cell under the set experimental conditions. For instance, the results show that at 200 W/m^2, the finned PV cell exhibits power and efficiency of 50.18 mW and 3.47%, respectively, while the PV cell without fins has power and efficiency of around 42.01 mW and 3.31%, respectively. Overall, the study shows an increase in power output of PV cell when employing cooling by 20%, at peak solar irradiance of 800 W/m^2. However, the maximum level of cooling is achieved at solar irradiance of 600 W/m^2. The increase of ambient temperature from 15 C to 35 C led to decrease in the efficiency of the PV with fins by 7.11% and the PV without fins by 7.7%. However, it is notable that the efficiency at 35 °C for the PV with fins and PV without fins is around 5.15% and 4.99%, respectively.

Natarajan et al. [32] performed a numerical study to investigate the effect of passive cooling on a PV with concentrator ratio of 10×. The focus of the study is the temperature of the PV, as authors developed a 2D thermal model which predicts the

Table 2.1 Electrical and thermal analysis

Reference summary		
Khelifa et al. [28] theoretically and experimentally studied a PV/T collector with sheet and tube absorber type. The mathematical model used in the study is based on the energy balance equations of different nodes across the system which are solved using finite difference method. The authors claim that using this collector led to PV cell temperature reduction which in turn led to increase of efficiency. Hence, the highest cell temperature is 53 °C, while the overall efficiency of the PV/T collector is around 69% which consists of 55% thermal efficiency and 14% electrical efficiency. The system was installed in a research facility in Algeria where the PV/T was mounted at a 32° tilt angle and mass flow rate fixed at 0.025 kg/s		
Remarks	**Parameter**	**Evaluation parameter**
The electrical power produced by the PV/T corresponds to the solar irradiance levels, where it peaks at around 40 W at 12 o'clock which is exact time of peak for direct solar irradiance of around 740 W/m^2. The electrical efficiency curve (over time of experiments) was similar to that of the thermal efficiency curve, in shape. The electrical efficiency of the PV/T system peaked at around 14.8%. From this, the authors claim that generated electrical energy is a function of absorber plate temperature and that lowering the absorber temperature leads to increase in PV/T energy generation	Solar irradiance	Climatic condition
The findings by Khelifa et al. [28] show that the thermal production of PV/T also corresponds to increase of solar irradiance where it peaks at around 350 W at 13 o'clock which is not where peak solar irradiance occurs which could be due to effects of other environmental components. The optical efficiency of the PV/T collector was found to be 64%. Furthermore, the overall trend of the thermal production shows increase with increase of solar irradiance. The outlet fluid peaks at 37 °C around 13 o'clock		
The thermal efficiency drops with increase in difference between inlet fluid temperature and the ambient temperature. Where at a difference of 0.04, the efficiency is at 64%, while for an increase difference of around 0.09, the thermal efficiency reaches around 26%	Inlet water temperature	Operating parameters
Khelifa et al. [28] numerically compared between a sheet and tube absorber PV/T and a serpentine absorber PV/T where the sheet and tube exhibits an optical efficiency of 60%, while serpentine's optical efficiency is around 65%. However, their behavior is quite different when increasing the difference between inlet and ambient temperature to 0.09, where sheet and tube absorber's thermal efficiency is around 37% which is higher than that of the serpentine's 30%	Absorber geometry and configuration	Design parameters

(continued)

Table 2.1 (continued)

Reference summary		
Sarhaddi et al. [29] investigated the performance of an air-based PV/T collector in terms of electrical and thermal aspects by developing mathematical models. Among parameters involved in the investigation are cell temperature, outlet air temperature, back surface temperature, maximum power point current and voltage, short circuit current, and voltage. The authors put much effort to improve the precision of the electrical and thermal models (e.g., corrections on heat loss coefficient, estimation of electrical parameters of PV module, etc.). The models were executed using computer simulation program. The main components of the PV/T air collector are glass (outer layer), solar cell, Tedlar, air passage, and insulation material. The simulation of the collector was experimentally validated using real data from an external reference, whom have measured all mentioned parameters. The authors declare a good agreement between experimental data and numerical simulation results. The results show that electrical, thermal, and overall efficiencies are around 10.01%, 17.18%, and 45%, respectively. Hence, the authors declare that electrical efficiency of the PV/T air collector discussed in the study has only slight change with respect to the design and operating parameters. Furthermore, the increase of either inlet air temperature, duct length, or wind speed causes decrease in thermal efficiency and overall PV/T efficiency. On the other hand, increase of inlet air velocity leads to increase of thermal and overall PV/T efficiencies. Also, increase in solar irradiance leads to initial increase in electrical and overall energy efficiency and then decrease		
Remarks	**Parameter**	**Evaluation parameter**
Increase of solar irradiance causes an increase in overall energy efficiency and electrical efficiency from 37% to 48.4% and from 6.5% to 11%, respectively. After solar irradiance reaches 160 W/m^2, both efficiencies then decrease to 42.7% and 9.3%, respectively. Still, the efficiency in this case is still higher than initial cases	Solar irradiance	Climatic condition
Sarhaddi et al. [29] state that increase of solar irradiance causes only a slight change, landing at 18%, in thermal efficiency		
Increase of the inlet air temperature did not cause much change in the electrical efficiency of the air-based PV/T system, where it remained around 10%, which is not the case for the thermal and overall energy efficiencies	Inlet air temperature	Operating parameters
According to Sarhaddi et al. [29], increase of the inlet air temperature was found to cause a decrease in the thermal and overall energy efficiencies. Raising the inlet air temperature from 300 k (26.85 °C) to 315 K (41.85 °C) caused thermal efficiency to drop from 17.2% to 9.2% and the overall energy efficiency to drop from 45% to 36.4%. This drop is mainly due to increased difference between inlet air temperature and ambient temperature		
As shown in the results, the electrical efficiency was not impacted by the change in inlet air velocity and remained at 10%	Inlet air velocity	Operating parameters
The air velocity increase is correlated to increase in the thermal and overall energy efficiency. Where raising the velocity from 0.001 m/s to 10 m/s leads to increasing thermal efficiency from 0% to 41% and overall efficiency from 26.5% to 71%		

(continued)

Table 2.1 (continued)

The electrical efficiency of an air-based PV/T collector experienced slight change with respect to increase of duct length form 1.2 m to 6 m to reach around 10.2%	Duct length	Design parameters
Increasing the duct length from 1.2 m to 6 m leads to decrease in overall energy efficiency and thermal efficiency of the air-based PV/T collector from 46% to 38.5% and from 17% to 10%, respectively		
Increase of wind speed led to increase in the electrical efficiency of the air-based PV/T collector, where wind speed rise of 10 m/s led to increase in efficiency from 9.5% to 11%	Wind speed	Climatic condition
Increase of wind speed from 0 to 10 m/s led the thermal and overall efficiencies to drop from 20.7% to 6.7% and 47.5% to 37.5%, respectively		
Reference summary		
Gang et al. [30] proposed a heat pipe-based PV/T system for simultaneous electricity and thermal energy generation. The system was compared to a conventional water-based PV/T system. The authors developed a dynamic model for systems' performance prediction and validated the model outcome with experiments. The collector is comprised of PV cell, heat pipe, black TPT layer, aluminum plate, and insulation layer. The system is composed of four collectors (of the proposed design), storage tank, circulation pump, and controller. The comparison between simulation and experiments shows satisfying predictions. The findings of the study show electrical and thermal efficiency of the heat pipe-based PV/T system to be around 41.9% and 9.4%, respectively. In addition to heat and electrical gains of 276.9 W/m^2 and 62.3 W/m^2, respectively. Furthermore, based on the second law of thermodynamics, the average total second law efficiency of the proposed is 6.8%		
Remarks	**Parameter**	**Evaluation parameter**
Overall, increase in solar irradiance is correlated with increase of the electricity gain and efficiency of the heat pipe-based PV/T system. However, ambient temperature affects the electrical performance as well. At 13:30 the maximum panel temperature is observed, while PV cell temperature is recorded around noon. At noon, the cell temperature was higher than that of the panel, whereas this changed during afternoon, as the panel temperature became higher. This is attributed to high solar irradiation in the morning and weak levels during afternoon. Hence, the PV temperature reached around 54 °C at approximately 12:09 which corresponds to beginning of drop in electrical efficiency and gain	Solar irradiance	Climatic condition
Increase in solar irradiation levels causes higher temperature for the outlet water from the collector; however, it is observed to fluctuate while increasing. This is mainly due to fluctuation in solar irradiance, which verifies solar irradiance effect on thermal component of PV/T. the average heat gain per unit area and total heat gain, for the tested period, are around 276.9 W/m^2 and 1272.8, respectively		
Increase of inlet water temperature across time led to decrease in electrical efficiency for the heat pipe-based PV/T collector. This is because increase of water temperature affected both panel and cell temperature, which resulted in heat thermal loss into the surroundings. In addition to the decrease in solar irradiation levels, this led to decrease in electrical efficiency	Inlet water temperature	Operating condition

temperature of PV with and without the proposed cooling arrangement. The proposed collector is composed of Fresnel lens, encapsulation, solar cell, backplate, and fins. The temperature distribution of seven cases was displayed. The investigated parameter includes number of fins, fin thickness and height, and spacing between two consecutive fins. The results of the simulation show that the highest reduction in cell temperature is achieved using four uniform fins with height and thickness of 5 mm and 1 mm, respectively.

Araki et al. [33] proposed the use of printed epoxy and copper sheet on aluminum plate as part of simple passive cooling method of a solar cell with 500× concentrator (Fresnel lens). Given that the incorporation of Fresnel lens could be helpful to increase solar irradiance, still drop in efficiency is observed due to rise of cell temperature. This prompts the use of a cooling mechanism for such systems and, hence, this study. Outdoor experiments were conducted to confirm the proposed cooling method. Heat analysis of cell temperature was conducted. The design is based on the heat spreading concept. The design is composed of the base plate which is made from aluminum. Printed on it is an epoxy that is filled with heat-conductive agents. The authors compared between a normal flat plate and a concentrated module. The results of the heat analysis showed a 10 ° C rise in surface cell temperature compared to conventional flat plane. Further heat dissipation was established using the metal frames. Temperature rise for the concentrated cell reached 18° to ambient and 10° to uniformly illuminated aluminum plate. Efficiency drop by 2% occurred when cell temperature increased and without help of heat sinks. Only rising 18° is in fact a good outcome of the cooling, especially under concentration ratio of 18×.

Tonui and Tripanagnostopoulos [34] presented a modified air-cooled PV/T with thin flat metallic sheet at the middle or fins at the back wall of the air absorber. In addition, they studied the difference in performance by forced and natural airflow. Figure 2.10 shows the schematic diagram showing temperature sensor positions. The electrical efficiency as a function of temperature is provided in Fig. 2.11. A comparison of steady-state thermal efficiencies between conventional air-based PV/T and the proposed type is presented as well. Different temperatures and energy yields are recorded for the proposed design to highlight its utility for PV systems. The results show temperature rise up to 12 ° C in sunny days for natural flow. As for forced airflow rate of 60 $m^3\ h^{-1}$, channel depth of 15 cm and utilizing fins produce efficiency of 30%. For the same airflow rate and channel depth, the thin metallic sheet leads to 28%, and the conventional air-based PV/T yields 25%. In addition, for all tested systems, the increase of airflow rate from 30 $m^3\ h^{-1}$ to 100 $m^3\ h^{-1}$ led to the increase in thermal efficiency. The authors conclude that the suggested modifications lead to better electrical and thermal performance for the overall system. A summary of studies carried out on passive cooling of PV modules is shown in Table 2.2.

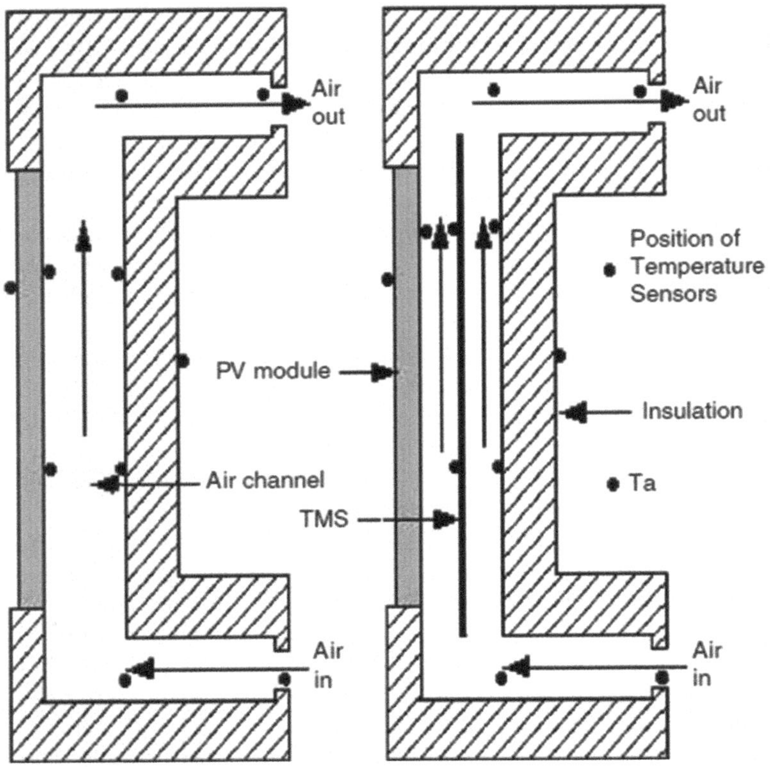

Fig. 2.10 Schematic diagrams showing temperature sensor position for REF (left) and TMS systems (right) [34]

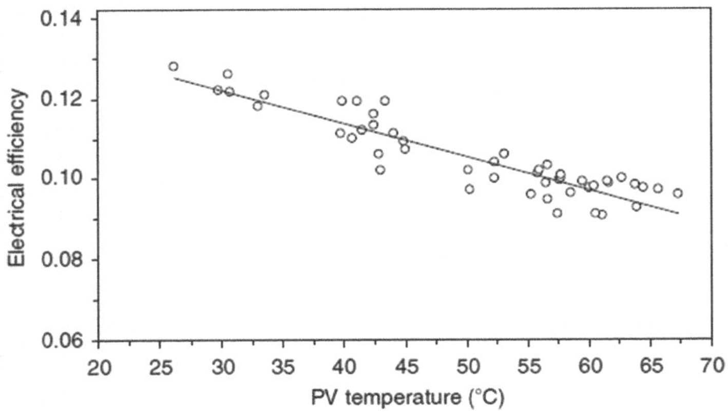

Fig. 2.11 Electrical efficiency as a function of PV temperature [34]

Table 2.2 Summary of research on passive cooling

Reference	Year	Location	Type of PV	Type of cooling method	Peak solar irradiance	Remarks
Cuce et al. [31]	2011	Turkey	Polycrystalline	Fins – aluminum heat sink	800 W/m^2	Electrical efficiency achieved 7.23%
Natarajan et al. [32]	2011	United Kingdom	Polycrystalline	Fins	1000 W/m^2	Minimum cell temperature ≈ 50 °C
Araki et al. [33]	2002	Japan	III–V multi-junction cell	Heat spreading structure	800 W/m^2	Temperature only rose 18° for under concentration ratio of 18×
Tonui and Tripanagnostopoulos [34]	2007	Greece	Polycrystalline	Fins and thin metal sheet	700 W/m^2	Electrical efficiency range [9–12.5%]

2.8 Theory and Literature Review of PV/T Systems with Various Working Fluids

This section displays a review of the design and analysis of PV/T systems according to their type of working fluid. A breakdown of main features and design considerations associated with each type is provided, along with a review of recent works in the literature. This section is further elaborated in Chap. 6.

2.8.1 Studies on Air-Based PV/T

Air is the circulating fluid or working fluid with these types of collectors. It is most advantageous for hot air demands such as drying applications, etc. The main advantage of using air as appose to water is its lower costs, lower risk of freezing or boiling, and furthermore, low risk for damage in case a leakage occurs. However, the disadvantage of this type of collectors is its lower thermal performance characteristics, relative to water-based or other liquid-based PV/T systems. This is because air exhibits lower heat capacity, and hence lower heat transfer occurs when circulating it through the tubes or pipes. Moreover, the low density of air leads to having significantly higher transfer volume than liquid- or water-based types. Given that higher volume is needed, the pipes or tubes would be bulkier and ultimately not suitable for applications in small areas, neither it is aesthetically pleasing. Despite these issues, it remains to be an appropriate choice for systems with air heating demands, and it does exhibit lower costs.

Ahn et al. [35] presented an air-type PV/T collector which is aimed to assist a heat recovery ventilation (HRV) system by using it, the PV/T, essentially for preheating the air and providing it as intake to the HRV system instead of outside air. The authors claim that coupling the two systems is expected to increase the efficiency of ventilation. The system was tested experimentally in building by employing a 1 kWp array consisting of four air-based PV/T collectors, installed at inclination of 30° and facing south. To view the difference between conventional HRV system and PV/T-assisted HRV system, the authors created the two modes, outside air fed directly into HRV and PV/T's outlet air fed into HRV and referred to them as OA mode and PV/T mode, respectively. The results of the study show that HRV system exhibited a heat transfer efficiency of 80%, where the heat transfer efficiency was improved by around 20%, the authors claim. Moreover, the PV/T collector showed thermal and electrical efficiencies amounting to 23% and 15%, respectively. Hence, the study concludes that higher efficiency is yielded when coupling an air-based PV/T collector into an HRV system.

Hu et al. [36] proposed and experimentally tested the performance of an ETFE (ethylene tetrafluoroethylene) cushion roof-integrated PV/T system which utilizes air as working fluid. Figure 2.12 shows the system configuration. The experiments were made for a period of 8 months, with around 6–7 hours of operation daily. The system was found to operate in a smooth and steady manner. Furthermore, 4 days were chosen for the performance investigation, including both days of summer and winter. The authors emphasize the importance of selecting clear sky days. The experimental procedure was subdivided into three main steps, which is to first connect batteries to the PV module and blow to the solar controller.

Second step was to connect the programmable logic controller (PLC) to the temperature patrol instrument and to the computer. Figure 2.13 shows photographs from the experiments. The third and last step is to check the pressure to confirm if system is in steady state. The results indicate that CIPV/T system produces an average total and net electricity of 54.5 Wh and 42.9 Wh, respectively. Furthermore, the system is found to achieve an overall or virtual system efficiency of 25.5%.

Rounis et al. [37] numerically introduced a multiple-inlet air-based building integrated (BIPV/T) system for improved electrical yield and heat extraction. The simulation was done to compare between single and multiple-inlet air-based BIPV/Ts and accounting for variations of wind conditions during summer and winter and at different airflow rates. The assumed PV module power is 300 W and forming an array sized 120 kW. The proposed systems are single-inlet with 0.1 m channel gap, single-inlet with 0.15 m channel gap, multiple-inlet with 0.1 m channel gap, and multiple-inlet with 0.15 m channel gap, where they were referred to as systems I, II, III, and IV, respectively. The results of the simulations show how single-inlet absorbers have lower overall performance for both electrical and thermal yields and efficiencies. Multiple-inlet system exhibits increase in electrical and thermal efficiencies of 1% and 14–25%, respectively. The authors declare that fan electricity consumption and net electricity production need to be investigated.

Kim et al. [38] designed, tested, and analyzed the performance of an air-based PV/T system. The experiments were conducted in outdoor environment, and the

Fig. 2.12 Schematic diagram of the proposed system

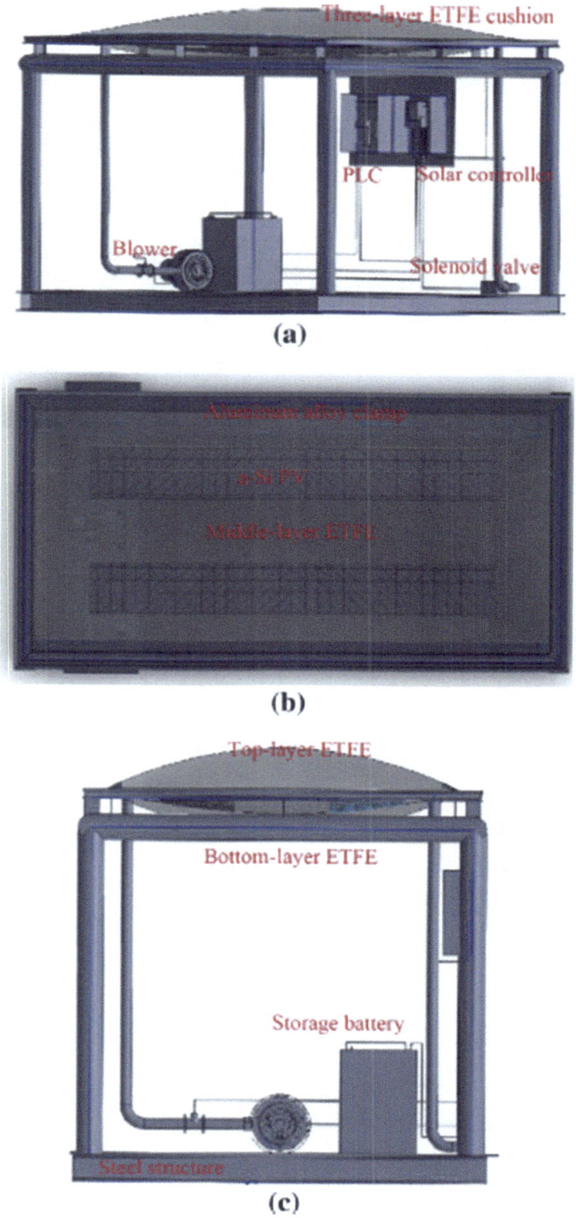

experimental setup was established and installed at 35° tilt angle, facing south. The temperature difference between the inlet air and outlet air is around 5 °C which makes the system suitable for preheating of fresh air for HRV system. The study shows that the average electrical and thermal efficiencies are around 15% and 22%, respectively. The electrical performance of the PV/T is described in terms of solar

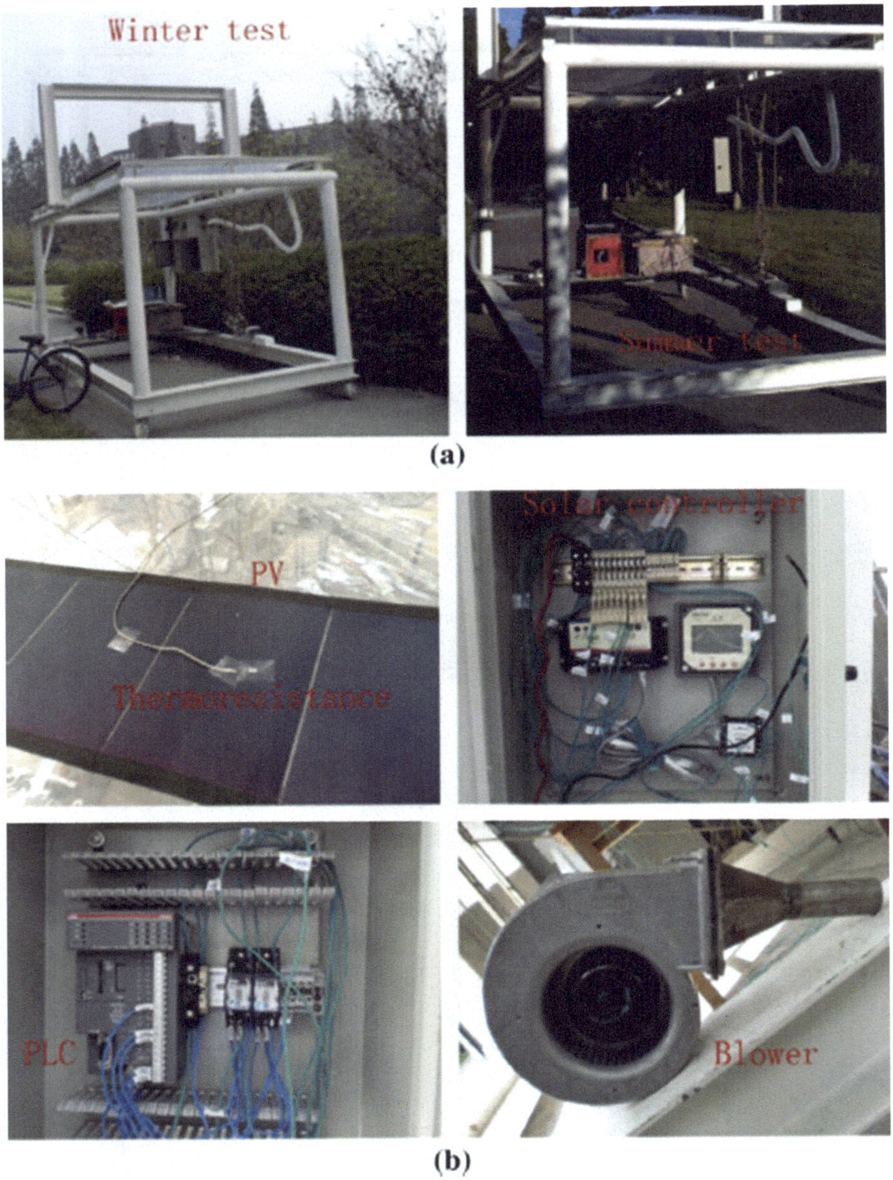

Fig. 2.13 Photographs from the experiments (**a**) test mockup (**b**) equipment [36]

irradiance, where increase of solar irradiance leads to increase in the electrical efficiency, yet at a certain point, it leads to increase of PV laminate temperature and hence decreases the electrical efficiency. Peak electrical efficiency was found to be 16% at solar irradiance levels of 750 W/m^2.

Ömeroğlu [39] designed and experimentally tested a forced circulation air-based PV/T panel and analyzed its performance using CFD software. Four array configurations were considered under constant solar irradiance of 1100 W/m^2 and variable air velocities. The study aimed to determine the critical threshold for parameters such as heat transfer area and velocity which are associated with fin arrangement and number of fins, in addition to airflow, etc. The heat transfer surface area and forced air circulation are known to positively affect the total heat transfer, and therefore, it is foreseeable that an increased number of fins and higher air velocities will help maintain the electrical efficiency of the panel at higher levels. The arrangements investigated were 54 pc shifted (type 1), 54 pc staggered (type 2), 108 pc double shifted (type 1), and 108 pc shifted (type 2). Testing the multiple arrangements allows for finding out the optimal design for the air-based PV/T, in addition to aiding in understanding parameters associated with fins and air velocity. In terms of electrical efficiency, the highest values were achieved by a 108pc double-shifted (type 1) arrangement which is around 12.02%. The reason different fin arrangements were considered is to create turbulence within the airflow. The testing period is done for 50 minutes for all cases, and after 15 minutes it can be clearly seen that the difference in efficiency between the conventional PV and proposed PV/T collectors increases immensely. By the end of the experimentation period, the electrical efficiency is around 7%, while electrical efficiency of the rival configurations is at a minimum of 11.2%. The CFD simulation was performed to validate the findings of the indoor experiments. The steps in the CFD simulation included meshing and defining material properties and model geometry. The simulation was done according to the validation and verification standards. According to the displayed results, increase of mass flow rate from 0.02 to 0.03 led to increase the thermal efficiency significantly for all cases.

Mojumder et al. [40] studied the impact of fin system and thin flat metallic sheet (TFMS) on the overall PV/T system. The collector is an air-based single-pass PV/T with rectangular fins. The energy balance for the proposed collector was made through an analytical expression. Temperature measurements were made for different locations on the PV/Ts under different fin numbers, mass flow rates, and solar irradiations of 0–4, 0.02 kg/s–0.14 kg/s, and 200 W/m^2–700 w/m^2. The results show maximum electrical and thermal efficiencies of 13.75% and 56.19%, respectively. The maximum electrical and thermal efficiency for the collector was achieved at a mass flow rate of 0.14 kg/s and a solar irradiance value of 700 w/m^2. In the discussion, the study presents an RMSP of deviation along with coefficient of correlation as parts of the statistical analysis. The results indicate that temperature increase of PV at a certain mass flow rate leads to drop in electrical efficiency, while raising the mass flow rate would lead to maintain that efficiency. This is due to heat being transferred to the base fluid. It was found that at a mass flow rate of 0.014 kg/s and a solar irradiance of 700 W/m^2 of the PV surface temperature, PV efficiency, thermal efficiency, and overall thermal efficiency are around 39.40 °C, 14.03%, 56.19%, and 93.11%, respectively.

Dubey et al. [41] prepared an analytical expression for the electrical efficiency of a PV module with and without flow. The expression is a function of climatic conditions and design parameters. The study considers four different cases/configurations of PV which are (a) glass-to-glass PV module with duct, (b) glass-to-glass PV module without duct, (c) glass-to-Tedlar PV module with duct, and (d) glass-to-Tedlar PV module without duct. In the results, glass-to-glass and glass-to-Tedlar are referred to as (GG) and (GT), respectively. The models are tested in outdoor conditions of India. The results of the study indicate that (GG) PV modules with duct achieve higher electrical efficiency as well as produce more outlet air temperature than the remaining configurations. In addition, the annual average efficiency of the (GG) PV module with and without duct is around 10.41% and 9.75%, respectively. Hence, optimum design is that of (GG) PV module with duct.

Jin et al. [42] developed a single-pass PV/T collector with rectangle tunnel absorber, located at the back of the PV module, and is connected in parallel. The collector was tested indoors using a solar simulator. The aim of the proposed collector is to raise electrical efficiency and use the hot air for drying purposes. The results, in terms of electrical and thermal efficiencies, of the study indicate that PV/T with rectangular absorber tunnel is better than without. The authors found that the optimal conditions are a solar irradiance of 817.4 W/m^2 and a mass flow rate of 0.0287 kg/s and an ambient temperature of 25 °C. This condition led to electrical, thermal, and overall PV/T efficiency of around 10.02%, 54.70%, and 64.72%, respectively. The results also show that increase of mass flow rate of air from 0.01 kg/s to 0.08 kg/s leads to drop in PV temperature from around 60 °C to 40 °C, respectively. Without tunnel, the PV temperature peaks at around 75, for 0.01 kg/s, and can be cooled down to around 58 °C, for 0.08 kg/s. It is important to note that the efficiency mentioned values are not average daily but peak efficiencies. The electrical and thermal efficiencies of selected referencing, from the literature, using air-based PV/T systems are shown in Figs. 2.14 and 2.15, respectively.

2.8.2 Studies on Water-Based PV/T

Water is the circulating fluid or working fluid with these types of collectors. The main advantage of water-based PV/Ts is for hot water demands. The hot water is usually stored in an external tank. Another advantage is its flat structure which makes it convenient for building integrated mode. However, the drawback of this type of collectors is (i) water leakage and (ii) freezing during extreme conditions. Still, certain measures for protection can be taken into consideration during the design and installation of these collectors.

Sultan and Tso [43] conducted experiments to compare between improved parallel flow and conventional direct flow absorbers. The evaluation parameter is the thermal efficiency of the collector. The mass flow rates range between 0.03 kg/s and 0.06 kg/s, while the solar irradiance range is between 400 W/m^2 and 850 W/m^2. The results indicate that better performance is achieved by the improved parallel flow absorber design for same operating conditions. The peak thermal efficiency for the

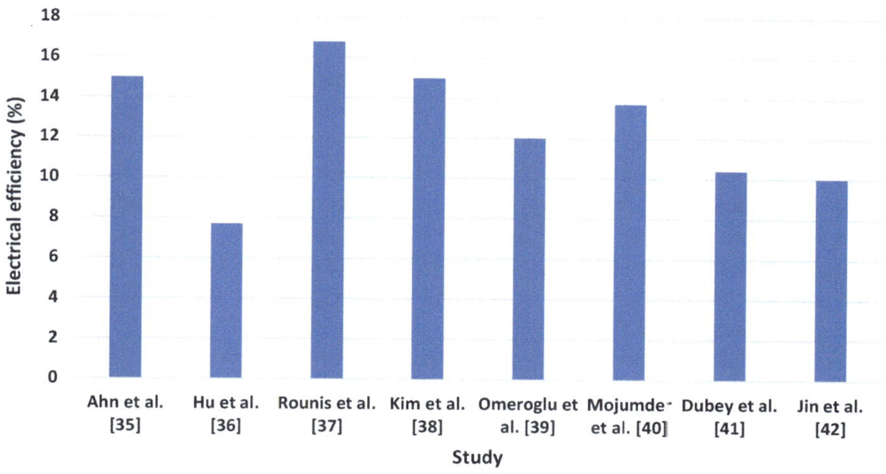

Fig. 2.14 The electrical efficiency of selected references using air-based PV/T

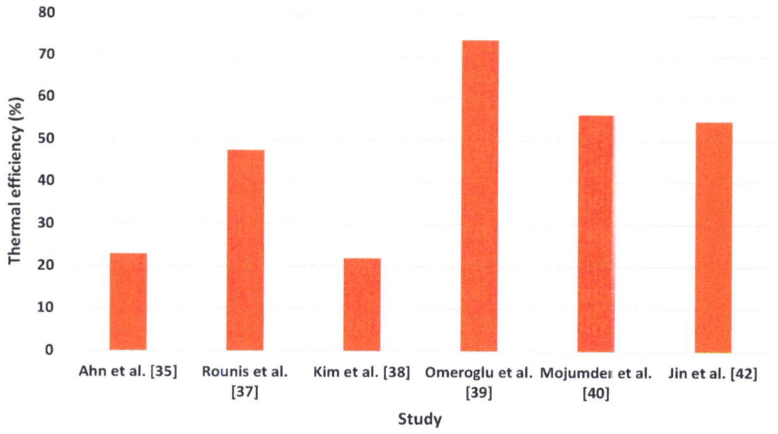

Fig. 2.15 The thermal efficiency of selected references using air-based PV/T

improved parallel flow and direct flow designs is 57.27% and 54.26%, respectively. Raising the mass flow rate from 0.03 kg/s to 0.057 kg/s led to increase of the thermal efficiency of proposed collector by 5.07%, while direct flow collector increased by 4.79%, for the same solar irradiance levels. The authors did not discuss the electrical performance of the PV/T collector, the components of the system, and the time period of the experiment.

Lu et al. [44] investigated the component of location arrangement of photovoltaic (PV) cells in the sheet and tube water-based flat-plate PV/T collectors. This is done by testing three arrangement configurations which are PVT-1, PVT-2, and PVT-3 which refer to conventional air gap type with cells stuck on the absorber, new air gap with cells stuck to bottom of glass cover, and zero air gap type, respectively.

Outdoor experiments were carried out, and numerical CFD analyses were conducted as well, to evaluate the performance of these collectors. In addition, the collector was compared to a conventional thermal collector, referred to as "pure thermal collector" in the study. The collectors were installed at 25° tilt angle facing normal south. The results of the study show that if the evaluation criteria are overall efficiency, then the best configuration is PVT-1, followed by PVT-3 and then PVT-2. However, if the evaluation criteria are in terms of electrical efficiency, then the highest is PVT-3, followed by PVT-2 and then PVT-1. The investigation also considers the covering factor (β) which is found to have significant effect on PVT-2. Hence, low value of the covering factor was found to be optimal for PVT-2, while higher values are better for PVT-1 and PVT-3.

Alzaabi et al. [45] fabricated and installed a hybrid water-based PV/T to improve the electrical performance of PV modules in UAE climate conditions. The authors performed experimental and numerical analysis for the PV/T performance. The evaluation parameters were studied for cooling and non-cooling PV modules. The results show that the PV panel temperature was reduced by 15–20%. Hence, the electrical efficiency of the PV module increased by 15–20% when cooling. The thermal efficiency of the proposed collector was found to be 60–70%. The electrical efficiency was found to increase between 12:30 PM and 3:30 PM by 6.7%, where at 3:30 the electrical power and efficiencies are around 52.8 W and 12.3%, respectively. The I–V curves for the cooled PV, non-cooled PV, and simulated PV are provided as well. The simulated PV produced the highest values for voltage, followed by the cooled PV. However, simulated PV exhibits the lowest value for current, as appose to cooled which is found to be the highest. The cooled having higher current than non-cooled is in disagreement with the theory, because increase of cell temperature leads to increase of current. However, the authors justified this outcome by stating that there was a slight difference in terms of solar irradiance value which led to increasing the current of cooled PV.

Nardi et al. [46] evaluated the thermal and electrical efficiency of a commercial PV/T panel along with infrared thermographic diagnosis during operation for both cooling and non-cooling modes. The tests were conducted during summer with the aim of environmental condition on system efficiency. The authors conducted seven tests under different conditions. Measurements were conducted for electrical energy production only and for simultaneous electrical and thermal energy production. Moreover, the authors employed a solar concentrator to enhance the overall efficiency. The PV and PV/T module efficiencies were maintained to the manufacturer's expectations. The authors conclude that electrical efficiency of the PV/T is superior to that of the conventional polycrystalline PVs. Test 2 shows how rapid solar irradiance variation affects the energy efficiency of the PV/T, although it seems to be stable for high water flow rates across time of experimentation. The results indicate that the setup with reflector and without cooling has the highest temperatures and is followed by that without cooling and without reflector. Furthermore, the study found a 28% increase of electrical efficiency due to employment of a simple reflector. However, for thermal performance, the authors claim that due to the set up being very simple, the thermal and overall efficiency was quite understated.

Liang et al. [47] tested the performance of a PV/T collector filled with graphite and compared its performance to a conventional PV. Measurements were recorded for temperatures of the backplane, inlet fluid, outlet fluid, and tank. As for PV/T production, the output power and thermal and electrical efficiency are used for evaluation. Cooling down the backplane temperature leads to enhancing electrical efficiency of the PV/T. In addition, having lower temperature for the inlet water leads to higher thermal efficiency, which is attributed to its proximity to ambient temperatures. Increasing tank temperature leads to increase of inlet temperature. The highest temperature difference observed between the two collectors was around 16.74 °C. The mean value of the proposed collector was found to be 6.46%, while that of conventional PV was found at 5.15%. The highest electrical and primary energy-saving efficiency are around 7.2% and 48%, respectively. Furthermore, the output power produced by the proposed collector and PV module are around 79 W and 70 W, respectively. Figures 2.16 and 2.17 show the output power and backplane temperatures for the PV module and PV/T collector across time of testing, respectively. The electrical and thermal efficiencies of selected referencing, from the literature, using water-based PV/T systems are shown in Figs. 2.18 and 2.19, respectively.

2.8.3 Studies on Air- and Water-Based PV/T

Air and water could be used simultaneously through separate channels (dual-channel) or pipes. Another method is to have the collector operate at either mode, water only or air only, separately according to the seasonal conditions. The air- and

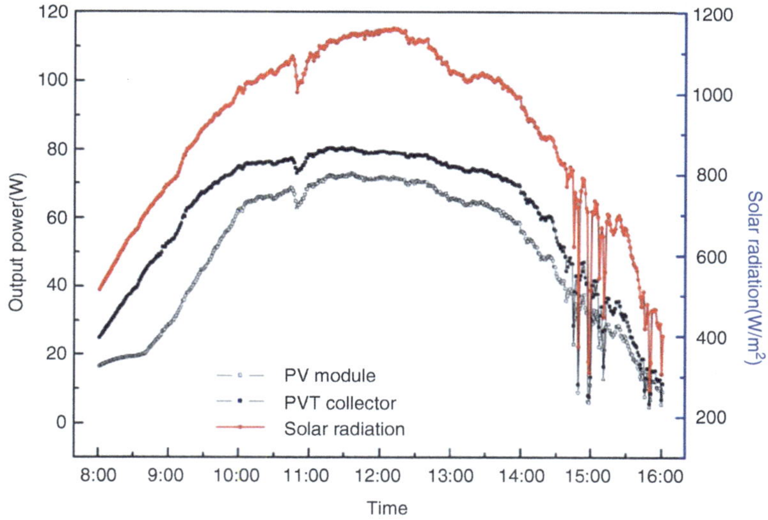

Fig. 2.16 The change of the output power with solar irradiance. (Refer to as solar radiation in the figure) [47]

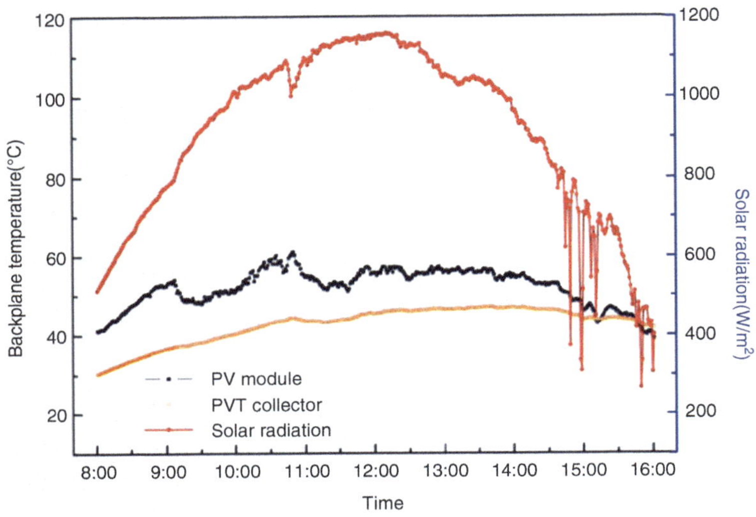

Fig. 2.17 The change of the backplane temperature with solar irradiance. (Referred to as solar radiation in the figure) [47]

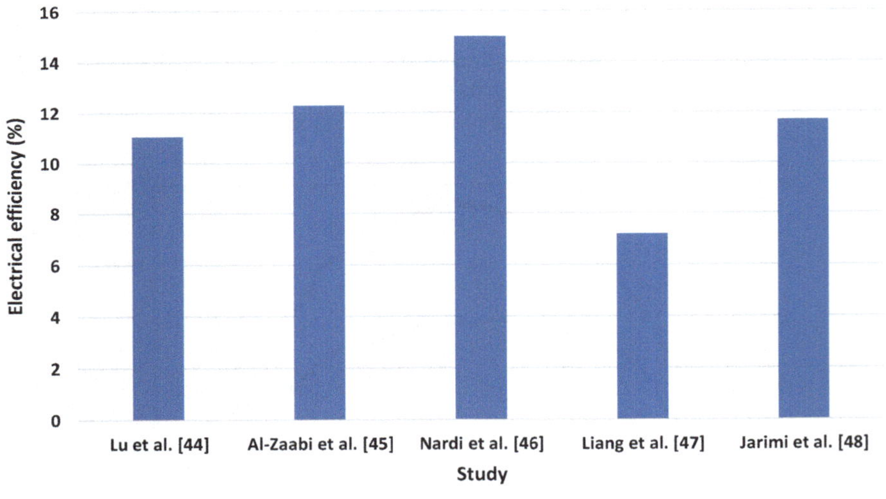

Fig. 2.18 The electrical efficiency of selected references using water-based PV/T

water-based PV/T collectors are very useful in that they exhibit better cooling, or heat transfer rate, and can meet both air heating and/or water heating demands. The main disadvantage of such system is its complexity and added costs and volume associated with having more than one channel or pipe. Furthermore, for active components, more spending must be made to purchase pump (for the water) and blowers/fans (for the air).

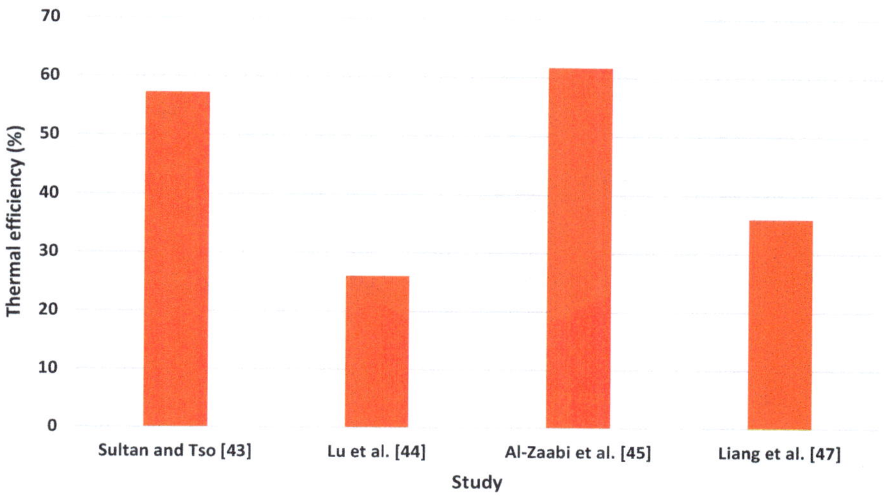

Fig. 2.19 The thermal efficiency of selected references using water-based PV/T

Jarimi et al. [48] carried out a 2D steady-state thermal modeling of a novel PV/T collector with both air and water used as working fluids. The collector is basically a combination of a finned air collector and tube and sheet water-based pipes. The pipes will convey the water while air will flow between the pipes (upper area) and the fins (lower area). The thermal model without working fluids is validated using the normal operating condition (NOCT) which is provided by manufacturer. However, for the cooled model, the findings are validated with results from prior research, and so the collector model is simulated. For simultaneous operation of water and air, only water, and only air, under irradiance levels of 500 W/m^2 to 800 W/m^2, the average temperature of the PV cells was predicted to be around 20, 16, and 14 °C, respectively. These findings are in agreement with theory and literature, meaning that air exhibits lowest cooling effect, followed by only water and both exceeded by the simultaneous use of air and water. Given that the PC cells are kept at low temperature, their efficiency was found to increase. Hence, the overall energy efficiency was predicted to be 40% higher which is attributed to the thermal component coving the entire area of PV. The authors claim that it is useful to utilize the produce hot air and water for preheating temperature for fish breeding and other applications such as clothe drying and hot water supply.

Su et al. [49] presented a study on PV/T collectors with dual channels to employ different fluids. The collector is shown in Fig. 2.20. The optimum fluid combination was investigated by comparing the use of (1) air-air-, (2) air-water-, (3) water-water-, and (4) water-air-based PV/T in terms of electrical and thermal parameters such as power generation, outlet fluid temperature, efficiency, etc.

This was established by building a collector that is composed of a glass cover (upper outer layer), flow passage (for either water or air), solar cells, backplane, flow passage (for either water or air), and shell (lower outer layer). The study

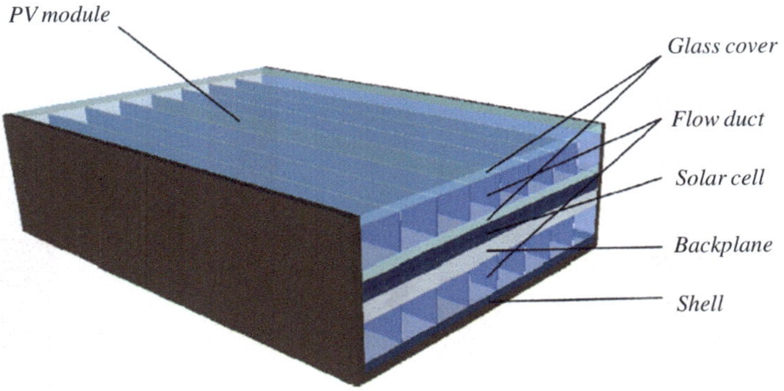

Fig. 2.20 The schematic diagram of the PV/T collector with dual channels [49]

forms the mathematical model for the system and then conducts simulation. The results of the simulation are validated and compared to experimental and numerical work of another study. The findings of the study show that using the water-water-cooled PV/T leads to better water heating. As for air-water, the temperature of the water is found to be the highest, while most amount of hot air was generated by the air-air-cooled PV/T. The observation of mass flow rate effect on overall efficiency of the water-water cooled PV/T is consistent with the literature, in that it raises the efficiency value. In addition, the authors found that increasing the higher between the upper and lower pipe causes efficiency enhancement as well. For a mass flow rate of 0.15 kg/s and height ratio of 3:1, the electrical and overall efficiency of the water-water-cooled PV/T reaches 7.8% and 83.4%, respectively. The thermal power corresponding to working fluid configuration is provided in Fig. 2.21.

Tiwari and Sodha [50] developed an analytical expression of overall thermal efficiency of a PV/T system with water-air that is a function of climatic and design parameters. The simulation uses weather data of India climate. The study compares four configurations where glazing/unglazing (G/UG) and Tedlar without Tedlar (T/WT) are considered (a) UGT, (b) GT, (c) UGWT, and (d) GWT. Furthermore, the authors compare between the use of air and water as working fluids. The results indicate that use of water led to higher daily efficiency of the proposed system than air for all considered configurations, except for GWT. For summer and winter conditions, the overall thermal efficiency was found to reach 65% and 77%, respectively. For the thermal model, the authors made a thermal resistance circuit diagram. The authors conclude that at lower operating temperatures, the unglazed without Tedlar (UGWT) PV/T showed best performance. Whereas for high operating temperature, the glazed with Tedlar (GT) PV/T exhibited the best performance. Furthermore, the study investigates aspects such as length of the PV module, mass of the water in storage tank, and mass flow rate on the thermal and overall thermal efficiencies. Increase of module length may lead to increased variation in water temperature, yet it does affect the overall thermal efficiency due to introducing more

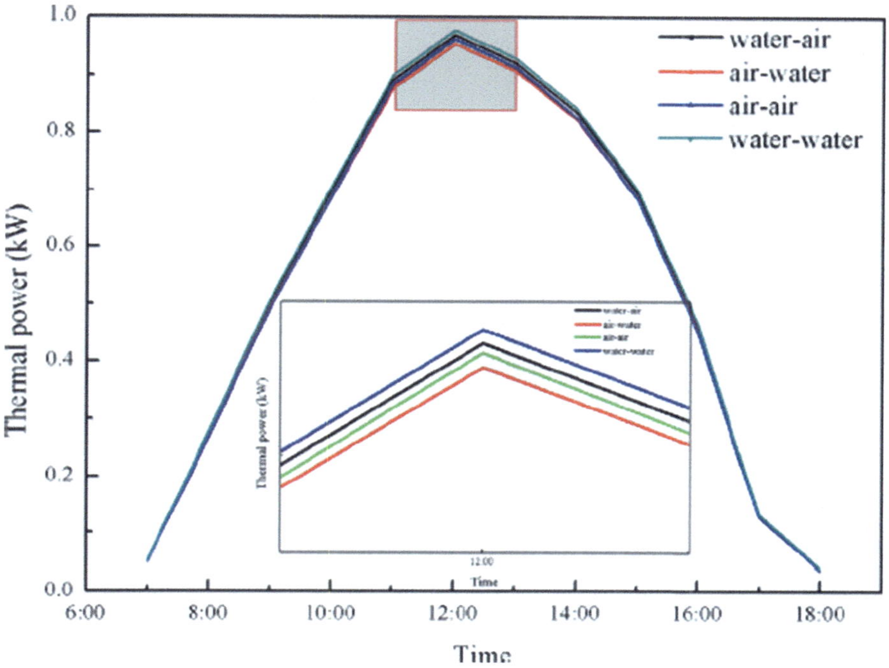

Fig. 2.21 The hourly variation of thermal power corresponding to different working fluid configurations [49]

heat losses to the system. According to the observed results, the water temperature peaks at 1 PM with a value of 55 °C when the PV module length is 8 meter and during winter conditions.

In 2016, Jarimi et al. [51] discussed the theoretical, indoor, and outdoor experimental study of a bi-fluid PV/T collector which can operate as (1) water-based, (2) air-based, and (3) water- and air-based. The authors conducted a 2D steady-state analysis for the collector and simulated its performance using MATLAB software. To validate the model, experiments were conducted indoors and outdoors (under Malaysian conditions). For indoor experiment, the average wind speed and solar irradiance were around 1 m/s and 700 W/m^2, respectively, while air and water mass flow rates ranged [0.0074–0.09 kg/s] and [0.0017–0.0265 kg/s], respectively. The results indicate that increase of mass flow rate leads to increase of thermal efficiency for all tested modes, up to a certain level. The optimal mass flow rates for air and water were found at 0.0262 kg/s and 0.0066 kg/s. For the simultaneous mode, the air and water mass flow rates were fixed at the optimal flow rates. The highest values achieved for thermal, electrical, and total PV/T efficiencies were around 4.37%, 66.32%, and 67.98%, respectively. The electrical and thermal efficiencies of selected referencing, from the literature, using air- and water- based PV/T systems are shown in Figs. 2.22 and 2.23, respectively.

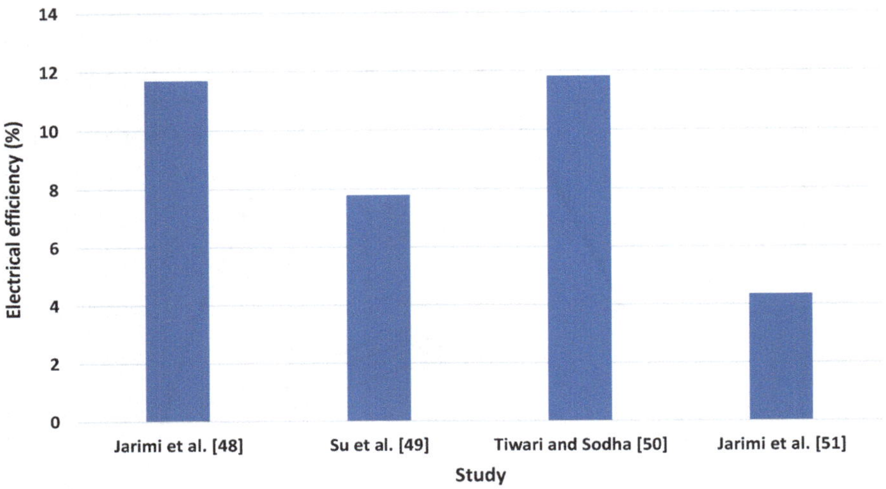

Fig. 2.22 The electrical efficiency of selected references using water- and air-based PV/T

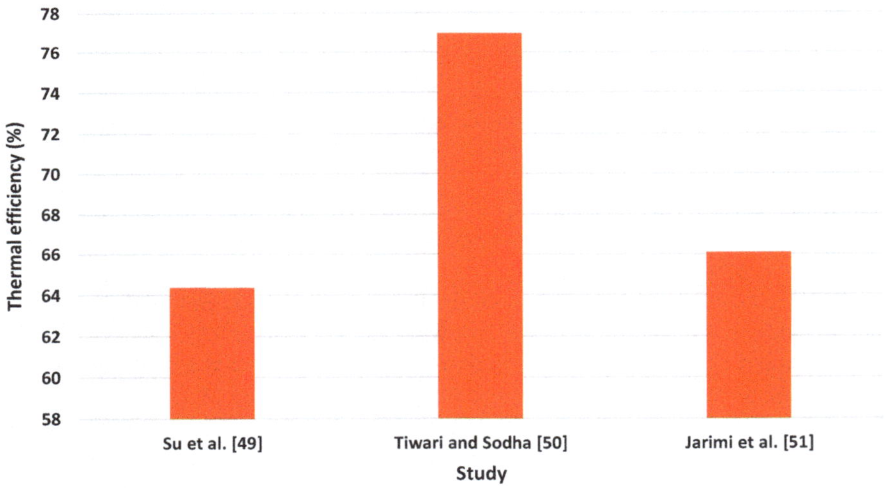

Fig. 2.23 The thermal efficiency of selected references using water- and air-based PV/T

2.8.4 Studies on Refrigerant-Based PV/T

Refrigerant-based PV/T has high heat transfer rate and thermal characteristics and can be used in extremely cold regions without fear of freezing, depending on the refrigerant used. The collectors can be designed as heat pipe-based PV/T where the direct expansion evaporation coils are placed underneath the PV module, and so when the refrigerant passes through, it will evaporate. Hence, the coils are considered the evaporation sector or element of the heat pump. The evaporation can occur at very low temperatures, e.g., 0–20 °C. Consequently, the PV module temperature will drop. The heat pump's compressor causes increase of pressure, from the vapor which is generator by the module, and delivers it to consider for heating.

Tsai [52] presented a mathematical model of a refrigerant-based PV/T-assisted heat pump water heater (HPWH) system and simulated the PV/T performance using MATLAB/Simulink software. The results of numerical study were compared to outdoor experimental data. The collector was installed at 23.5° tilt angle. The authors used a model-based control methodology in order to feed the HPWH system with power produced from the PV. The accuracy of the model is examined by comparing the numerical and experimental findings. The authors claim that the model-based predication control methodology allowed the PV to successfully power the compressor of the HPWH system. Finally, the PV/T efficiency and COP of PVTA-HPWH were found to reach 86% and 7.09, respectively. As for the average electrical and thermal efficiency, they reached 12.37% and 73.90%, respectively.

Ji et al. [53] presented a novel PV/T-solar-assisted heat pump (PV/T-SAHP) with a special design of direct expansion evaporator with PV cells used as lamination (hence, PV evaporator) and a thermal absorber. The configuration of the system is provided in Fig. 2.24. The distributed model approach was used to create a dynamic model of the system. The model inputs are solar irradiance and temperature values,

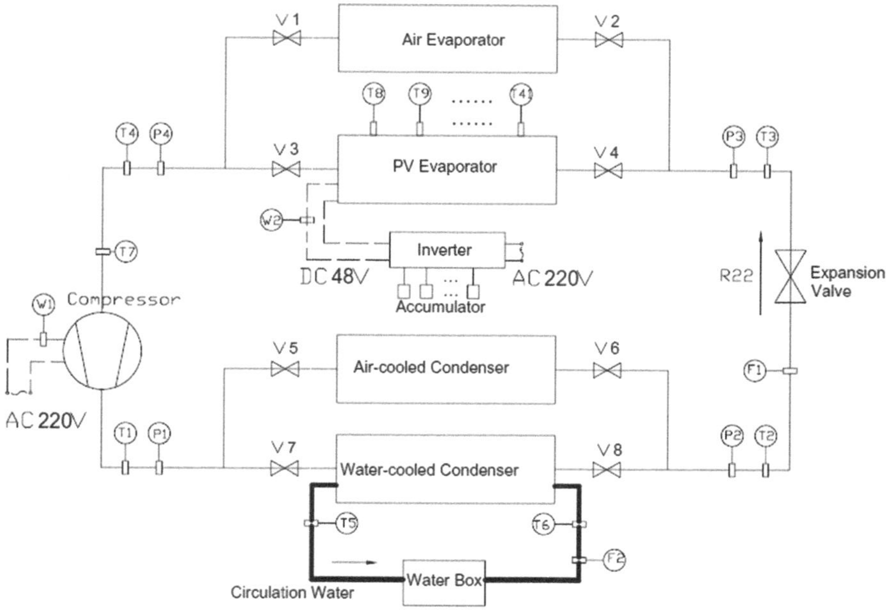

T1-T41, Thermal couple; V1-V8, Cut-off valve; P1-P4, Pressure sensor;
F1-F2, Flowmeter; W1-W2, Wattmeter

Fig. 2.24 Schematic diagram of the PV-SAHP experimental setup [53]

while its output is the spatial distributions of refrigerant conditions. The inputs of the model are shown in Fig. 2.25. Experimental work was also done, to compare between the measured and predicted data and ultimately to validate the model. The experiments were conducted for China's climate condition. The results of the experiments show that the heat gain exceeds 2500 W from 11:00 to 13:00 o'clock. The comparison results between measured and predicted values are in agreement, within ±8% deviation between predicted and measured output electricity and PV efficiency. Finally, the PV and thermal efficiencies were found to exceed 12% and 50%, respectively.

Zhao et al. [54] designed a novel PV/e roof module to function as roof element, electricity generation, and evaporation for a heat pump system. Conceptually, the model is composed of flat-plate glazing, PV cells, copper plate, evaporation coil, insulation, and hose connectors to inlet and outlet. The study analyzes the energy profile of the proposed hybrid system and investigates the temperature distribution across its layers. Mathematical model describes the energy transfer and conversion process and setup using computer software. The model was validated using experimental data from another research work. The authors investigated parameters such as type of glazing, type of solar cell (mono- or polycrystalline), and system operation under UK climate condition. The system prototype is composed of 10 PV/e modules coupled to a 5 kW heat pump. The results of the study indicate the following:

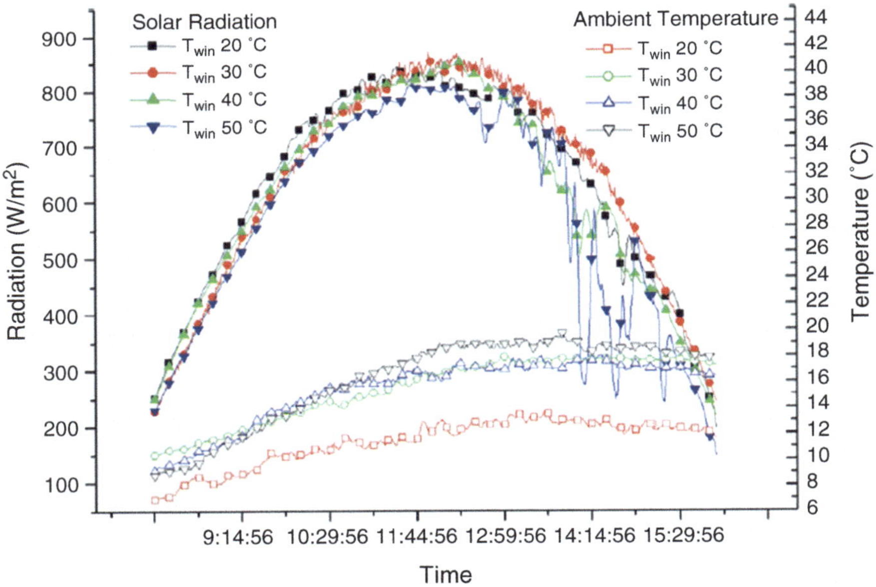

Fig. 2.25 Daily variation of solar irradiance (referred to as solar radiation) and ambient temperature during the experiment period [53]

1. The proposed system should operate at temperature of 10 °C and 60 °C of evaporation and condensation, respectively.
2. Monocrystalline cells achieved better efficiency than polycrystalline ones.
3. For UK operating condition, the PV/T collector achieves thermal and electrical efficiencies of 55% and 19%, respectively, although the value of electrical efficiency is quite high. Under the same conditions, the module-based heat pump system achieves an overall efficiency exceeding 70%. Table 2.3 provides a summary of the discussed studies in terms of year of publication, location, type of working fluid, peak solar irradiance and peak mass flow rate, and electrical, thermal, and total PV/T efficiencies. The electrical and thermal efficiencies of selected referencing, from the literature, using refrigerant-based PV/T systems are shown in Figs. 2.26 and 2.27, respectively.

It is noteworthy to mention that the algebraic addition of thermal and electrical efficiencies, to produce total PV/T efficiency, may not be valid for peak values if those values occur at different times. It is preferable to obtain peak efficiency values occurring in the same time or to add average values of electrical and thermal efficiencies.

2.9 Theory and Literature Review of PV/T Systems with Various Passage Flow Configurations

This section displays a review of the design and analysis of PV/T systems according to their type of passage flow. A breakdown of main features and design considerations associated with each type is provided, along with a review of recent works in the literature.

Rosli et al. [55] performed a CFD simulation using ANSYS Fluent on different absorber-shaped designs for water-based PV/T collectors. The designs are spiral, serpentine, and u-flow shaped which are all designed using CATIA V5R20 software. The authors compared the simulation results with experimental data from previous works and found good agreement, root mean square of 1.29 °C. This is compared to root mean square between experiments and simulation of the previous research which was found at 2.08 °C. The results for effect of mass flow rate show that increase of mass flow rate led to the highest thermal efficiency. At mass flow rate of 0.005 kg/s, the spiral, u-flow, and serpentine absorber PV/Ts achieved thermal efficiencies of 22.96%, 21.02%, and 22.62%, respectively. However, at 0.005 kg/s, the highest electrical efficiency was around 11.78% and was achieved when using the u-flow absorber. The spiral designed absorber showed the highest overall PV/T efficiency of around 34.63%. The least overall efficiency was found for the u-flow design with 32.8%.

Sopian et al. [56] numerically investigated the energy performance of three different types of absorber configurations (direct flow, parallel flow, and split flow) of PV/T. The focus of the work was to examine the thermal efficiency of the systems. The split flow PV/T design was found to achieve the highest thermal efficiency,

Table 2.3 Summary of studies in working fluid-based PV/T collectors

Reference	Year	Location	Working fluid	Type of PV	Type of system	Peak mass flow rate (Kg/s)	Peak solar irradiance (W/m^2)	Electrical efficiency (%)	Thermal efficiency (%)	Total efficiency (%)
Ahn et al. [35]	2015	South Korea	Air	Monocrystalline silicon	Direct-active	0.0277	1050	15	23	*38
Hu et al. [36]	2016	China	Air	Amorphous silicon	Direct-active	–	1158.3	7.7	–	25.5
Rounis et al. [37]	2016	Canada	Air	–	Direct-active	0.22	–	16.8	47.5	–
Kim et al. [38]	2014	Korea	Air	Monocrystalline	Direct-active	0.07801	910	15	22	*37
Ömeroğlu [39]	2018	Turkey	Air	Polycrystalline	Direct-active	0.032	1100	12.02	74	
Mojumder et al. [40]	2016	Malaysia	Air	Polycrystalline	Direct-active	0.14	700	13.75	56.19	*69.94
Dubey et al. [41]	2009	India	Air	–	Direct-active	2 m/s	820	10.41	–	–
Jin et al. [42]	2010	Malaysia	Air	–	Direct-active	0.08	817.4	10.02	54.7	64.72
Sultan and Tso [43]	2018	Malaysia	Water	Monocrystalline	Direct-active	0.06	850	–	57.27	–
Lu et al. [44]	2017	China	Water	Polycrystalline	Direct-passive	0.002	1000	11.07	26.07	37.14
Al-Zaabi et al. [45]	2014	UAE	Water	Polycrystalline	Direct-active	0.09	899.7	12.3	61.7	74
Nardi et al. [46]	2017	Italy	Water	Polycrystalline	Direct-active	0.212	1000	15–18	–	–
Liang et al. [47]	2014	China	Water	Polycrystalline	Direct-active	0.012	1000	7.2	36	43.2

Reference	Year	Location	Working fluid	Type of PV	Type of system	Peak mass flow rate (Kg/s)	Peak solar irradiance (W/m^2)	Electrical efficiency (%)	Thermal efficiency (%)	Total efficiency (%)
Jarimi et al. [48]	2013	Malaysia	Combi	Monocrystalline	Direct-active	Air: 0.034 Water: 0.0079	800	11.7	–	55
Su et al. [49]	2016	China	Combi	Polycrystalline	Direct-active	0.15	677	7.8	64.4	84.2
Tiwari and Sodha [50]	2006	India	Combi	–	Direct-active	0.02	704	11.83	65–77	–
Jarimi et al. [51]	2016	Malaysia	Combi	–	Direct-active	Air: 0.0262 Water: 0.0066	1016	4.37	66.12	78.98
Tsai [52]	2014	Taiwan	Refrigerant	Polycrystalline	Indirect-active	–	1000	12.5	74	86.5
Ji et al. [53]	2009	China	Refrigerant	Single-crystalline	Indirect-active	–	840	12	50	–
Zhao et al. [54]	2011	United Kingdom	Refrigerant	Monocrystalline	Indirect-active	0.006	–	19.02–19.07	55.58	56.40–74.13

(−) stands for nowhere found in the article or not mentioned. (∗) not calculated in the article, (~) approximately

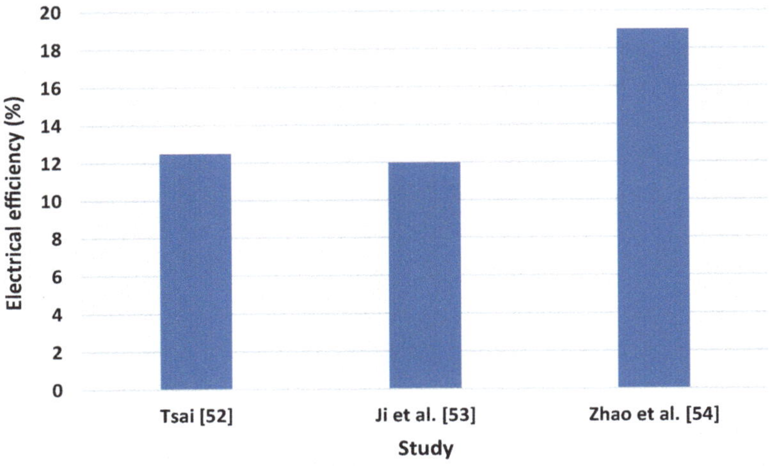

Fig. 2.26 The electrical efficiency of selected references using refrigerant-based PV/T

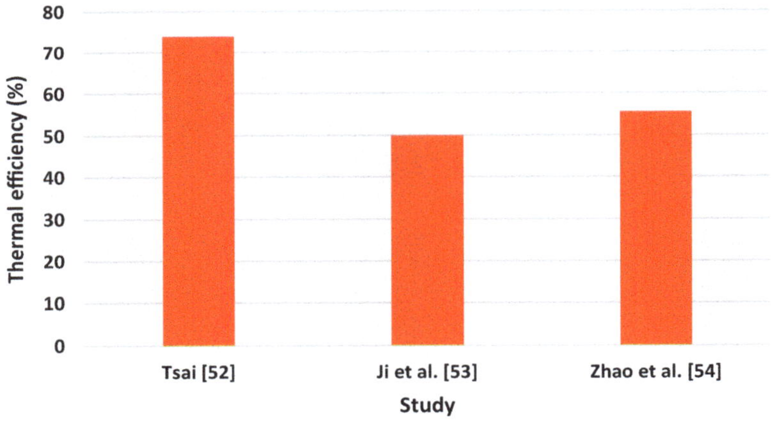

Fig. 2.27 The thermal efficiency of selected references using refrigerant-based PV/T

about 51.4%, compared to the other two types. The results also show proximity in thermal efficiency of parallel and direct flow configurations. Consistently, the parallel flow absorber was found to be the least efficient thermally. Furthermore, the increase of module numbers (as in the size of the panel) was found to cause an increase in outlet fluid temperature. However, reducing the mass flow rate for increased number of collectors was found to lead to higher outlet fluid temperature. Other studies involving different flow configurations were discussed in the literature [57–59]. Table 2.4 shows the different types of flow passage configurations. Figures 2.28 and 2.29 show PV module and PV/T layout and 3D view of the absorber used by Hossain et al. [59], respectively.

Table 2.4 Flow passage configurations used in PV/T systems

Passage flow configuration	Drawing	Refs.
Direct flow	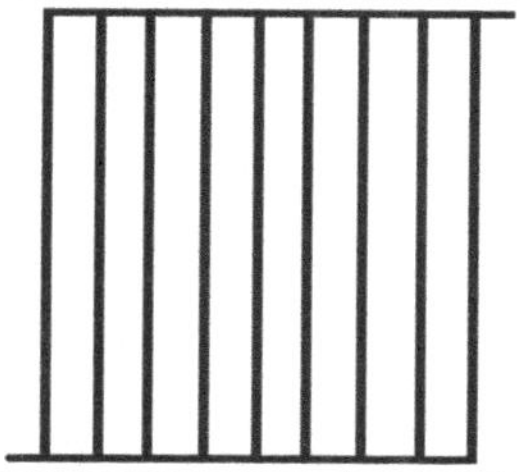	[56]
Serpentine flow		[55]
Parallel flow	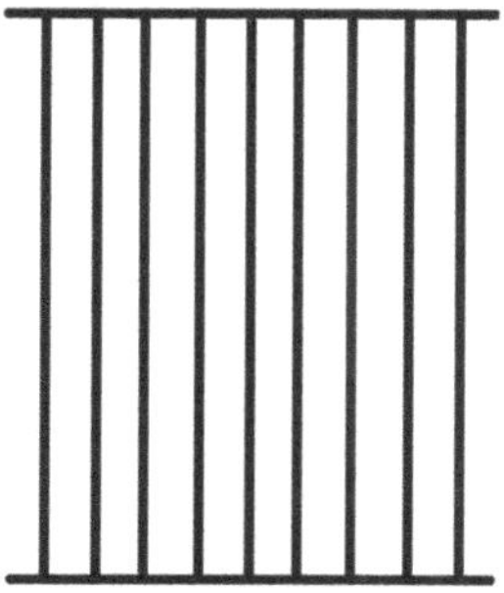	[56]
Web flow		[7]

(continued)

Table 2.4 (continued)

Passage flow configuration	Drawing	Refs.
Split flow	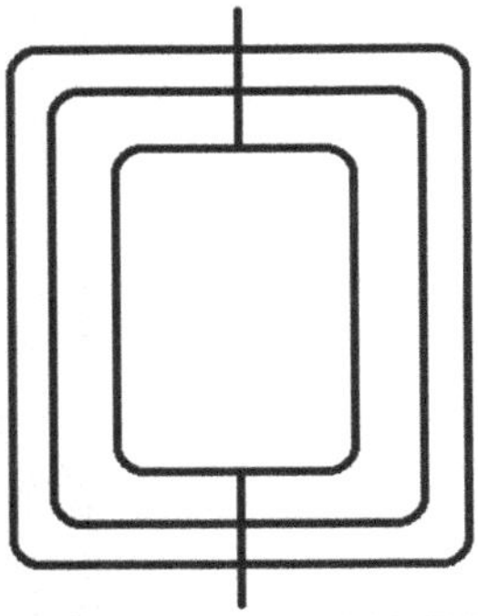	[56]
Oscillating flow	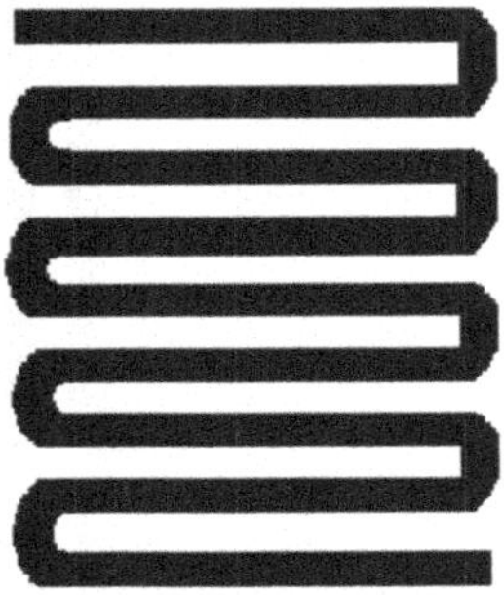	[57]
Spiral flow	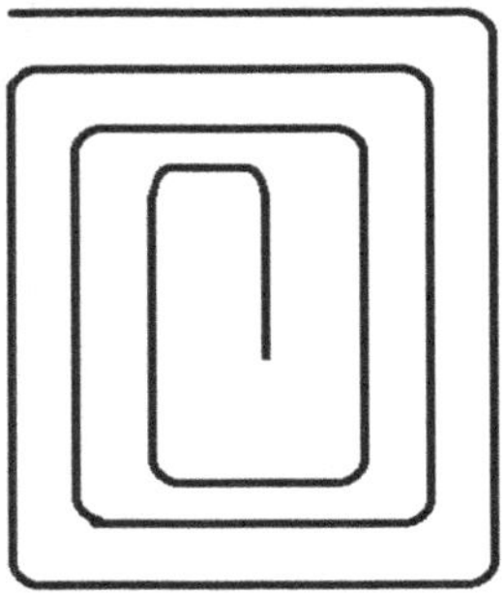	[55]

2.10 Theory and Literature Review of PV/T Systems with Various Geometry Types

Various types of absorber geometries are used for absorber designs of PV/T collectors [60]. The dimensions, geometry, and material of the absorber all affect its thermal performance parameters such as the heat removal factor (F_R). The fluid gains heat as its passes inside the absorber tubes, attributed to contact underneath the PV module. The fluid will gain the heat and is then fed into the water tank and/or heat exchanger. PV/T absorbers have points of inlet and outlet (in some cases multiple

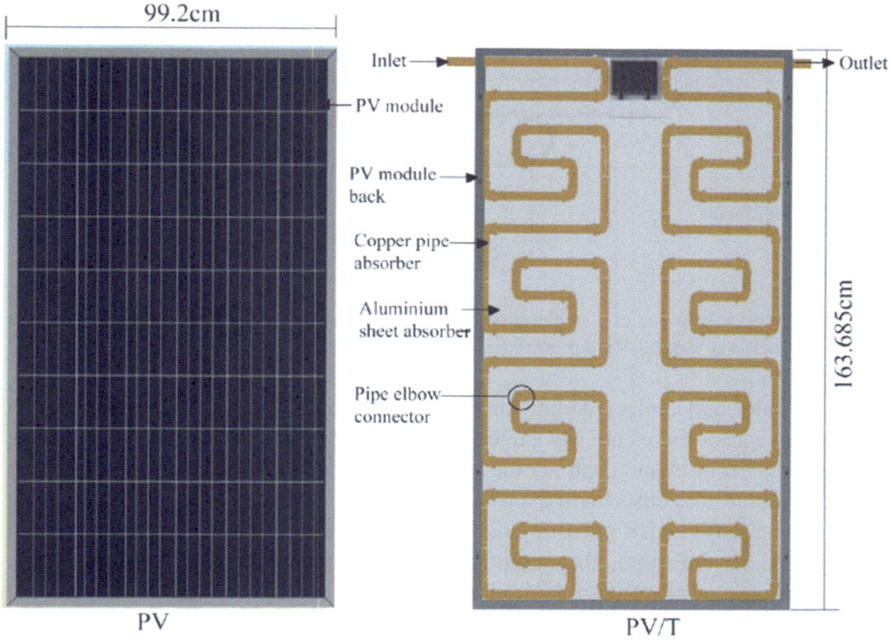

Fig. 2.28 PV module and layout of PV/T thermal collector [59]

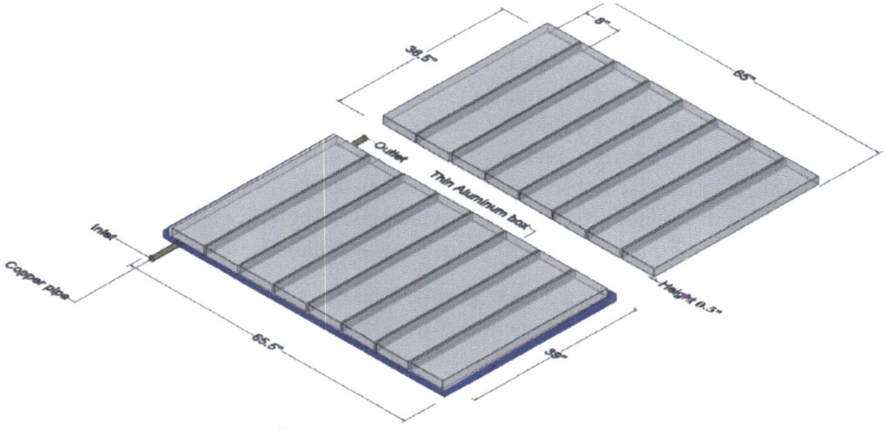

Fig. 2.29 3D view of flow channel and PCM packets [59]

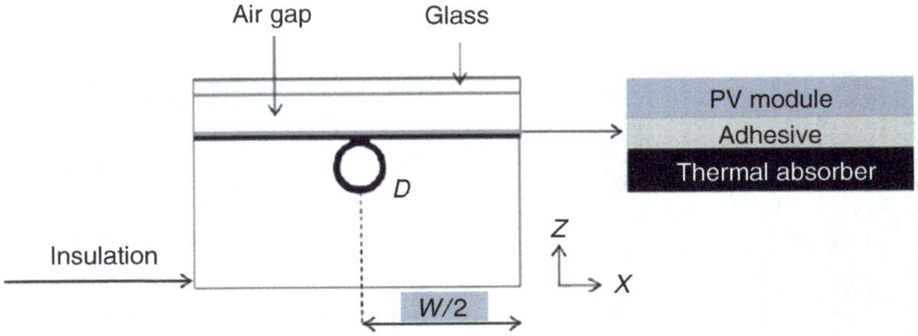

Fig. 2.30 Cross section x-z of a single-glazed PV/T module showing the pipe diameter D and the distance 14 between two adjacent pipes W [66]

points of inlet and outlet). Major energy losses are caused by friction within the pipe, while minor losses are associated, caused by change in velocity, both in magnitude and direction, and each geometry has a loss coefficient. Figure 2.30 shows a typical circular tube absorber used for a PV/T by Guarracino et al. [66].

The thermodynamics must be taken into consideration when examining these geometries. Furthermore, the manufacturing aspects should also be considered, for instance, it is more complicated to weld round (circular) tubes than it is to weld square tubes. Table 2.5 shows some of the geometries discussed in the literature.

2.11 Research Schemes and Methodologies in PV/T Field

Much information and ideas can be surmised from the aforementioned studies. Hence, Table 2.6 provides a summary of critical details from different studies from the literature review, while Table 2.7 shows the lessons learned from each study with regard to theory, design, and implementation of PV/T systems. By obtaining such critical details, research can be directed to drawing the commonalities and inconsistencies throughout the literature, which could help in enriching the field.

From Table 2.6 the following observation can be made:

1. Most of the studies are based on experiments, specifically outdoor experiments. Some studies only focus on theoretical, either analytical or numerical, while fewer studies combine both experiments and numerical work for validations.
2. Studies with both experimental and numerical results typically employ methods to estimate the difference in results and/or errors.
3. The experiments conducted use different experimentation time. In some studies, the time of the experiments is only 4 hours, while in others it is 8 hours. Most conducted experiments are found to exceed 1 hour of testing. Mostly, length of the experiment is 8 hours, starting from 8 AM and stopping at 4 PM. The shortest and longest periods of experiments are 12 hours and 32 minutes, respectively.

Table 2.5 Summary of geometries used in PV/T systems

Geometry	Explanation	Drawing	Refs.
Square-shaped	The tube exhibits four equal straight sides. Covers more area than circular-shaped tube	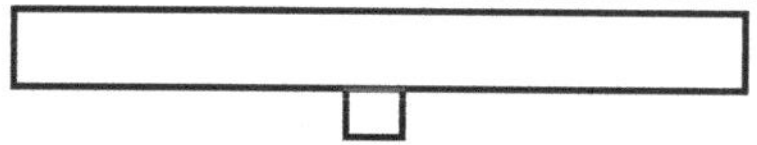	[61, 62]
Rectangular-shaped	More contact area than square-shaped tube	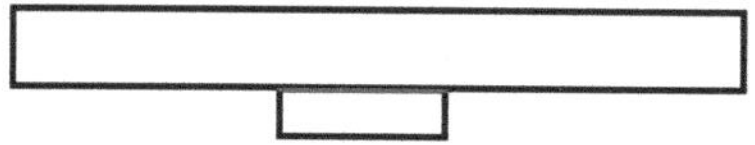	[63]
Circular-shaped	Higher heat transfer than square; better contact around the walls	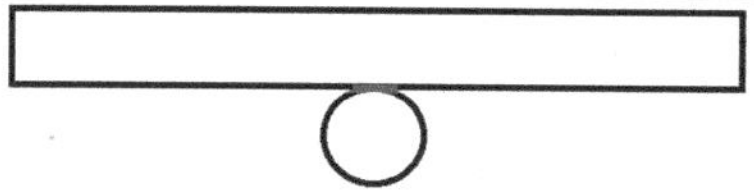	[64–68]
Ellipse-shaped	Lesser pressure drops and higher heat transfer rate than circular	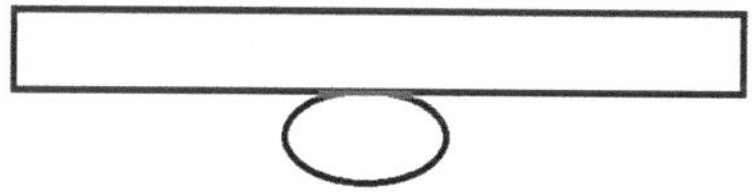	[69, 70]

Table 2.6 Summary description of research studies throughout the literature

Reference	Type of fluid	Type of study	Parameters of evaluation	Time of experiments	Type of flow	Range of mass flow rates (kg/s)	Range of solar irradiance (W/m^2)
Ahn et al. [35]	Air	Experimental (outdoors)	Enthalpy, electrical and thermal efficiency	5 hours [9:40–15:40]	Forced circulation	0.0277	700~1050
Hu et al. [36]	Air	Experimental (outdoors)	Energy yield, electrical and virtual efficiency	6–7 hours [10:00–17:00]	Forced circulation	–	574.2–1158.3
Rounis et al. [37]	Air	Numerical simulation	Electrical and thermal efficiencies	–	Forced circulation	[0.11–0.22]	–
Kim et al. [38]	Air	Experimental (outdoor)	Electrical and thermal efficiencies	9 hours [08:00–17:00]	Forced circulation	0.078	~60–910
Ömeroğlu [39]	Air	Experimental (indoor) and numerical (CFD ANSYS fluent)	Electrical and thermal efficiency	0–50 minutes	Forced circulation	0.018–0.032	1100
Mojumder et al. [40]	Air	Experimental (outdoor)	Electrical and thermal efficiency	0–32 minutes	Forced circulation	0.02–0.14	100–700
Dubey et al. [41]	Air	Experimental (outdoor) and analytical	Electrical efficiency and daily average electrical efficiency	8 hours [08:00–16:00]	Forced circulation	0.0058	820
Jin et al. [42]	Air	Experimental (indoor)	Electrical, thermal, and overall PV/T efficiency	–	Forced circulation	0.01–0.08	0–817.4
Sultan and Tso [43]	Water	Experimental (outdoor)	Thermal efficiency and outlet water temperature	–	Forced circulation	0.03–0.06	400–850
Lu et al. [44]	Water	Experimental (outdoor) and numerical (CFD software Fluent 14.0)	Daily electrical (of PV/T and solar cell), thermal, and overall efficiencies	8 hours [9:21–17:21]	Natural circulation	0.002	~90–1000
Al-Zaabi et al. [45]	Water	Experimental (outdoor) and numerical (Simulink)	Voltage, current, power, and efficiency	5 hours [12:00–17:00]	Forced circulation	0.09	~110–912

Nardi et al. [46]	Water	Experimental (outdoor)	Thermal, electrical, and overall PV/T efficiencies	6 hours [09:00–15:00]	Forced circulation	0.106–0.212	~ 500–1067
Liang et al. [47]	Water	Experimental (outdoor)	Produced power, electrical efficiency, thermal and overall efficiency	8 hours [08:00–16:00]	Forced circulation	0.012	~ 210–1000
Jarimi et al. [48]	Combi	Numerical modeling and simulation	Electrical power and efficiency, average PV temperature, and overall energy efficiency	12 hours [06:00–18:00]	Forced circulation	Air: 0.034, water: 0.0079	500–800
Su et al. [49]	Combi	Experimental data (outdoor) and numerical simulation (MATLAB)	Electrical efficiency, thermal efficiency, thermal power, and overall efficiency	9 hours [8:00 to 17:00]	Forced circulation	0.05–0.15	0–1000
Tiwari and Sodha [50]	Combi	Numerical (analytical expression)	Thermal efficiency, overall efficiency, outlet fluid temperature	7 hours [09:00–16:00]	Forced circulation	0.01–0.5	167–704
Jarimi et al. [51]	Combi	Experimental (indoor and outdoor) and 2D steady-state analysis	Thermal and electrical efficiencies, overall thermal efficiency, and total PV/T efficiency	6 hours [10:30–16:30]	Forced circulation	Air [0.0074–0.09] and water [0.0017–0.0265]	400–1016
Tsai [52]	Refrigerant	Experimental (outdoor) and numerical (MATLAB/Simulink)	Coefficient of performance (COP) of HPHW system, electrical, thermal, and total PV/T efficiencies. Outlet water temperature and generated PV power	1 hour [12:00–13:00]	Forced circulation	–	0–1000
Ji et al. [53]	Refrigerant	Experimental (outdoor)	Yield electricity and heat gain, vapor quality, enthalpy, electrical and thermal efficiencies	8 hours [08:00–16:00]	Forced circulation	0.217	228–830
Zhao et al. [54]	Refrigerant	Numerical and experimental (prototype)	Electrical and thermal efficiency	4 hours [09:00–13:00]	Forced circulation	0.006	–

(−) stands for nowhere found in the article or not mentioned. (∗) not calculated in the article, (~) approximately

Table 2.7 Lessons learned from different studies throughout the literature

Reference	Lessons learned
Ahn et al. [35]	The utility of air-based PV/T systems for preheating inlet air of a heat recovery ventilation (HRV) system; increase of the HRV heat transfer efficiency by 20%
Hu et al. [36]	Using pressure sensor to detect pressure data. If in case it detects certain pressure value (300 pa) and does not decrease dramatically to certain pressure levels (240 pa), then the system is under steady state
Rounis et al. [37]	1. Lower PV temperatures are achieved by increase of the total air mass flow. At conditions with no wind, this difference in PV temperatures (between low air mass flow rate and high air mass flow rate) is more prominent. Higher wind velocities reduce the differences due to it being dominated by external convection. The increase of wind velocity causes PV temperature variations to drop, across the same string 2. Multiple-inlet air-based systems have better uniformity of PV temperature distribution. Even at low mass flow rates, this type, multiple-inlet, keeps PV temperatures low
Kim et al. [38]	Mitigation of the increase and decrease of electrical efficiency with respect to solar irradiance levels and concerns of counterproductive conditions. Hence, although solar irradiance levels reach 910 W/m^2, the optimum efficiency was found at solar irradiance of around 750 W/m^2
Ömeroğlu [39]	1. Increasing the number of fins and the air velocities will help maintaining the electrical efficiency of PV/T panel 2. The operating temperature of the PV module reaches a value as high as 120 °C, and the electrical efficiency drops about to 7% from 12.5% 3. The CFD phase was done to validate the conceptual design. The most important step is the step where defining and establishing the material properties and model geometry are done, in addition to performing the meshing 4. After 15 minutes of testing, the difference in electrical efficiency between the conventional PV and rival fined configurations becomes significant
Mojumder et al. [40]	Increase of mass flow rate led to significant increase in heat gain, which is found to reach 121.94 W at a mass flow rate of 0.014 kg/s
Dubey et al. [41]	1. The electrical efficiency of the PV/T drops with rise of solar irradiance and ambient temperature which means the rise of PV temperature is causing reduction in electrical efficiency. This is displayed as the cell temperature has the exact opposite curve to that of the electrical efficiency, for both types of PV modules 2. The electrical efficiency of PV module with duct is higher than without duct 3. Evaluated the correlation coefficient and root mean square percent deviation
Jin et al. [42]	Higher change in thermal efficiency is observed with and without the rectangle tunnel, while moderate difference is observed for the electrical efficiency
Sultan and Tso [43]	Increase of the mass flow rate led to decrease in the outlet water temperature. Hence, increase of mass flow rate from 0.03 lg/s and 0.057 kg/s that led to decrease in outlet air temperature of direct flow and parallel flow absorber by 7.12% and 8.47%, respectively

(continued)

Table 2.7 (continued)

Reference	Lessons learned
Lu et al. [44]	1. Given that natural circulation was used (thermosyphon principle), the velocity of water was low in the tubes which makes CFD assumption of the fluid to be considered laminar and incompressible 2. Hexahedral mesh was used and no PV cell with negative volume appeared 3. The number of cell produced is around 1,000,000 and mass flow rate was assumed at 0.002 kg/s 4. The heat generation rate value for PVT-1 is around 720 W/m^2, which was made by assuming the reduction in solar irradiance due to losses associated with reflection to ambient by glass cover (10%) and that polycrystalline silicon absorbers 70% of incident irradiance and converts one tenth of it to electrical power 5. The customer needs are very important and set the evaluation criteria for the PV/T collector. Customer electrical and thermal needs allow for judgment to electrical and/or thermal bias of the PV/T
Al-Zaabi et al. [45]	Although the solar irradiance levels decrease after 12:30, the electrical, thermal, and overall efficiency was found to increase which is partially because of PV panel temperature drop and decrease in heat losses. At 12:30 PM, the solar irradiance levels are at around 899.7 W/m^2, yet the electrical, thermal, and overall efficiencies are around 11.5%, 58.8%, and 70.3%. However, those values are higher at 3:30 PM, where solar irradiance is around 545.4 W/m^2, the efficiency are around 12.3%, 61.7%, and 74%, respectively
Nardi et al. [46]	1. Impact of a nonuniform temperature distribution can be investigated with infrared thermographic diagnosis. As well as, measurement of electrical and thermal parameters 2. The authors conclude the importance of thermal images to identify possible damage and inoperative cells and to observe visually the gains attributed to operative conditions 3. Applying simple reflector can enhance the electrical efficiency by 28% 4. Lowering the mass flow rate causes the electrical efficiency trend to recall the solar irradiance increase
Liang et al. [47]	1. Backplane temperature is reverse proportional to the electrical efficiency 2. Decrease in fluid inlet temperature leads to increase of thermal efficiency
Jarimi et al. [48]	Further enhancement of heat transfer is achieved by establishing fins. In addition to using both water and air simultaneously
Su et al. [49]	1. Air- and water-cooled PV/T is more efficient than separate air- or water-cooled PV/T. however, dual-channel water-based (water and water) produced better performance than (air-water/water-air) in this study. The peak electrical power produced was around 110 W. while, thermal power peaked at 978 W at noon 2. For the case of air-air-cooled PV/T, the lowest value of electrical efficiency reached around 4.8% for the solar cell 3. Despite variation in solar irradiance, the overall efficiency remains stable relative to changes in electrical and thermal efficiencies. The overall efficiency achieved is between 80% and 84%

(continued)

Table 2.7 (continued)

Reference	Lessons learned
Tiwari and Sodha [50]	1. Water-based PV/T exhibits better thermal characteristics than air-based 2. Variation of water temperature was observed, on hourly basis, to be proportional to the length of the PV module. The opposite can be said about overall thermal efficiencies' relationship with length of PV module 3. Variation of water temperature was observed, on hourly basis, to be inverse proportional to mass of water in storage tank 4. Higher thermal efficiency is achieved in the winter season than summer season which is attributed to lower heat losses 5. For high operating temperature, glazed with Tedlar water-based PV/T has better thermal performance than unglazed with or without Tedlar. For low operating temperature, the unglazed without Tedlar exhibits better performance
Jarimi et al. [51]	Higher efficiencies are achieved using simultaneous cooling than separate air- and water-cooling configurations
Tsai [52]	1. The use of refrigerant led to increase in PV/T efficiency and improved coefficient of performance (COP) 2. From a dynamic point of view, the small change of environmental conditions does not affect the operation or overall HPWH system
Ji et al. [53]	Increase of pipe (copper coil) length leads to increase of average specific enthalpy and vapor quality. Increase of pipe length from 0 to 16 caused an increase in average specific enthalpy from 450 to around 580 kJ
Zhao et al. [54]	The integration of PV cells and evaporation coil into a prefabricated roof introduces significant savings in capital and running costs over individual, separate, arrangements of PV, heat pump, and roof structure

4. The, overwhelming, majority of the studies employ forced circulation mechanism for cooling of the PV module or operation of the PV/T collector. Most of them conducted in outdoor environments.
5. Variable mass flow rates were used for the different proposed collectors. The range of mass flow rates, in total, is from 0.0017 kg/s to 0.5 kg/s. The mass flow rates (kg/s) of working fluids range [0.0058–0.22], [0.002–0.212], [0.0017–0.5], and [0.006–0.217] for air-, water-, air- and water-, and refrigerant-based PV/Ts, respectively.
6. Variable solar irradiance values are detected through measurements; the lowest and highest solar irradiance levels are around 0–1158.3 W/m^2. Although the value 1158.3 W/m^2 exceeds the 1 sun value, of 1000 W/m^2, it is possible to obtain such measurement.
7. Most of the studies aim to evaluate the performance of the proposed PV/T by calculating the electrical, thermal, and total PV/T efficiencies. Refrigerant-based or heat pipe-based PV/T systems are also evaluated with coefficient of performance (COP) and vapor quality in some cases.
8. Numerical studies often use MATLAB software, and CFD analysis is often done using ANSYS Fluent software.

It is important to note that this is a mere sample of the work conducted in the field of PV/T and does not reflect the entirety of studies and articles conducted.

2.12 Energy and Exergy of PV/T

Energy analysis of the PV/T's performance combines both efficiency expressions for thermal and electrical efficiencies, given that both provide an indication of ratio of energy gain (thermal and electrical) to incident solar irradiation on a collector over a period. Hence, in other term the total efficiency, also referred to as combined efficiency, is provided in Eq. 2.2. Another method of performance evaluation is proposed, which is to consider the primary energy-saving efficiency (η_f) given that electrical energy is a high-grade form of energy gain and hence, comparing the energy grade differences between electrical and thermal aspect, which is illustrated in Eq. 2.6 [71]. This indicator takes into consideration both quality and quantity of the converted solar energy by PV/T.

$$\eta_f = \frac{\eta_e}{\eta_{\text{power}}} + \eta_{\text{th}} \tag{2.6}$$

Exergy analysis is useful in determining the PV/T collector efficiency and degradation of energy during thermal and electrical conversion processes and establishing strategies to design and operate PV/Ts for optimal use of energy. The first law of thermodynamics is used to define the thermal efficiency of a system, including concepts of energy balance equations and accounting for input, desired output, and lost energies. The exergy efficiency on the other hand is based on the second law of thermodynamics by accounting for total exergy inflow, exergy outflow, and exergy destructed from the system [72]. Exergy efficiency is defined as the ratio of product exergy, which is the desired output exergy, to the exergy inflow. The general exergy balance of water-based PV/T is provided in Eqs. (2.7) and (2.8) [73, 74].

$$\Sigma E_{x_{\text{in}}} - \Sigma E_{x_o} = \Sigma E_{x_d} \tag{2.7}$$

or it can be written as:

$$\Sigma E_{x_{\text{in}}} - \Sigma\left(E_{x_{\text{th}}} + E_{x_{\text{PV}}}\right) = \Sigma E_{x_d} \tag{2.8}$$

where the input exergy is $E_{x_{\text{in}}}$, the output exergy is E_{x_o}, and the thermal exergy is $E_{x_{\text{th}}}$. The equations describing each are illustrated bellow [61]:

$$\Sigma E_{x_{\text{in}}} = A_C N_C I\left[1 - \frac{4}{3}\left(\frac{T_a}{T_s} + \frac{1}{3}\left(\frac{T_a}{T_s}\right)^4\right)\right] \tag{2.9}$$

$$\Sigma E_{x_{\text{th}}} = Q_u\left[1 - \frac{T_a + 273}{T_o + 273}\right] \tag{2.10}$$

$$\Sigma E_{x_{\text{PV}}} = \eta_c A_C N_C I \tag{2.11}$$

where A_C, N_C, and I represent the collector area, number of collectors, and solar radiation, respectively. The ambient and sun temperatures are represented by T_a and T_s, respectively. Hence, the PV/T exergy is provided in Eq. 2.12 [61].

$$E_{x_{\mathrm{PV/T}}} = E_{x_{\mathrm{th}}} + E_{x_{\mathrm{PV}}} \tag{2.12}$$

Finally, the exergy destruction, or irreversibility, and exergy efficiency are provided in Eqs. 2.12 [75–77] and 2.13 [61].

$$E_{x_d} = T_{a,\mathrm{ref}} .\dot{S}_{\mathrm{gen}} \tag{2.13}$$

where $T_{a,\mathrm{ref}}$ is the reference ambient temperature and S_{gen} is the entropy generation rate.

$$\eta_{\mathrm{ex}} = 1 - \frac{E_{x_d}}{E_{x_{\mathrm{in}}}} \tag{2.14}$$

Ceylan and Gurel [78] experimentally analyzed a new design of water-cooled PV/T system with forced circulation. Exergy analysis was performed for the experimental data to calculate exergy variations of thermal and electrical exergies. Moreover, the overall exergy efficiency was found to reach 17% at 45 °C and 21% at 55 °C. While Fayaz et al. [79] performed a 3D numerical analysis of PV/T with water and MWCNT-water nanofluid using COMSOL Multiphysics software, Chap. 3 will introduce nanofluid fundamentals and concepts. The numerical model accounted for incident energy, generated electrical power, extracted thermal energy, and PV/T efficiencies. Different methods of energy and exergy analysis are presented in the literature [80–82].

2.13 Conclusions and Recommendations

In this chapter, the chronology, rationale, theory, concepts, and literature associated with photovoltaic thermal (PV/T) collectors have been discussed. Moreover, the aims, design considerations, and evaluation criteria are presented. The literature review work was done to further elaborate and introduce the research studies on working fluid-based PV/T designs. This was done due to working fluids being the main way to classify PV/T systems. Another aspect of PV/T design were also discussed, such as the passage flow and tube geometry of the thermal absorber. This section will revisit some of the main topics, of this chapter, as well as present the conclusions and recommendations.

Conclusions

1. PV/T systems exhibit better all-around thermal and electrical efficiencies compared to conventional photovoltaic modules. Refer to Sect. 2.3.1.
2. Higher overall energy yields can be established by installing PV/T collectors on rooftops instead of separate photovoltaic modules and solar thermal collectors. Refer to Sect. 2.3.1.
3. Across the literature, high efficiencies are observed for PV/T collectors during indoor experiments [42] than those of outdoor experiments [35–37].
4. Higher overall total efficiencies are observed for refrigerant-based [52, 54] and air- and water-based PV/T systems [48–51], followed by water-based PV/Ts, and the lowest are air-based PV/Ts.
5. In most cases, the increase of solar irradiance levels leads to increase in electrical, thermal, and overall efficiencies. However, when the heat transfer is insufficient to cool the PV module temperature or any disruption in the process, then even at high solar irradiance, the efficiencies of the collector will drop. In that case, increase of solar irradiance will most probably lead to opposite effects.
6. In most cases, the increase of mass flow rate leads to better overall heat transfer between the PV module and solar thermal collector which leads to enhanced electrical, thermal, and overall PV/T efficiencies.

Recommendations

1. The lack of consensus in period of experiments, method of testing, and range of mass flow rates calls for the need for establishing a global standard for designing, testing, evaluating, and installing of photovoltaic-thermal (PV/T) collectors.
2. To perform more long-term tests on PV/T collectors to study factor such as the lifetime of the PV module. Given that degradation in PV modules occurs due to thermal stress and that PV/Ts serve in reducing that stress, it is recommended to analyze and study the power yield, exergy, and efficiency of the PV module for long stretches of time.
3. To investigate the performance of different geometries and different base fluids for the same geometries to further understand the effect of PV/T design parameters.
4. Most of the studies discussed in the literature are of direct systems (for direct use) and with active components. The mechanism to regulate those components or integrate such systems to an actual thermal load is usually not discussed. Types of valves, interconnecting elements and their material, and controller (if any) should also be discussed in terms of power requirements and design aspects.
5. Majority of research in the field of PV/T does not examine pressure drop and pumping power requirements. Hence, it is recommended to produce work which employs those metrics which can help in the complete evaluation of the system.
6. The type and efficiency of the heat exchanger in indirect systems should be discussed as well. This is because it can help to view the temperature value of the water for use and it can also help other researchers obtain most suitable type for studies, to enrich the field.

References

1. P.G. Charalambous, G.G. Maidment, S.A. Kalogirou, K. Yiakoumetti, Photovoltaic thermal (PV/T) collectors: A review. Appl. Therm. Eng. **27**(2–3), 275–286 (2007)
2. E. Skoplaki, J.A. Palyvos, On the temperature dependence of photovoltaic module electrical performance: A review of efficiency/power correlations. Sol. Energy **83**(5), 614–624 (2009)
3. S. Dubey, J.N. Sarvaiya, B. Seshadri, Temperature dependent photovoltaic (PV) efficiency and its effect on PV production in the world–a review. Energy Procedia **33**, 311–321 (2013)
4. A. Royne, C.J. Dey, D.R. Mills, Cooling of photovoltaic cells under concentrated illumination: A critical review. Sol. Energy Mater. Sol. Cells **86**(4), 451–483 (2005)
5. V.L. Dalal, A.R. Moore, Design considerations for high-intensity solar cells. J. Appl. Phys. **48**(3), 1244–1251 (1977)
6. R. Daghigh, M.H. Ruslan, K. Sopian, Advances in liquid based photovoltaic/thermal (PV/T) collectors. Renew. Sust. Energ. Rev. **15**(8), 4156–4170 (2011)
7. A. Fudholi, K. Sopian, M.H. Yazdi, M.H. Ruslan, A. Ibrahim, H.A. Kazem, Performance analysis of photovoltaic thermal (PVT) water collectors. Energy Convers. Manag. **78**, 641–651 (2014)
8. F. Yazdanifard, M. Ameri, Exergetic advancement of photovoltaic/thermal systems (PV/T): A review. Renew. Sust. Energ. Rev. **97**, 529–553 (2018)
9. H.A. Kazem, Evaluation and analysis of water-based photovoltaic/thermal (PV/T) system. Case Stud. Therm. Eng. **13**, 100401 (2019)
10. M.S. Hossain, A.K. Pandey, J. Selvaraj, N.A. Rahim, A. Rivai, V.V. Tyagi, Thermal performance analysis of parallel serpentine flow based photovoltaic/thermal (PV/T) system under composite climate of Malaysia. Appl. Therm. Eng. **153**, 861–871 (2019)
11. E. Lorenzo, *Solar Electricity: Engineering of Photovoltaic Systems* (Earthscan/James & James, Spain, 1994)
12. E. Radziemska, Performance analysis of a photovoltaic-thermal integrated system. Int. J. Photoenergy (2009)
13. T.T. Chow, J.W. Hand, P.A. Strachan, Building-integrated photovoltaic and thermal applications in a subtropical hotel building. Appl. Therm. Eng. **23**(16), 2035–2049 (2003)
14. E. Radziemska, The effect of temperature on the power drop in crystalline silicon solar cells. Renew. Energy **28**(1), 1–12 (2003)
15. M. Lämmle, A. Oliva, M. Hermann, K. Kramer, W. Kramer, PVT collector technologies in solar thermal systems: A systematic assessment of electrical and thermal yields with the novel characteristic temperature approach. Sol. Energy **155**, 867–879 (2017)
16. N.J. Regnier, L.W. Schmidt, D.E. Kefes; Hoffman Electronics Corp, Solar energy converting apparatus or the like. U.S. Patent 2,946,945 (1960)
17. R.H. Dean, L.S. Napoli, S.G. Liu; RCA Corp, Method of fabricating a photovoltaic device. U.S. Patent 3,999,283 (1976)
18. P.F. Varadi; Solarex Corp, Panel for solar energy cells. U.S. Patent 4,056,405 (1977)
19. A. Kirpich, *Conceptual Design and Systems Analysis of Photovoltaic Power Systems* (Energy Research and Development Administration, 1977)
20. E.C. Kern Jr, M.C. Russell, Combined photovoltaic and thermal hybrid collector systems (No. COO-4577-3; CONF-780619-24). Massachusetts Inst. of Tech., Lexington (USA). Lincoln Lab (1978)
21. L.W. Florschuetz, Extension of the Hottel-Whillier model to the analysis of combined photovoltaic/thermal flat plate collectors. Sol. Energy **22**(4), 361–366 (1979)
22. P. Raghuraman, Analytical predictions of liquid and air photovoltaic/thermal, flat-plate collector performance. J. Sol. Energy Eng. **103**(4), 291–298 (1981)
23. B. Lalović, Z. Kiss, H. Weakliem, A hybrid amorphous silicon photovoltaic and thermal solar collector. Sol. Cells **19**(2), 131–138 (1986)
24. H.P. Garg, R.K. Agarwal, Some aspects of a PV/T collector/forced circulation flat plate solar water heater with solar cells. Energy Convers. Manag. **36**(2), 87–99 (1995)

25. C.F. Johnson, Solar concentrator for heat and electricity. U.S. Patent 6,080,927 (2000)
26. T. Bergene, O.M. Løvvik, Model calculations on a flat-plate solar heat collector with integrated solar cells. Sol. Energy **55**(6), 453–462 (1995)
27. J. Coventry, Simulation of a concentrating PV/thermal collector using TRNSYS. Conference paper (ANU Research Publications, 2002). http://hdl.handle.net/1885/40836, http://digitalcollections.anu.edu.au/handle/1885/40836
28. A. Khelifa, K. Touafek, H.B. Moussa, Approach for the modelling of hybrid photovoltaic–thermal solar collector. IET Renew. Power Gener. **9**(3), 207–217 (2014)
29. F. Sarhaddi, S. Farahat, H. Ajam, A.M.I.N. Behzadmehr, M.M. Adeli, An improved thermal and electrical model for a solar photovoltaic thermal (PV/T) air collector. Appl. Energy **87**(7), 2328–2339 (2010)
30. P. Gang, F. Huide, Z. Tao, J. Jie, A numerical and experimental study on a heat pipe PV/T system. Sol. Energy **85**(5), 911–921 (2011)
31. E. Cuce, T. Bali, S.A. Sekucoglu, Effects of passive cooling on performance of silicon photovoltaic cells. Int. J. Low Carbon Technol. **6**(4), 299–308 (2011)
32. S.K. Natarajan, T.K. Mallick, M. Katz, S. Weingaertner, Numerical investigations of solar cell temperature for photovoltaic concentrator system with and without passive cooling arrangements. Int. J. Therm. Sci. **50**(12), 2514–2521 (2011)
33. K. Araki, H. Uozumi, M. Yamaguchi, A simple passive cooling structure and its heat analysis for 500/spl times/concentrator PV module. In Conference Record of the Twenty-Ninth IEEE Photovoltaic Specialists Conference (IEEE, May 2002), pp. 1568–1571
34. J.K. Tonui, Y. Tripanagnostopoulos, Improved PV/T solar collectors with heat extraction by forced or natural air circulation. Renew. Energy **32**(4), 623–637 (2007)
35. J.G. Ahn, J.H. Kim, J.T. Kim, A study on experimental performance of air-type PV/T collector with HRV. Energy Procedia **78**, 3007–3012 (2015)
36. J. Hu, W. Chen, D. Yang, B. Zhao, H. Song, B. Ge, Energy performance of ETFE cushion roof integrated photovoltaic/thermal system on hot and cold days. Appl. Energy **173**, 40–51 (2016)
37. E.D. Rounis, A.K. Athienitis, T. Stathopoulos, Multiple-inlet Building Integrated Photovoltaic/Thermal system modelling under varying wind and temperature conditions. Sol. Energy **139**, 157–170 (2016)
38. J.H. Kim, S.H. Park, J.T. Kim, Experimental performance of a photovoltaic-thermal air collector. Energy Procedia **48**, 888–894 (2014)
39. G. Ömeroğlu, CFD analysis and electrical efficiency improvement of a hybrid PV/T panel cooled by forced air circulation. Int. J. Photoenergy **2018**(12), 1–11 (2018)
40. J.C. Mojumder, W.T. Chong, H.C. Ong, K.Y. Leong, An experimental investigation on performance analysis of air type photovoltaic thermal collector system integrated with cooling fins design. Energ. Buildings **130**, 272–285 (2016)
41. S. Dubey, G.S. Sandhu, G.N. Tiwari, Analytical expression for electrical efficiency of PV/T hybrid air collector. Appl. Energy **86**(5), 697–705 (2009)
42. G. Jin, H. Ruslan, S. Mat, M.Y. Othman, A. Zaharim, K. Sopian, Experiment study on single-pass photovoltaic-thermal (PV/T) air collector with absorber. In 9th WSEAS International Conference on System Science and Simulation in Engineering, ICOSSSE'10 (October 2010), pp. 435–438
43. S.M. Sultan, C.P. Tso, A thermal performance study for different glazed water based photovoltaic thermal collectors. In AIP Conference Proceedings, Vol. 2030, No. 1 (AIP Publishing, November 2018), p. 020307
44. L. Lu, X. Wang, S. Wang, X. Liu, Analysis of three different sheet-and-tube water-based flat-plate PVT collectors. J. Energy Eng. **143**(5), 04017022 (2017)
45. A.A. Alzaabi, N.K. Badawiyeh, H.O. Hantoush, A.K. Hamid, Electrical/thermal performance of hybrid PV/T system in Sharjah, UAE. Int. J. Smart Grid Clean Energy **3**(4), 385–389 (2014)
46. I. Nardi, D. Ambrosini, T. de Rubeis, D. Paoletti, M. Muttillo, S. Sfarra, Energetic performance analysis of a commercial water-based photovoltaic thermal system (PV/T) under summer conditions. In Journal of Physics: Conference Series, Vol. 923, No. 1. (IOP Publishing, November 2017), p. 012040

47. R. Liang, J. Zhang, L. Ma, Y. Li, Performance evaluation of new type hybrid photovoltaic/thermal solar collector by experimental study. Appl. Therm. Eng. **75**, 487–492 (2015)
48. H. Jarimi, M.N.A. Bakar, N.A. Manaf, M. Othman, M. Din, Mathematical modelling of a finned bi-fluid type photovoltaic/thermal (PV/T) solar collector. In 2013 IEEE Conference on Clean Energy and Technology (CEAT) (IEEE, November 2013), pp. 163–168
49. D. Su, Y. Jia, X. Huang, G. Alva, Y. Tang, G. Fang, Dynamic performance analysis of photovoltaic–thermal solar collector with dual channels for different fluids. Energy Convers. Manag. **120**, 13–24 (2016)
50. A. Tiwari, M.S. Sodha, Performance evaluation of hybrid PV/thermal water/air heating system: A parametric study. Renew. Energy **31**(15), 2460–2474 (2006)
51. H. Jarimi, M.N.A. Bakar, M. Othman, M. Din, Bi-fluid photovoltaic/thermal PV/T solar collector with three modes of operation: Experimental validation of a theoretical model, in *Mediterranean green buildings & renewable energy*, (Springer, Cham, 2017), pp. 445–464
52. H.L. Tsai, Design and evaluation of a photovoltaic/thermal-assisted heat pump water heating system. Energies **7**(5), 3319–3338 (2014)
53. J. Ji, H. He, T. Chow, G. Pei, W. He, K. Liu, Distributed dynamic modeling and experimental study of PV evaporator in a PV/T solar-assisted heat pump. Int. J. Heat Mass Transf. **52**(5–6), 1365–1373 (2009)
54. X. Zhao, X. Zhang, S.B. Riffat, Y. Su, Theoretical study of the performance of a novel PV/e roof module for heat pump operation. Energy Convers. Manag. **52**(1), 603–614 (2011)
55. M.A.M. Rosli, Y.J. Ping, S. Misha, M.Z. Akop, K. Sopian, S. Mat, A.N. Al-Shamani, M.A. Saruni, Simulation study of computational fluid dynamics on photovoltaic thermal water collector with different designs of absorber tube. J. Adv. Res. Fluid Mech. Therm. Sci. **52**(1), 12–22 (2018)
56. K. Sopian, G.L. Jin, M.Y Othman, S.H. Zaidi, M.H. Ruslan, Advanced absorber design for photovoltaic thermal (PV/T) collectors. Recent Researches in Energy, Environment, and Landscape Architecture (2011)
57. M.M. Sardouei, H. Mortezapour, K.J. Naeimi, Temperature distribution and efficiency assessment of different PVT water collector designs. Sādhanā **43**(6), 84 (2018)
58. N.A. Manaf, A. Bakar, M. Nazari, H. Jarimi, S. Muhamed, M. Othman, Design of a single-pass bi-fluid photovoltaic/thermal (PV/T) solar collector. Int. J. Chem. Environ. Eng. **4** (2013)
59. M.S. Hossain, A.K. Pandey, J. Selvaraj, N.A. Rahim, M.M. Islam, V.V. Tyagi, Two side serpentine flow based photovoltaic-thermal-phase change materials (PVT-PCM) system: Energy, exergy and economic analysis. Renew. Energy **136**, 1320–1336 (2019)
60. M.A.M. Rosli, S. Misha, K. Sopian, S. Mat, M.Y. Sulaiman, E. Salleh, Parametric analysis on heat removal factor for a flat plate solar collector of serpentine tube. World Appl. Sci. J. **29**(2), 184–187 (2014)
61. A.H.M.A.D. Fudholi, A. Ibrahim, M.Y. Othman, M. Hafidz, Energy and exergy analyses on water based photovoltaic thermal (PVT) collector with spiral flow absorber. In 2nd International Conference on Energy Systems, Environment, Antalya, Turkey (October 2013), pp. 70–74
62. V.N. Palaskar, S.P. DESHMUKH, Study of oscillatory flow heat exchanger used in hybrid solar system fitted with fixed reflectors. Int. J. Ren. Energy Res. **4**(4), 893–900 (2014)
63. A. Ibrahim, G.L. Jin, Hybrid Photovoltaic Thermal (PV/T) Air and Water Based Solar Collectors Suitable for Building Integrated Applications Adnan Ibrahim, Goh Li Jin, Roonak Daghigh, Mohd Huzmin Mohamed Salleh, Mohd Yusof Othman, Mohd Hafidz Ruslan, Sohif Mat and Kamaruzzaman Sopian Solar Energy Research Institute, University Kebangsaan Malaysia, 43600, Bangi, Selangor, Malaysia. Am. J. Environ. Sci. **5**(5), 618–624 (2009)
64. M.R. Karim, M.A.R. Akhanda, Study of a Hybrid Photovoltaic Thermal (PVT) Solar Systems Using Different Ribbed Surfaces Opposite to Absorber Plate (2011)
65. D.İ.L.Ş.A.D. Engin, M.U.S.T.A.F.A. Engin, Simulation modelling of a photovoltaic and thermal collector (PV/T) hybrid system. In 6th International Ege Energy Symposium and Exhibition (June 2012), pp. 28–30

66. I. Guarracino, A. Mellor, N.J. Ekins-Daukes, C.N. Markides, Dynamic coupled thermal-and-electrical modelling of sheet-and-tube hybrid photovoltaic/thermal (PVT) collectors. Appl. Therm. Eng. **101**, 778–795 (2016)
67. M. Alobaid, B. Hughes, D. O'Connor, J. Calautit, A. Heyes, Improving thermal and electrical efficiency in photovoltaic thermal systems for sustainable cooling system integration. J. Sustain. Dev. Energy Water Environ. Syst. **6**(2), 305–322 (2018)
68. M.A.M. Rosli, S. Mat, H. Ruslan, K. Sopian, H.T. Jaya, Parametric study on water based photovoltaic thermal collector. In 7th International Conference on Renewable Energy Sources (RES'13) (2013), pp. 135–140
69. A.N. Al-Shamani, S. Mat, M.H. Ruslan, A.M. Abed, K. Sopian, Effect of new ellipse design on the performance enhancement of PV/T collector: CDF approach. Int. J. Environ. Sustain. **5**(2) (2016)
70. L. Sun, L. Yang, L.L. Shao, C.L. Zhang, Overall thermal performance oriented numerical comparison between elliptical and circular finned-tube condensers. Int. J. Therm. Sci. **89**, 234–244 (2015)
71. W. He, Y. Zhang, J. Ji, Comparative experiment study on photovoltaic and thermal solar system under natural circulation of water. Appl. Therm. Eng. **31**(16), 3369–3376 (2011)
72. A. Hepbasli, Exergetic modeling and assessment of solar assisted domestic hot water tank integrated ground-source heat pump systems for residences. Energ. Buildings **39**(12), 1211–1217 (2007)
73. A. Tiwari, S. Dubey, G.S. Sandhu, M.S. Sodha, S.I. Anwar, Exergy analysis of integrated photovoltaic thermal solar water heater under constant flow rate and constant collection temperature modes. Appl. Energy **86**(12), 2592–2597 (2009)
74. S. Dubey, G.N. Tiwari, Analysis of PV/T flat plate water collectors connected in series. Sol. Energy **83**(9), 1485–1498 (2009)
75. A.S. Joshi, I. Dincer, B.V. Reddy, Energetic and exergetic analyses of a photovoltaic system. In Canadian Society for Mechanical Engineering (CSME) Forum, Ottawa, Canada (June 2008)
76. A. Hepbasli, A key review on exergetic analysis and assessment of renewable energy resources for a sustainable future. Renew. Sust. Energ. Rev. **12**(3), 593–661 (2008)
77. F. Sarhaddi, S. Farahat, H. Ajam, A. Behzadmehr, Exergetic performance evaluation of a solar photovoltaic (PV) array. Aust. J. Basic Appl. Sci. **4**(3), 502–519 (2010)
78. İ. Ceylan, A.E. Gürel, Exergetic analysis of a new design photovoltaic and thermal (PV/T) System. Environ. Prog. Sustain. Energy **34**(4), 1249–1253 (2015)
79. H. Fayaz, R. Nasrin, N.A. Rahim, M. Hasanuzzaman, Energy and exergy analysis of the PVT system: Effect of nanofluid flow rate. Sol. Energy **169**, 217–230 (2018)
80. B.J. Huang, T.H. Lin, W.C. Hung, F.S. Sun, Performance evaluation of solar photovoltaic/thermal systems. Sol. Energy **70**(5), 443–448 (2001)
81. T.T. Chow, J. Ji, W. He, Photovoltaic-thermal collector system for domestic application. J. Solar Energy Eng. **129**(2), 205–209 (2007)
82. X. Zhang, X. Zhao, S. Smith, J. Xu, X. Yu, Review of R&D progress and practical application of the solar photovoltaic/thermal (PV/T) technologies. Renew. Sust. Energ. Rev. **16**(1), 599–617 (2012)

Chapter 3
Advanced PV/T Systems

3.1 Background

In this chapter, recent developments and advanced designs of photovoltaic thermal (PV/T) systems are introduced by first addressing their concepts and then providing recently published works to support their explanation. The theory and concepts are first explained for building-integrated, nanofluid-based, and nano-phase change material (nano-PCM)-based PV/T systems. Deeper analysis of performance in terms of energy and exergy are provided as well. Then, discussion on various types of novel PV/T systems is carried out to cover a broader scope on PV/T. The developments in this field include the applications of PV/T systems in industrial, commercial, and residential settings. Hence, the final section of the chapter presents a discussion on the various proposed applications for PV/T systems.

3.2 Introduction

The field of photovoltaic thermal (PV/T) collectors can be multidisciplinary, given that many dimensions, from different disciplines, are discussed and considered for the evaluation of PV/T. For instance, it is necessary for the evaluation of the system to consider the electrical, thermal, economic, etc., aspects. The type of studies or processes within PV/T includes design, manufacturing or building prototype, installation, testing, and evaluation. The associated aspects, with those studies and processes, include electrical, thermal, economic, environmental, and manufacturing. Hence, the disciplines that are tied to this topic include electrical engineering, mechanical engineering (heat transfer, thermodynamics and fluid dynamics, etc.), physics, economics, environmental science and technology, and manufacturing engineering. Each discipline aims to point out or highlight parameters that help describe the aspects. For instance, electrical engineering is used to describe the

A. H. A. Al-Waeli et al., *Photovoltaic/Thermal (PV/T) Systems*,
https://doi.org/10.1007/978-3-030-27824-3_3

electrical behavior of the PV/T with parameters such as electrical power, energy, and exergy. In addition, power conversion efficiency and I-V, P-V, and P-I curves. Therefore, given the multitude of ways to configure and evaluate PV/T systems, the scope for development is quite broad. Hence, many studies have been presented with novel methods to improve the performance, sustainability, and reliability of PV/T systems.

This chapter aims to explain the theory and concepts of these developments and provide a review of associated studies. The focus of the review will be to present the technique and its outcomes to rate the current status of the technique and potential for industry integration. In addition, conclusions and recommendations are made in the chapter summary section.

3.3 Trends

As displayed in the history and chronology of research in PV/T systems, Sect. 2.4.1, the focus of studies was mainly to verify the utility of combining photovoltaic modules and solar thermal collectors to form this technology. Moreover, the aims developed to viewing the performance of this hybrid collector considering different design and operating conditions. Lately, more focus has been paid to employ this technology into industrial, commercial, and residential settings. Hence, more efforts have been made to investigate special and novel material for enhanced heat transfer process and better thermal regulation, such as nanofluids and nano-phase change material (nano-PCM), instead of the conventional working fluids and design material. Moreover, research to study the integration of these collectors into buildings, referred to as building-integrated PV/T (BIPV/T), and modes of grid-connected configurations, grid-connected PV/T (GCPV/T), has been carried out. Figure 3.1a–d shows examples of nanofluid-based, nano-PCM-based, building-integrated, and grid-connected PV/Ts discussed by different studies, respectively.

The figure provides examples of nanofluid samples, nano-PCM samples and preparation, building-integrated PV/T view, and the grid-connected PV/T view. Each one of these methods and/or systems is explained in detail in Sects. 3.4, 3.5, and 3.6, along with review of recent works and state-of-the-art experimental facilities.

3.4 Nanofluid-Based PV/T System

Nanofluids are fluids that contain nanometer-sized particles (solid) in a base fluid (liquid). There are two methods to prepare nanofluids which are (i) the one-step method and (ii) the two-step method. Most commonly used is the two-step method for its simplicity and low cost. The two-step method is done by first forming the particles and then dispersing them in a base fluid. Nanofluids are used instead of

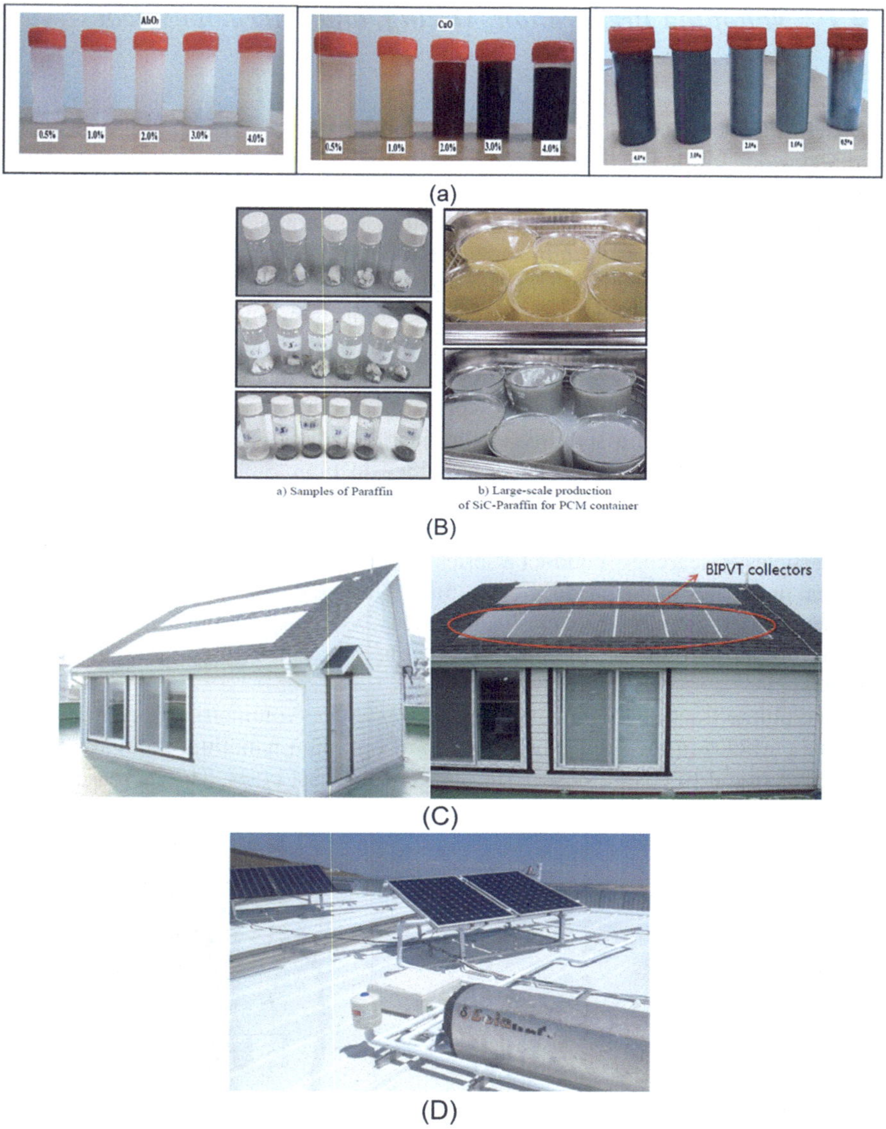

Fig. 3.1 (**a**) Nanofluid samples from PV/T system – study by Alwaeli et al. [1]. (**b**) Nano-PCM sample from PV/T system – study by Alwaeli et al. [2]. (**c**) Building-integrated PV/T system by Kim et al. [3] and (**d**) grid-connected Fuentes et al. [4]

water in a nanofluid-based PV/T system. This is because these fluids exhibit better thermophysical properties [5], which is beneficial to enhancing the heat transfer of a PV/T collector. Ultimately, this leads to achieving a better thermal and electrical performance.

3.4.1 Nanofluid Preparation, Mixing, and Thermophysical Properties

The two-step method is conducted by first obtaining the solid nanoparticles, usually in dry powder form. The second step is to mix the nanoparticles with a base fluid using a sonication device or magnetic stirrer. Because of the high surface activity of nanoparticles made by this method, they tend to agglomerate. Hence, mixing using sonicator device, such as an ultrasonic probe, is highly recommended to create a uniform dispersion of the nanoparticles and eliminate side effects such as agglomeration and sedimentation. The nanofluid solution then must be sonicated using ultrasonic shakers or vibrates which provide sonication sessions, typically 80–100 minutes, to achieve a stable, uniform, and continuous suspension [6]. The use of surfactant, also referred to as dispersant, is highly recommended to improve thermal conductivity and stability of the nanofluid [7]. A photograph of a probe sonicator and an ultrasonic shaker is provided in Figs. 3.2 and 3.3, respectively, while Fig. 3.3 shows the methodology of two-step method in published research.

In order to view the changes that occur to the base fluid as a result of adding nanoparticles, researchers conduct a series of tests aimed to investigate the thermophysical properties of the produced nanofluid. These tests are carried out to measure the thermal conductivity, specific heat capacity, viscosity, and density. Figure 3.4 shows this as part of the methodology being performed throughout the literature [9–12].

The volume fraction of the nanofluids expresses the volumetric concentration of the nanoparticles in the fluid. The selection of optimum nanofluid volume fraction is highly critical to ensure best performance with least costs. The increase of the

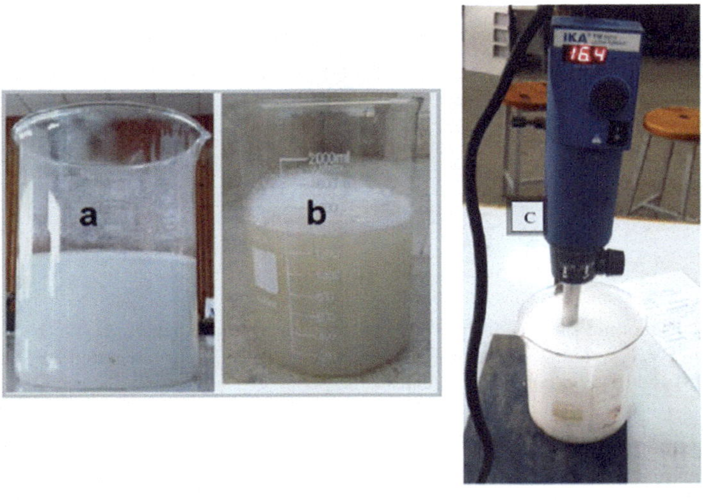

Fig. 3.2 Silica nanofluid (**a**) prior to experiment, (**b**) post-experiment, and (**c**) during experiment; preparation with a probe sonicator [8]

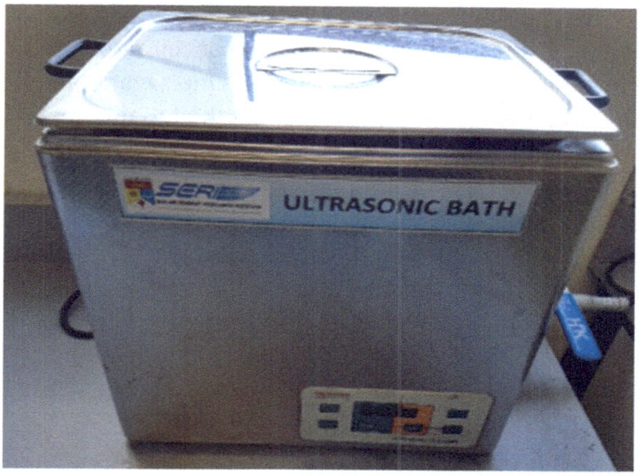

Fig. 3.3 Ultrasonic bath shaker used for two-step method

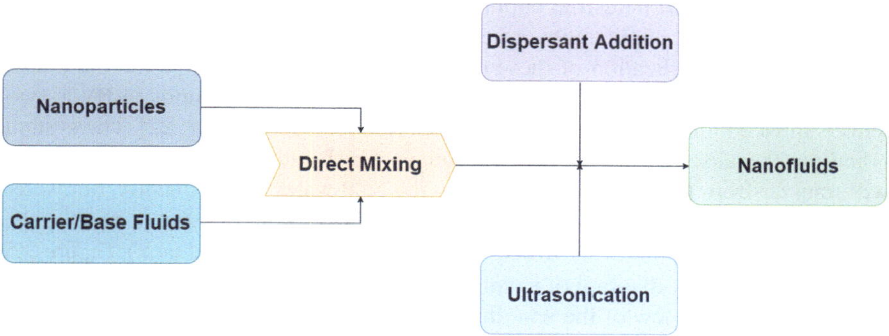

Fig. 3.4 Methodology of the two-step method

volume fraction means that more nanoparticles are required to form the desired suspension, and hence it is more costly. However, the decrease of the volume fraction means that less nanoparticles are employed in the desired suspension, and hence its thermal conductivity is reduced [13]. This means that examining the cost-effectiveness of nanoparticles leads to finding the optimum nanofluid volume fraction. Equation 3.1 [14–16] shows how to calculate the volume fraction:

$$\left(\frac{m_{np}/\rho_{np}}{m_{np}/\rho_{np} + m_{bf}/\rho_{bf}} \right) \tag{3.1}$$

where ∅ is the nanoparticle volume fraction in percentage; m_{np} and m_{bf} are the weight of nanoparticles and base fluid, respectively; while ρ_{np} and ρ_{bf} are the densities of nanoparticles and base fluid, respectively.

As Fig. 3.4 shows, different samples with different concentration ratios are produced to examine the optimum concentration ratio for the desired application (working fluid in a PV/T system). Drazazga et al. [17] conducted an experimental investigation using the two-step method to mix metal oxide (CuO and Al_2O_3) water nanofluids in order to study the effect of sonication time, stabilizer type, and suspension pH on stability of these nanofluids. The study presented a brief discussion on effectiveness of preparation method on nanofluid stability. A visual test was conducted for all samples after the mixing and 3 days after the process. The author's claim that larger nanoparticle diameter leads to worse suspension stability of nanofluids. Moreover, worse stability can also be achieved by having higher mass fraction, the author's claim. The study recommends employing SDBS as dispersant and employing the physical dispersion method of stirring and oscillation to make the nanofluid uniform and stable.

Given that reduction of the size of a heat exchange system can be done by employing highly efficient heat transfer fluids, this increase in thermal conductivity will attribute to miniature devices, which provide many advantages that are highly desirable in advanced industrial applications. In view of the advantages nanofluids hold in heat transfer applications, research studies in solar energy have been made to investigate the utilization of these nanofluids in solar thermal, photovoltaic and photovoltaic thermal (PV/T) systems [18–20]. The use of nanofluids in PV/T systems is most commonly done for indirect-active PV/T systems [21, 22]. The system is indirect because it employs heat exchanger which intakes cold water. The heat will transfer from the heater nanofluid into the cold water. Consequently, hot water will be produced for use, while nanofluids will be cooled to start a new cycle. Commonly, the nanofluids are stored at a separate tank (nanofluid tank). Figure 3.5 show a typical block diagram of a nanofluid-based PV/T system.

The literature review of the nanofluid-based PV/T systems is presented in Sect. 3.4.2 which encompasses a variety of proposed ideas from the literature of established frame works.

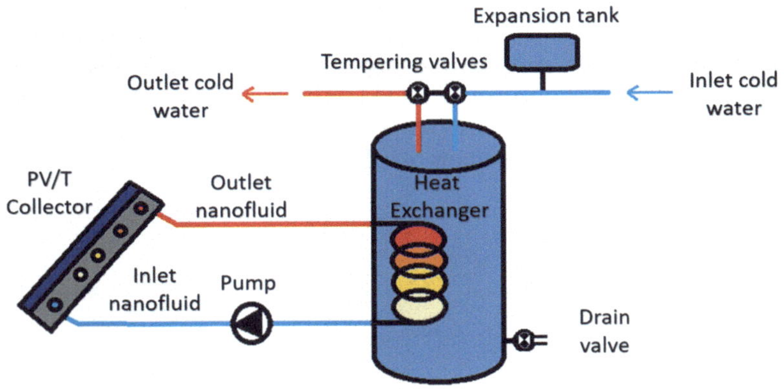

Fig. 3.5 Schematic diagram of a nanofluid-based PV/T system

3.4.2 Literature Review on Nanofluid-Based PV/T Systems

This section provides a detailed review of recent studies in nanofluid-based PV/T systems. The section is divided into four subsections: (i) nanofluids in flat-plate type (PV/T) collectors, (ii) nanofluids in spectral splitting (PV/T) collectors, (iii) nanofluids in Jet impingement (PV/T) collectors, and (iv) advanced nanofluid-based (PV/T).

Nanofluids in Flat-Plate (PV/T) Collectors

As explained in Chaps. 1 and 2, flat-plate (PV/T) collectors are composed of photovoltaic modules attached to absorbers and/or tubes which carry fluids, either liquid or air, for heat transfer between them. In this section, studies that investigate the performance of PV/T collectors with nanofluids as working fluids are reviewed.

Michael and Iniyan [22] constructed a nanofluid-based PV/T collector using CuO/H2O nanofluid and experimentally tested its performance to evaluate the design and nanofluid for PV/T compared to conventional water-based PV/T collector. The volume concentration of nanofluid was set for 0.05%, and the mass flow rate was kept at 0.01 kg/s (laminar flow). The findings of the study show an increase in thermal efficiency up to 45.76%. Authors recommend the use of heat exchangers with higher effectiveness for improved performance. For the mixing procedure, surfactant was used to improve the stability of the nanofluid. Increase of heat transfer coefficient of around 10.03% and 9.91% was observed as resultant of using CuO/water nanofluid compared to water with and without glazing, respectively.

Purohit et al. [23] numerically investigated the use of alumina-water nanofluid as working fluid for a flat-plate PV/T under laminar flow condition. The study offers comparison between nanofluid and other working fluids in terms of Reynold number and pumping power. Different volume fractions are used 1%, 4%, and 6% of nanoparticles (dimension of 20 nm) with Reynolds number ranging between 300 and 1800. The findings show an average improvement in heat transfer coefficient of 25.2% for equal Reynolds number and 13.8% for equal pumping power. The entropy generation, under equal Reynolds number, for nanofluid is reduced significantly with around 31%. The results show increase of average heat transfer rate with increase of nanofluid volume fraction and increase of Reynolds numbers. Figures 3.6 and 3.7 show the discussed comparisons.

Khanjari et al. [24] numerically investigated the effect of two environmental parameters on performance of water-based and nanofluid (Al2O3-water)-based PV/T and compared the two. The study employs CFD simulation using ANSYS Fluent with consideration of conduction and convection heat transfer. Solar irradiance and fluid inlet temperature are modelled for the proposed system. The results of the simulation show decrease of electrical efficiency with increase of solar irradiance and also with increase of fluid inlet temperature. The comparison results showcase the utility of nanofluid as working fluid where it exhibits higher heat transfer

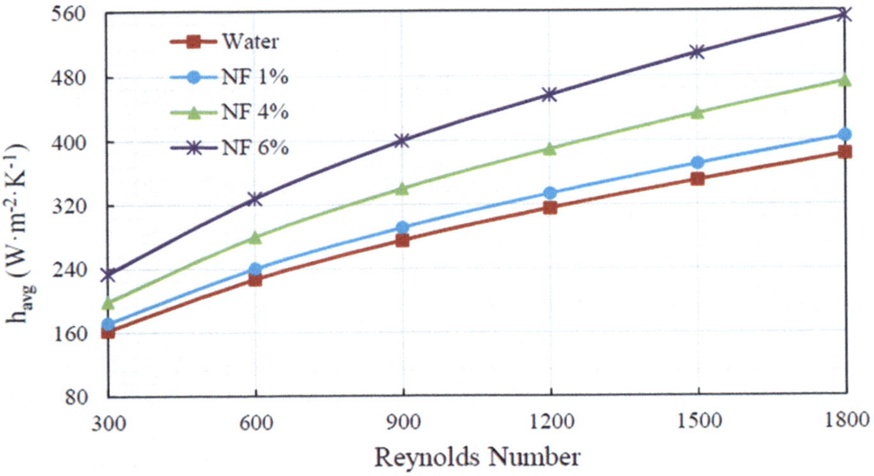

Fig. 3.6 The heat transfer coefficient comparison between base fluid and different volumes of nanofluid (1%, 4%, and 6%) under equal Reynolds number [23]

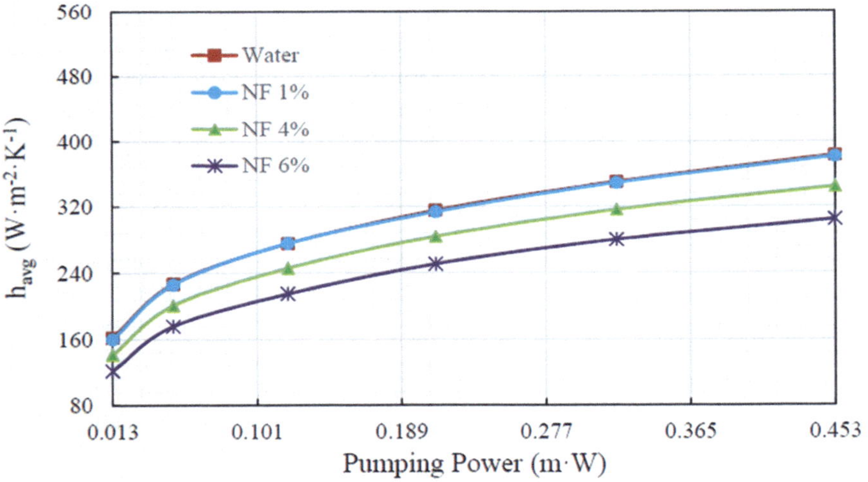

Fig. 3.7 The heat transfer coefficient comparison between base fluid and different volumes of nanofluid (1%, 4%, and 6%) under equal Reynolds number [23]

coefficient and efficiency over pure water. Experimental data from the literature was used to validate the simulation results. The simulation shows higher thermal efficiency for the nanofluid-based PV/T with around 81%, compared to around 80.1% of pure water. However, increase of inlet fluid temperature did not cause significant effect on thermal efficiency; this could be due to proximity ambient temperature.

Hussien et al. [25] conducted an indoor experimental investigation of nanofluid-based PV/T system using Al2O3-water nanofluid with different concentration ratios to view the corresponding enhancement in electrical and thermal efficiencies. The concentration ratio was varied from 0.1% to 0.5% with a step of 0.1% at a fixed mass flow rate of 0.2 liter per second. The findings show that more cell temperature drop is achieved under a concentration ratio of 0.3% which led to rise in electrical efficiency to 12.1%. The author's claim that further increase of concentration ratio (above 0.3%) had a reverse effect and led to decrease in electrical efficiency. The range and observed behavior in the study is not consistent with the literature. Overall, it is observed that use of Al2O3 nanofluid at 0.3% concentration ratio for cooling of PV/T led to drop the temperature from 79.1 °C, for conventional PV, to 42.2 C°.

Nanofluids in Spectral Splitting (PV/T) Collectors and Spectral Selective Nanofluids

Solar beam splitting is a technique used to improve the performance of a single-junction solar photovoltaic. The solar spectrum is split into different wavelength bands and employed in separate PV and thermal systems. The concept of the spectrum beam splitting of PV/T system is provided in Fig. 3.8.

Throughout the literature [26, 27], studies conducted used different selective glasses as spectrum filters. However, the use of these glasses is not economical due to their high costs, which makes this technology hard to introduce to photovoltaic market. However, selective liquids are suggested as spectrum filters to replace

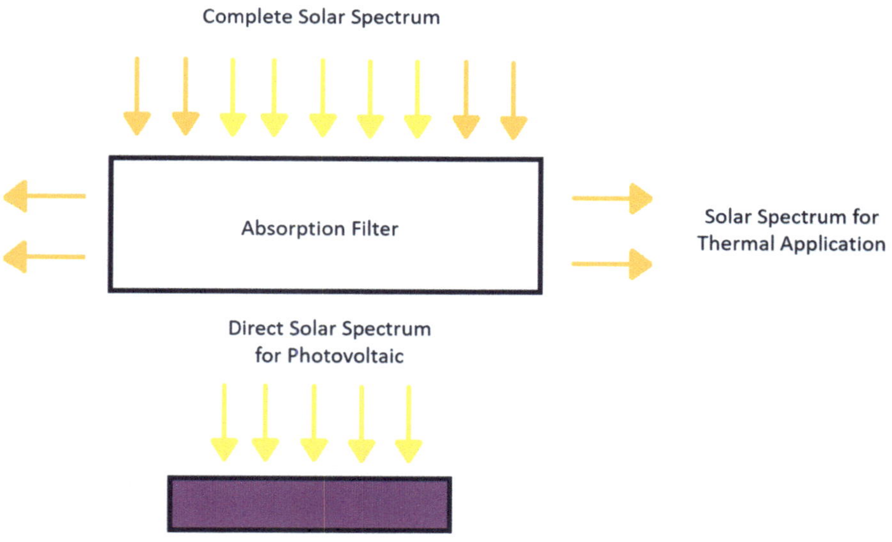

Fig. 3.8 Concept of spectrum beam splitting PV/T system

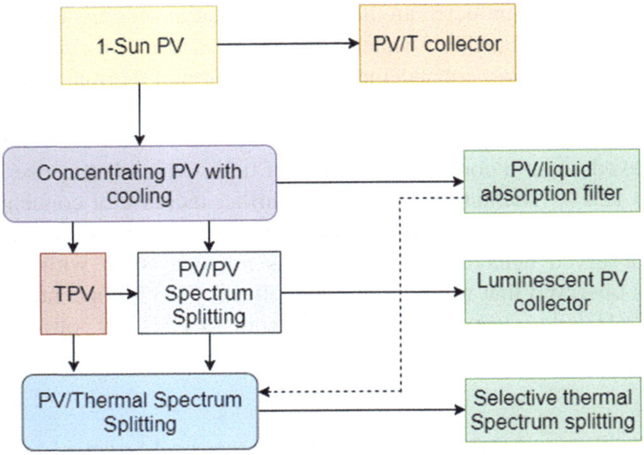

Fig. 3.9 Development in solar concentrating beam splitting systems, as described by [27]

glasses for use with photovoltaics [28–31]. As explained in the concept, unnecessary parts of the solar spectrum are filtered and hence separated to be stored as heat in the liquids, which can be used in thermal applications. These liquids include water, heat transfer oils, ethylene glycol, and nanofluids. Many of which have been are reported as feasible liquids for spectrum filtration and heat absorber suitable for PV/T applications. Imenes and Mills [27] explained the outline of development in solar concentrating beam splitting systems, as shown in Fig. 3.9.

Hjerrild et al. [32] investigated properties, optical and thermal, and characteristics of core-shell silver-silica nano-discs in a glycerol base fluid for use in PV/T applications such as spectrum splitting nanofluid PV/T. The study focuses on the stability of these nanofluids. The experiments include thermal testing and intensive UV irradiation.

Du et al. [33] presented a plasmonic nanofluid-based PV/T system with silica aerogel glazing and designed an optical filter for the nanofluid. The purpose of using silica aerogel is for thermal insulation as well as being an optically transparent layer, making it suitable for reduction of thermal losses. The concept of this system is to utilize plasmonic nanofluids in cooling PV cell through use in absorber channel under the PV, and then use the nanofluid to flow in photothermal channel placed above PV cell to absorber the parts of photons which cannot be utilized by PV cell efficiently. Hence, absorbing high-energy photons and low-energy photons of the solar spectrum. Numerical simulation was used to study the electrical and thermal behavior of energy and exergy efficiencies, as well as comparing the system's performance with that of conventional PV and PV/T under different conditions. Electrical, optical, and thermal modes were considered for the modelling. To solve simultaneously the electrical and thermal equations, the authors used a FORTRAN code. The obtained results show a clear enhancement in performance of the proposed PV/T system relative to its counterparts in terms of electrical and high-grade thermal exergy. An enhancement of 13.3% in exergy output is observed when applying concentrated light of $C = 10$, compared to conventional PV. This is mainly to PV's voltage drop due to increase of cell temperature. The proposed nanofluid-based PV/T with aerogel performs less, in terms of electrical efficiency, than a conventional PV/T. However, nanofluid-based PV/T with aerogel outperforms a conventional PV/T in terms of thermal efficiency. This is mainly due to the design of the collector in this study. The same can be said about high-grade thermal exergy and efficiency. Even at concentration of 1, the proposed system achieves a thermal efficiency above 60%.

Nanofluids in Jet Impingement (PV/T) Collectors

Different cooling technologies have been investigated for improvement of PV performance, such as spray cooling, jet impingement, and microchannels. These techniques have shown promising results [34–36]. Jet impingement of a liquid (such as nanofluids) or gas (such as air) onto a surface (such as photovoltaic module) on a continuous basis is explained in this section. This mechanism of heat transfer is common and ongoing. Figure 3.10 shows a schematic drawing of the jet impingement cooling for photovoltaic collectors. The cooling of the PV is expected to cause increase in its electrical efficiency. Furthermore, the gained heat by the impinged fluid can be collected and utilized for thermal applications, if a heat exchanger is used. Barrau et al. [37] conducted experimental tests on a novel hybrid-cooling scheme to study the influence of jet impingement and microchannel cooling to improve temperature uniformity of the cooled object or heat sink.

Hasan et al. [38] investigated the performance of different jet impinged nanofluids in terms of cooling and thermal energy generation as part of a PV/T system. The tested were conducted indoors and under controlled conditions. The nanofluids employed were formed of nanoparticles (SiC, TiO2, and SiO2) and base fluid (water). The system consisted of 4 parallel tubes and 36 nozzles that directly, with

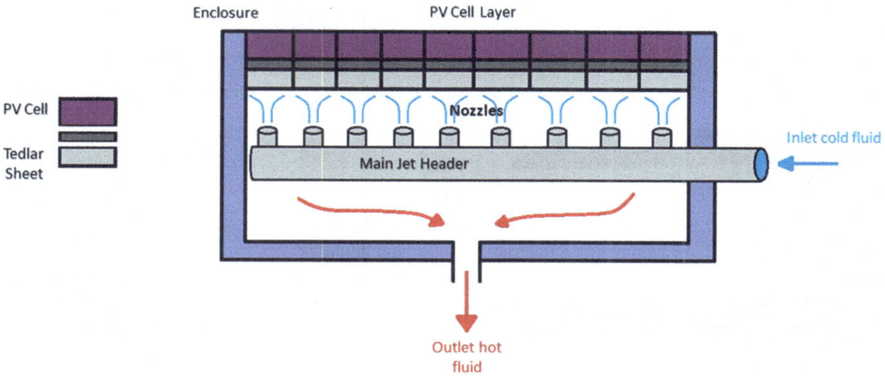

Fig. 3.10 Concept of PV/T jet impingement

direct injection onto PV/T collector. The results show that the collector achieved peak electrical, thermal, and overall efficiencies of 12.75%, 85%, and 97.75%, respectively, when using SiC/water nanofluid at solar irradiance, ambient temperature, and mass flow rate of 1000 W/m^2, 30 °C, and 0.167 W/m^2, respectively. In addition, to an increase of maximum power by 62.5% over the conventional PV module, for the same conditions, the thermal efficiency is quite high in this study; this could be attributed to different reasons:

1. The nature of the experiments which are done indoors with control operational parameters.
2. No layers between the PV (absorber) and impinged nanofluids.
3. The number of nozzles and type of jet impingement.
4. The thermal conductivity of nanofluids.

Jaaz et al. [39] used a similar experimental setup with water (without implementing nanofluids) and also added a compound parabolic concentrator (CPC) to the system. The system was designed, fabricated, and tested under climate conditions of Malaysia. The improvement in electrical efficiency was observed throughout the day. The correlation between output power of PV and mass flow rate was observed in the results. The findings surmise that using jet impingement cooling with CPC and without CPC led to increase in photovoltaic power output by 31% and 16%, respectively.

Haitham [40] conducted experimental and numerical performance investigation on PV panel cooled with jet impingement to assess its viability for hot climate conditions. The experimental setup is provided in Fig. 3.11. The author proposed a comprehensive model or the numerical work. The results indicate that in June, the cooling leads to drop of cell temperature from 69.7 °C to 36.6 °C, which is significant. Five thermocouples were used to measure the surface temperature of the PV

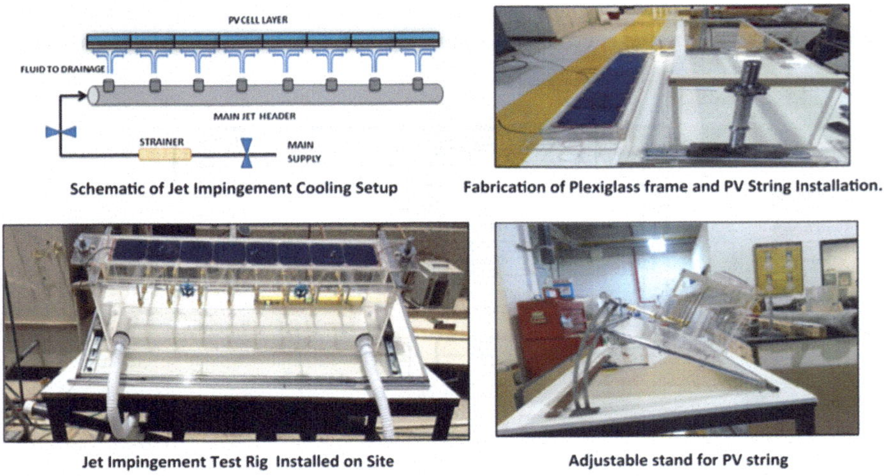

Fig. 3.11 Experimental setup of jet impinged PV by Haitham [40]

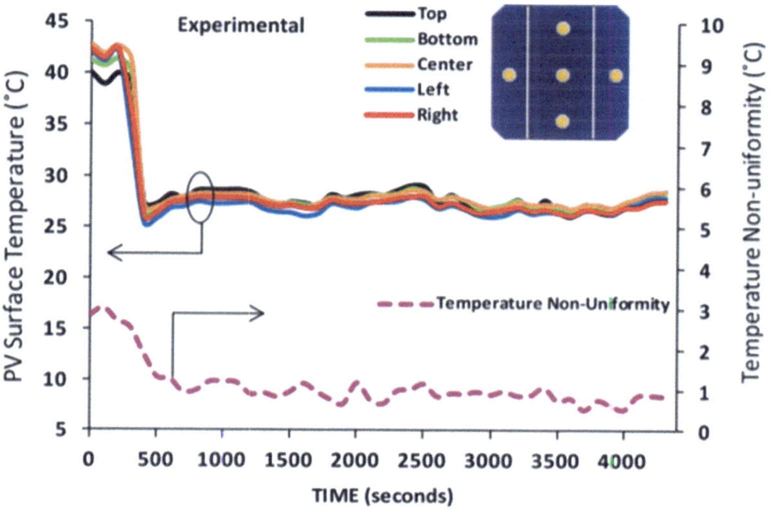

Fig. 3.12 Single-cell surface temperature profile variation with temperature non-uniformity for PV cooling (test conducted on December at 11:00 AM) [40]

cell, while jet cooling is applied, for a period of 4000 s. Figure 3.12 shows the cell surface temperature profile. In June, cooling the PV led to increase in its power output and efficiency by 51.6% and 66.6%, respectively. The numerical simulation highlights the advantage of not only cooling the cell temperature but having uniform temperature for each cell.

In Fig. 3.12, temperature non-uniformity is provided on the right axis. It was observed by the author of the study that cell temperature drops from 46 °C to 26 °C within 500 s.

Various Advanced Nanofluid-Based PV/T

The development in types of nanofluids and designs of PV/T collectors which employ these nanofluids is ongoing. Several works have been made enhance the heat transfer and consequently improve the cooling of PV module. This section provides some of the advance types of nanofluid-based PV/T.

Carbon Nanotubes (CNTs) Based PV/T

Carbon nanotubes (CNTs) are novel nanocarrier systems which can be used in various applications of science and engineering. Two classifications for the shape of CNT are made, namely, SWCNT and MWCNT. Conceptually, SWCNT is made by rolling a graphene sheet into a seamless cylinder, while more rolled-up graphene sheets formed MWCNT. These materials exhibit large surface area and can possibly be manipulated to serve specific thermal conductivity requirements of some applications. Abdallah et al. [41] presented an outdoor experimental study of multiwalls carbon nanotubes (MWCNT)-based PV/T system. The authors investigated the effect of nanofluid volume concentration ratio ranging between 0% and 0.3% and under a fixed circulation of 1.2 liter per minute. The remarkable advantage of carbon nanotubes is their clearly high thermal conductivities, which is demonstrated in this study. The results show that lowest module temperature is achieved when employing volume fraction of 0.075% of MWCNT with a 12 ° C reduction in temperature during peak solar irradiance and leading to an overall efficiency of 83.26%.

Low-Concentrated Photovoltaic Thermal (LCPV/T) with Nanofluids

Concentrated photovoltaic (CPV) is a highly useful technology to overcoming high running costs and environmental negative side effects of conventional energy production systems. This is because the use of low-cost concentrators allows for reducing the cell area needed and raise the efficiency to high of the multi-junction solar cell to high amounts. Radwan et al. [42] proposed a low-concentrated photovoltaic thermal (LCPV/T) system with a microchannel heat sink using aluminum oxide (Al2O3)-water and silicon carbide (SiC)-water nanofluids. The study also used different volume fractions and mass flow rates. In addition to investigation of different concentration ratios of the concentrator, the concept of the proposed design is illustrated in Fig. 3.13.

Numerical simulation of the thermo-fluid dynamics of the system was conducted in order to assess performance parameters, such as solar cell temperature and energy efficiencies.

The findings show that use of nanofluids allows for reduction in cell temperature at high concentration ratios, relative to using water. Also, silicon carbide (SiC)-water nanofluids led to, relatively, higher reduction in cell temperature than aluminum oxide (Al2O3)-water nanofluids. Also increase of nanoparticle volume fraction for both nanofluids causes higher reduction in cell temperature. This is mainly due to increase of nanofluid's thermal conductivity and hence overall heat transfer between the PV and the heat sink. Hence, higher electrical efficiency is achieved when employing nanofluids, especially at lower Reynolds number (Re) and higher concentration ratio. The concentration ratio is found to affect the dynamic between nanofluids and thermal efficiency. However, the increase of Reynolds number and nanoparticle volume fraction was found to increase the friction power. The author's claim this technique leads to decrease in cell temperature to 38 °C and increase electrical efficiency up to 19%.

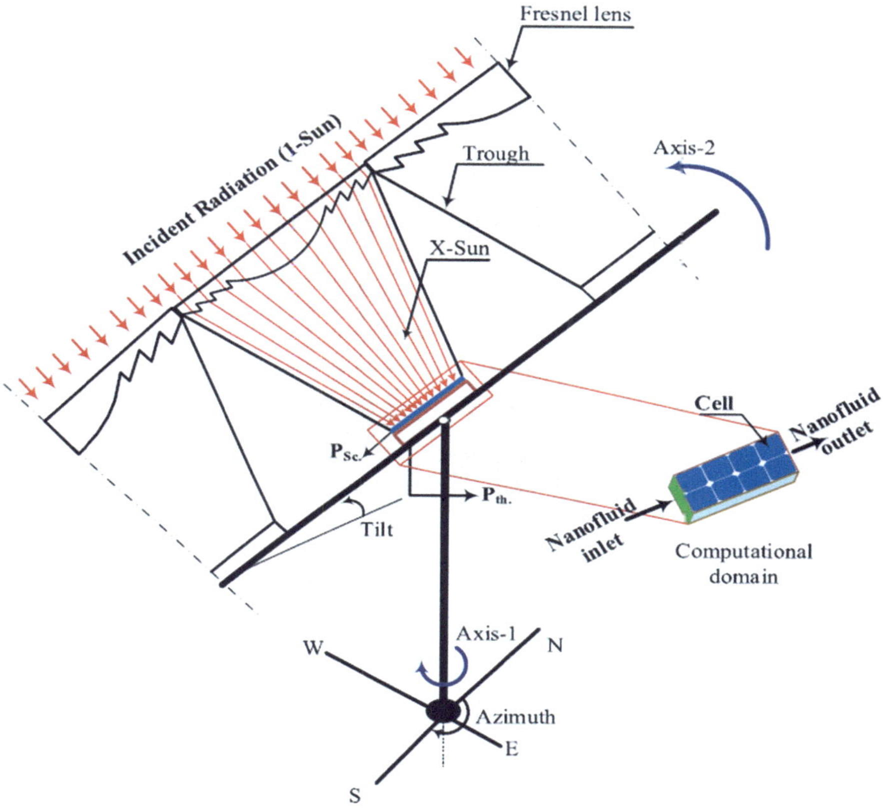

Fig. 3.13 Proposed physical model and the coordinate system by Radwan et al. [42]

Hybrid Photovoltaic/Thermoelectric System with Nanofluid

The concept of photovoltaic/thermoelectric system is to utilize a thermoelectric module (TEM) to capitalize on heat generated, from losses, from the photovoltaic. The thermoelectric device is capable of directly converting heat into electrical yield. The TEM is attached to back side of the cell. The dissipated heat is considered heat source for TEM. Working fluids such as liquids or air are the heat sink for the TEM. This system was proposed by Soltani et al. [43] using nanofluids as working fluids. Figure 3.14 shows a schematic diagram of proposed system.

The authors compared between nanofluid-based cooling method and conventional cooling methods of water- and air-based cooling, in an experimental investigation. Among the methods of cooling, the authors compared between natural and

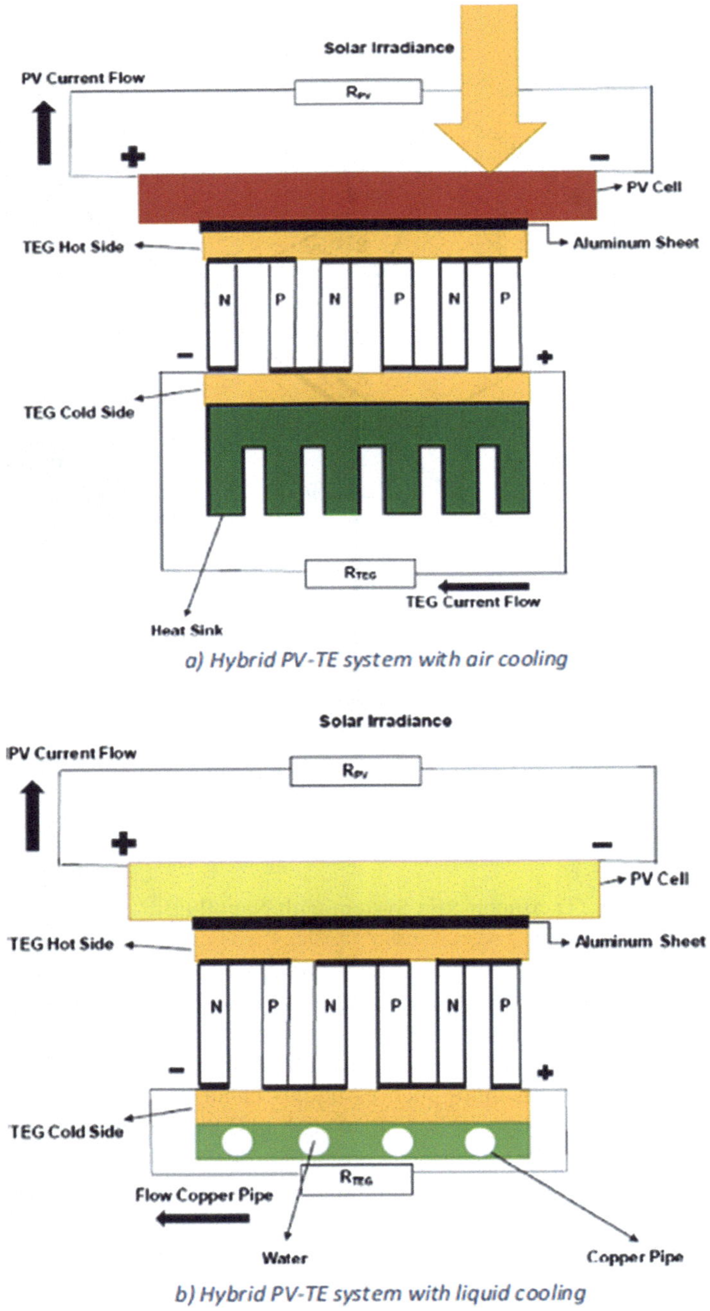

Fig. 3.14 Drawing of the proposed hybrid PV/TE system with (**a**) air cooling, (**b**) liquid cooling by Soltani et al. [43]

forced cooling. The nanofluids used were silica (SiO2)-water nanofluid and Iron(II,III) oxide (Fe3O4)-water nanofluid. The lowest PV power is produced when natural cooling is used, while highest power and efficiency are achieved when employing SiO2-water nanofluid for cooling, where the author's claims led to improvement of 54.29% and 3.35%, respectively. The improvement in Fe3O4-water nanofluid is around 52.40% and 3.13%, respectively. For natural cooling, the TEM voltage peaked at around 12:30 (noon) with a peak voltage of 1.64 volts, while the highest voltage for TEM is achieved when using SiO2-water nanofluid cooling with around 2.5 volts. The output of TEM is clearly governed by temperature gradient of the system (refer to Figs. 3.10 and 3.11 in the study [43]).

3.5 PCM and Nano-PCM-Based PV/T Systems

It is necessary to employ means of storage for solar energy technologies, given the discrepancy between supply and demand for solar energy during nighttime where heat demand in many cases is maximum and solar irradiance is virtually zero. Batteries can be used to store electrical energy during daytime and meet the demands at night. While phase change material (PCM) offers a suitable mean of storing thermal energy given they exhibit properties such as quick charging/discharging and high efficiency, moreover, they are economically feasible and readily available in the market. Different melting points for PCM exist which allows them to fit the desired application. However, these materials should be implemented for systems with high temperature during daytime and low temperature during nighttime to achieve high heat extraction rate (and temperature drop) and high solidification rate, for regeneration or recovery, respectively. Factors such as thermal conductivity, heat capacity, and mass of the PCM material affect the phase change of PCM. PCMs exhibit high latent heat of fusion and suitable phase-transition temperature range. Furthermore, they are nonhazardous and nontoxic [44]. These materials can be classified into organic material, such as paraffin and non-paraffin, and nonorganic material, such as salt hydrates, etc. However, the former classification is considering the most commonly used. The main issue of paraffin is its low thermal conductivity [45, 46] and exhibits high volume change upon melting. Hence, many authors proposed encapsulating these materials with nanoparticles, forming nano-PCM or nano-enhanced PCM material. These materials are highly useful and lead to increase in recovery of PCM and its thermal conductivity for heat transfer applications. PCM has been used by authors to thermally regulate photovoltaic modules. Stropnik and Stritih et al. [47] investigated the performance of a PV with PCM (type RT28HC) both experimentally and numerically, using TRNSYS. The results show how PV with PCM is lower in temperature by 35.6 °C than the PV without PCM. The simulation results, for 1 year, show an increase of 7.3% in electrical efficiency when using PCM, in Ljubljana, Slovenia. As for nano-PCMs, Karunamurthy et al. [48] dispersed cupric oxide (CuO) nanoparticles and paraffin, through two-step method, to improve the PCM's thermal conductivity. The authors used an ultrasonic stirrer for the mixing and produced seven samples of different volume fractions, ranging from 0.01% to 0.15%. This material was used for low temperature energy storage (LTES) system for a solar pond. The increase of volume fraction was found to decrease charging time of the LTES.

Sardarabadi et al. [49] used a PV/T collector employing PCM and nanofluids, while Alwaeli et al. [2] used a PV/T collector employing nano-PCMs and nanofluids. Both studies report increase in electrical and thermal efficiency with respect to bare PV modules during testing time. It is important to note that cooling maintains the efficiency of PV module to a value approximating efficiency standard testing condition. However, authors state "increase efficiency" as in compared to conventional PV module under same operating conditions. This study is revisited in Chap. 6.

3.6 Grid-Connected (PV/T) System

It is highly important to understand performance metrics of grid-connected photovoltaic thermal (GCPV/T) systems and their performance on local level. Given that power supplied to utility grid has gained more attention and hence is an option for meeting the increasing power demands [50]. However, high costs of this technology compared to traditional energy sources are a disadvantage. In many cases, government programs to support grid-connected PV configurations are put into effect such as feed-in tariffs (FiT) to encourage the public invest in this technology. However, more development in efficiency and productivity of GCPV systems are necessary. Improving different aspects of the system such as inverter efficiency is critical. However, more success can be achieved by provide cooling of these GCPV systems. Hence, GCPV/T offers an interesting solution that can help popularize this technology among end consumers. Hence, the investigation of GCPV/T must include the performance of the module in terms of power and efficiency as well as technical evaluation parameters or factors. Hence, techno-economic evaluations are conducted to assess the performance of GCPV systems. The technical aspect of GCPV/T is provided in this section, while the economic aspect is presented in Chap. 4. Equation (3.2) is used to calculate the energy produced by the system [51]:

$$P_{PV}(t) = P_{peak}\left(\frac{G(t)}{G_{stc}}\right) - \alpha_T\left[T_c(t) - T_{stc}\right] \tag{3.2}$$

where the cell temperature (T_c) is calculated by [51]:

$$T_c(t) - T_{amb} = \left(\frac{NOCT - 20}{800}\right)G(t) \tag{3.3}$$

Also, the PV/T electrical power generated is via the alternating current (AC) measurement, taken from the inverter output in the form of set of time periods (hour, day, or month) [51]:

$$E_{AC,t1} = \sum_{t=1}^{N} E_{AC,t2}. \tag{3.4}$$

Also, $N = 60, 24,$ and 30 for hour, day, and month, respectively. The photovoltaic unit efficiency is calculated on a DC current, while efficiency is a function of AC power. The PV module and system efficiencies are specified in Eqs. 3.5 and 3.6, respectively [51]:

$$\eta_{PV} = \frac{E_{DC}}{G(t) \times A_c} \times 100\% \tag{3.5}$$

$$\eta_{sys} = \frac{E_{AC}}{G(t) \times A_c} \times 100\% \tag{3.6}$$

Specific yield or simply "yield," abbreviated (SY), with unit of kWh/kWp, is a very important performance metric of PV systems of all sizes. It is useful for comparison between different locations and analysis of different designs for health assessment of PV array. Specific yield, also referred to as yield factor (YF) [54], is an indicator of the amount of produced energy (kWh) for every (kWp) of module capacity over the course of a typical or actual year. Refer to Eqs. 3.7 and 3.8 [51, 52]:

$$YF_d = \frac{E_{PV}\left(kWh / year\right)}{PV_{WP}\left(kWp\right)} \tag{3.7}$$

$$YF_F = \frac{E_{AC}\left(kWh / year\right)}{PV_{WP}\left(kWp\right)} \tag{3.8}$$

Capacity factor is defined as "the ratio of the annual average energy production (kWh_{AC}) of an energy generation plant divided by the theoretical maximum annual energy production of a plant assuming it operates at its peak rated capacity every hour of the year." Equation (3.9) shows the formula for calculating the capacity factor [53]:

$$CF = \frac{YF}{8760} = \frac{E_{PVannual}}{\left(P_R \times 8760\right)} \tag{3.9}$$

Finally, the performance ratio (P_R) describes the ratio of specific yield (or yield factor – YF) to reference yield, which is defined as solar irradiation received for a period. Equation (3.10) is used to calculated performance ratio [54]:

$$PR = \frac{YF}{Y_R} \tag{3.10}$$

Finally, the thermal design quality is of great importance for maximizing and cooling the GCPV. It is important to ensure uniform temperature distribution and limit the occurrence of hot spots. In the case of GCPV array, it is highly critical to make sure optimum configuration of the thermal system is made so that the right number of cascaded thermal absorbers is selected. Still, more research in this area is needed for better designed of PV/T system.

3.7 The Role of Artificial Neural Networks (ANN) in PV/T System Prediction

Artificial Neural Networks (ANN) are defined as computing systems which are inspired by the biological neural networks that constitute animal brains, mainly human. Neural networks are not algorithms of themselves, they are frameworks for various machine learning algorithms to function together and process complex data inputs. The ANN can be used for various applications due to their characteristics, e.g., function of classification, which is useful digitally for face recognition, and function of prediction which is useful for reliability assessment and weather forecasting [55]. The implementation of these neural networks in the solar energy field is described in Chap. 1. The simulation method highly affects the prediction of PV/T performance [56]. The purpose of ANN is the mimicry of humans' brain functions and/or processes in computers, specifically the ability to learning [57]. This is accomplished by depending on pre-specified functions by the programmer and a real data input to achieve the desired results. To improve the performance and adaptability, the algorithms of these networks can change their own instructions [58]. Meanwhile, the implementations of "Artificial Neural Networks" (ANN) as tools to predict the performance of PV and PV/T systems have been studied by several researchers [59–61]. These methods are classified as experimental mathematical models, regression, artificial intelligence based on networks, and finally statistical models based on a time series of data.

Alwaeli et al. [62] used different Artificial Neural Networks through NeuroSolution software to predict the performance of nanofluid and nano-PCM-based PV/T and to compare between the findings of these neural networks. The ANN networks used were support vector machine (SVM), multilayer perceptron (MLP), and self-organizing feature map (SOFM). The flowchart of the methodology of the study is provided in Fig. 3.15.

Ammar et al. [63] conducted a study to layout ground work for tracking of maximum power generation of PV/T systems using control algorithm based on Artificial Neural Networks (ANN). This ANN is used to detect the optimal power operating point (OPOP) with considerations of PV/T behavior. The calculation depends on

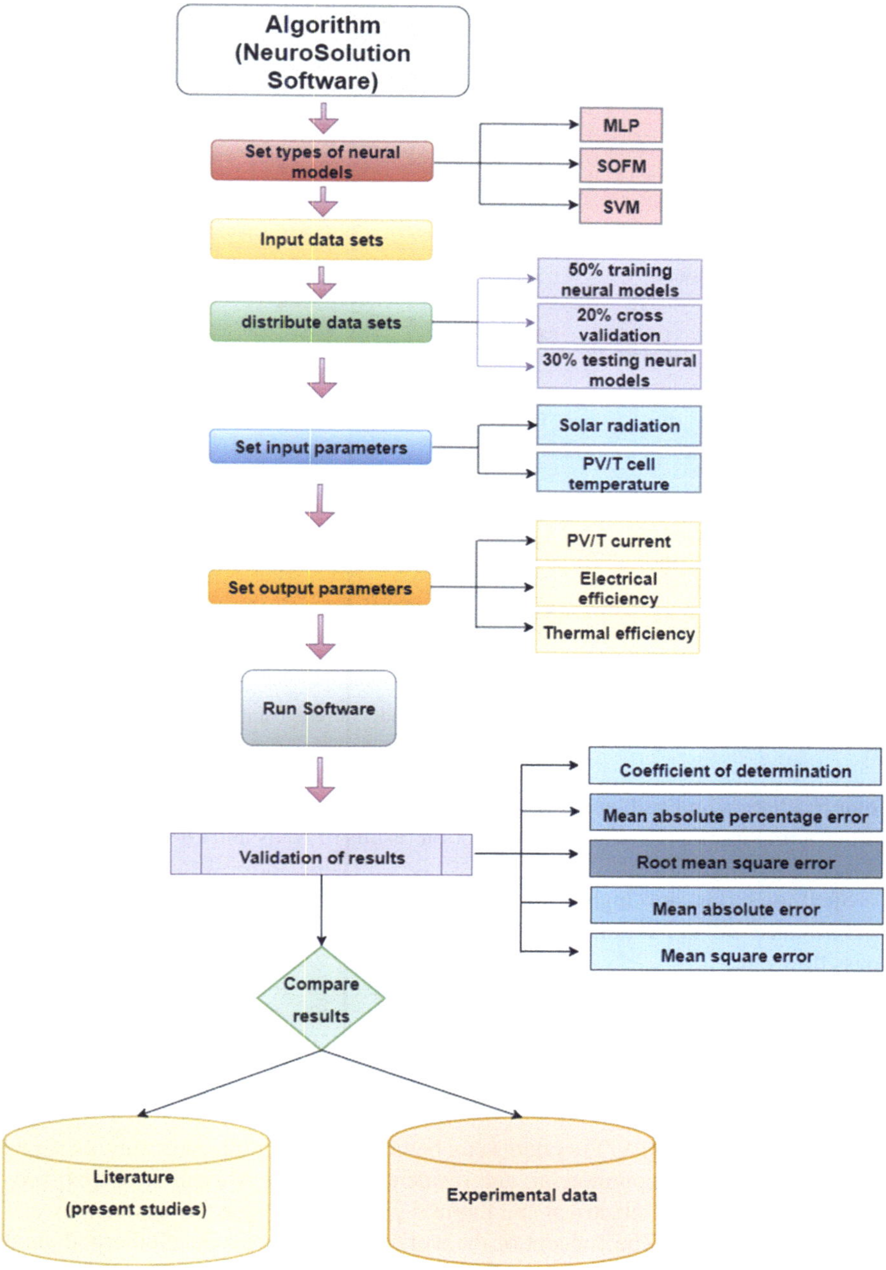

Fig. 3.15 ANN methodology flowchart

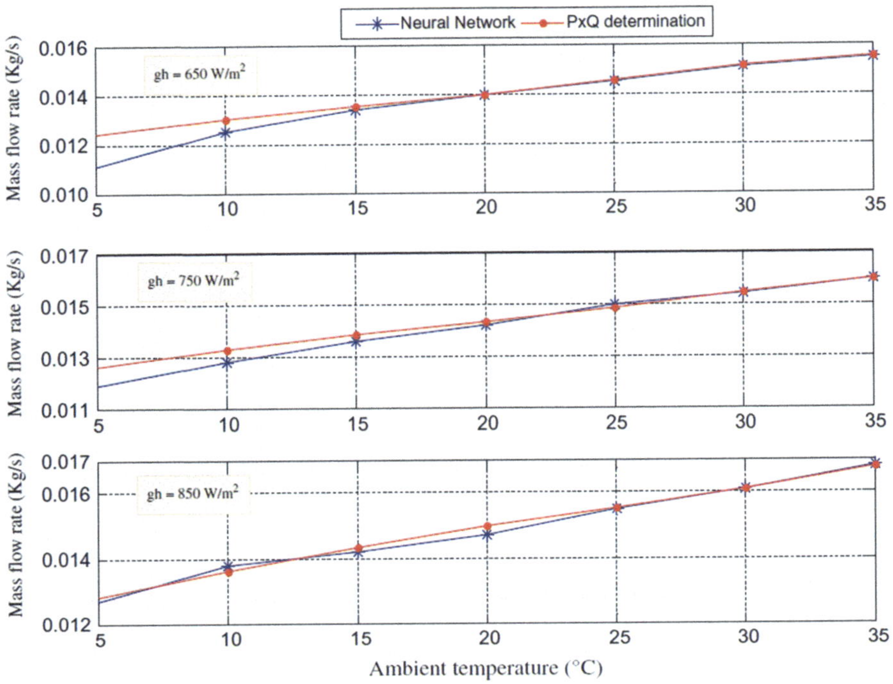

Fig. 3.16 Estimated and prediction OPOP via ANN under different ambient temperature and constant solar irradiance by Ammar et al. [63]

solar irradiation and ambient temperature to provide the optimum mass flow rate for the system. Figure 3.16 shows the prediction results of ANN and OPOP with respect to solar irradiance.

Such methods are highly useful to understand and predict the long-term performance of PV/T for different location and the optimum operating conditions of mass flow rate.

3.8 Applications of PV/T Systems

Photovoltaic thermal (PV/T) systems can be implemented/utilized in various applications in industrial, commercial, and residential settings. Alwaeli et al. [64] proposed the use of PV/T electrical and thermal yields to increase the sustainability of the palm oil process. The concept of the work is to use the produced thermal yield for preheating and to be employed in the sterilization process. Fudholi et al. [65] pointed out the use of PV/T technology for drying purposes of agricultural and marine products. Kern and Russell [66] used PV/T for a heat pump application.

3.9 Conclusions and Recommendations

This chapter discussed advanced design and concepts of PV/T systems from recent works in the literature. A comprehensive literature survey was conducted to display the research and development (R&D) in the field and to link the various components of these systems to the concept laid out in Chap. 2. Nanofluid-based PV/T systems of different types were discussed in terms of concept, operation and research findings. The use of nano-PCMs for thermal regulation of PV modules has been discussed as well. In addition, the major aspects of design and evaluation of grid-connected PV/T systems were illustrated. Finally, some applications of PV/T systems were mentioned briefly.

Conclusions

1. Nanofluids exhibit better thermophysical properties for heat transfer applications than pure water or conventional fluids all together.
2. Increase of nanoparticle volume fraction leads to increase in thermal conductivity of the nanofluid and hence the overall heat transfer between the PV module and thermal absorber.
3. For accurate comparison of performance of different PV/T systems, two methods of comparison can be used, in terms of equal Reynolds number basis or equal pumping power basis.
4. Dispersing nanoparticles into phase change material (PCM) leads to enhancing the thermal conductivity of PCM and improve its recovery time.
5. The nanofluid and nano-PCM-based PV/T system exhibits superior performance of its conventional PV and nanofluid-based PV/T counterpart.

Among the observations form the literature is the following:

1. Lack of consensus on optimum nanofluid mass fraction or concentration ratio.
2. Lack of consensus on stability and sonication time.

Recommendations

1. To further investigate advanced nanofluids to enhance heat transfer process of the PV/T collector. This is to include magnetic nanofluids, Boehmite nanofluid, etc.
2. To investigate the long-term stability of nanofluids used as working fluids in PV/T systems.
3. To assess the number of cycles, nano-PCM can withstand melting and solidifications.
4. To evaluate the feasibility and life cycle assessment of the advanced type of PV/T systems in the literature. Although they exhibit high efficiencies, it is necessary to view them in terms of costs and environmental impact.

References

1. A.H. Al-Waeli, K. Sopian, M.T. Chaichan, H.A. Kazem, H.A. Hasan, A.N. Al-Shamani, An experimental investigation of SiC nanofluid as a base-fluid for a photovoltaic thermal PV/T system. Energy Conver. Manage. **142**, 547–558 (2017)
2. A.H. Al-Waeli, K. Sopian, M.T. Chaichan, H.A. Kazem, A. Ibrahim, S. Mat, M.H. Ruslan, Evaluation of the nanofluid and nano-PCM based photovoltaic thermal (PVT) system: an experimental study. Energy Conver. Manage. **151**, 693–708 (2017)
3. J.H. Kim, S.H. Park, J.G. Kang, J.T. Kim, Experimental performance of heating system with building-integrated PVT (BIPVT) collector. Energy Procedia **48**, 1374–1384 (2014)
4. M. Fuentes, M. Vivar, J. de la Casa, J. Aguilera, An experimental comparison between commercial hybrid PV-T and simple PV systems intended for BIPV. Renew. Sustain. Energy Rev. **93**, 110–120 (2018)
5. N.A.C. Sidik, I.M. Adamu, M.M. Jamil, G.H.R. Kefayati, R. Mamat, G. Najafi, Recent progress on hybrid nanofluids in heat transfer applications: a comprehensive review. Int. Commun. Heat Mass Transfer **78**, 68–79 (2016)
6. W. Yu, H. Xie, A review on nanofluids: preparation, stability mechanisms, and applications. J. Nanomater. **2012**, 1 (2012)
7. X.F. Li, D.S. Zhu, X.J. Wang, N. Wang, J.W. Gao, H. Li, Thermal conductivity enhancement dependent pH and chemical surfactant for Cu-H2O nanofluids. Thermochimica Acta **469**(1–2), 98–103 (2008)
8. A.A. Hawwash, A.K.A. Rahman, S.A. Nada, S. Ookawara, Numerical investigation and experimental verification of performance enhancement of flat plate solar collector using nanofluids. Appl. Therm. Eng. **130**, 363–374 (2018)
9. R.A. El-Salamony, R.E. Morsi, A.M. Alsabagh, Preparation, stability and photocatalytic activity of titania nanofluid using gamma irradiated titania nanoparticles by two-step method. J Nanofluids **4**, 442–448 (2015)
10. S. Mukherjee, S. Paria, Preparation and stability of nanofluids-a review. IOSR J. Mech. Civil Eng. **9**(2), 63–69 (2013)
11. N.A. Ali, Preparation and characterisation of physicochemical properties of Aluminium Oxide (Al2O3)-Water Nanofluids Using Two Step Methods (Doctoral dissertation, UMP) (2010)
12. S. Suresh, K.P. Venkitaraj, P. Selvakumar, M. Chandrasekar, Synthesis of Al2O3–Cu/water hybrid nanofluids using two step method and its thermo physical properties. Colloids Surf. A Physicochem. Eng. Asp. **388**(1–3), 41–48 (2011)
13. J.B. Castellanos, *Thermal conductivity of alumina and silica nanofluids* (Minnesota State University, Mankato, 2014)
14. H.P. Lehmann, X. Fuentes-Arderiu, L.F. Bertello, Glossary of terms in quantities and units in clinical chemistry (IUPAC-IFCC recommendations 1996). Pure Appl. Chem. **68**(4), 957–1000 (1996)
15. S.K. Das, S.U. Choi, W. Yu, T. Pradeep, *Nanofluids: Science and Technology* (Wiley, 2007)
16. S.K. Das, N. Putra, P. Thiesen, W. Roetzel, Temperature dependence of thermal conductivity enhancement for nanofluids. J. Heat Transfer **125**(4), 567–574 (2003)
17. M. Drzazga, M. Lemanowicz, G. Dzido, A. Gierczycki, Preparation of metal oxide-water nanofluids by two-step method. Inż. Ap. Chem. **51**(5), 213–215 (2012)
18. Y. Khanjari, F. Pourfayaz, A.B. Kasaeian, Numerical investigation on using of nanofluid in a water-cooled photovoltaic thermal system. Energy Conver. Manage. **122**, 263–278 (2016)
19. T. Yousefi, F. Veysi, E. Shojaeizadeh, S. Zinadini, An experimental investigation on the effect of Al2O3–H2O nanofluid on the efficiency of flat-plate solar collectors. Renew. Energy **39**(1), 293–298 (2012)
20. D. Jing, Y. Hu, M. Liu, J. Wei, L. Guo, Preparation of highly dispersed nanofluid and CFD study of its utilization in a concentrating PV/T system. Solar Energy **112**, 30–40 (2015)
21. A.H. Al-Waeli, K. Sopian, H.A. Kazem, M.T. Chaichan, Photovoltaic/Thermal (PV/T) systems: status and future prospects. Renew. Sustain. Energy Rev. **77**, 109–130 (2017)

22. J.J. Michael, S. Iniyan, Performance analysis of a copper sheet laminated photovoltaic thermal collector using copper oxide–water nanofluid. Solar Energy **119**, 439–451 (2015)
23. N. Purohit, S. Jakhar, P. Gullo, M.S. Dasgupta, Heat transfer and entropy generation analysis of alumina/water nanofluid in a flat plate PV/T collector under equal pumping power comparison criterion. Renew. Energy **120**, 14–22 (2018)
24. Y. Khanjari, A.B. Kasaeian, F. Pourfayaz, Evaluating the environmental parameters affecting the performance of photovoltaic thermal system using nanofluid. Appl. Therm. Eng. **115**, 178–187 (2017)
25. H.A. Hussien, A.H. Noman, A.R. Abdulmunem, Indoor investigation for improving the hybrid photovoltaic/thermal system performance using nanofluid (Al2O3-water). Eng. Technol. J. **33**(4 Part (A) Engineering), 889–901 (2015)
26. A. Mojiri, R. Taylor, E. Thomsen, G. Rosengarten, Spectral beam splitting for efficient conversion of solar energy—a review. Renew. Sustain. Energy Rev. **28**, 654–663 (2013)
27. A.G. Imenes, D.R. Mills, Spectral beam splitting technology for increased conversion efficiency in solar concentrating systems: a review. Solar Energy Mater. Solar Cells **84**(1–4), 19–69 (2004)
28. M.A.C. Chendo, D.E. Osborn, R. Swenson, Analysis of spectrally selective liquid absorption filters for hybrid solar energy conversion. In *Optical Materials Technology for Energy Efficiency and Solar Energy Conversion IV* (Vol. 562, pp. 160–166). International Society for Optics and Photonics (1985, December)
29. R. Looser, M. Vivar, V. Everett, Spectral characterisation and long-term performance analysis of various commercial Heat Transfer Fluids (HTF) as direct-absorption filters for CPV-T beam-splitting applications. Appl. Energy **113**, 1496–1511 (2014)
30. J. Kaluza, K.H. Funken, U. Groer, A. Neumann, K.J. Riffelmann, Properties of an optical fluid filter: theoretical evaluations and measurement results. Le Journal de Physique IV **9**(PR3), Pr3-655 (1999)
31. S.S. Joshi, A.S. Dhoble, P.R. Jiwanapurkar, Investigations of different liquid based spectrum beam splitters for combined solar photovoltaic thermal systems. J. Solar Energy Eng. **138**(2), 021003 (2016)
32. N.E. Hjerrild, J.A. Scott, R. Amal, R.A. Taylor, Exploring the effects of heat and UV exposure on glycerol-based Ag-SiO2 nanofluids for PV/T applications. Renew. Energy **120**, 266–274 (2018)
33. M. Du, G.H. Tang, T.M. Wang, Exergy analysis of a hybrid PV/T system based on plasmonic nanofluids and silica aerogel glazing. Solar Energy **183**, 501–511 (2019)
34. M. Abdolzadeh, M. Ameri, Improving the effectiveness of a photovoltaic water pumping system by spraying water over the front of photovoltaic cells. Renew. Energy **34**(1), 91–96 (2009)
35. N. Karami, M. Rahimi, Heat transfer enhancement in a PV cell using Boehmite nanofluid. Energy Conver. Manage. **86**, 275–285 (2014)
36. P. Valeh-e-Sheyda, M. Rahimi, E. Karimi, M. Asadi, Application of two-phase flow for cooling of hybrid microchannel PV cells: a comparative study. Energy Conver. Manage. **69**, 122–130 (2013)
37. J. Barrau, M. Omri, D. Chemisana, J. Rosell, M. Ibañez, L. Tadrist, Numerical study of a hybrid jet impingement/micro-channel cooling scheme. Appl. Therm. Eng. **33**, 237–245 (2012)
38. H.A. Hasan, K. Sopian, A.H. Jaaz, A.N. Al-Shamani, Experimental investigation of jet array nanofluids impingement in photovoltaic/thermal collector. Solar Energy **144**, 321–334 (2017)
39. A. Jaaz, H. Hasan, K. Sopian, A. Kadhum, T. Gaaz, A. Al-Amiery, Outdoor performance analysis of a photovoltaic thermal (PVT) collector with jet impingement and compound parabolic concentrator (CPC). Materials **10**(8), 888 (2017)
40. H.M. Bahaidarah, Experimental performance evaluation and modeling of jet impingement cooling for thermal management of photovoltaics. Solar Energy **135**, 605–617 (2016)

41. S.R. Abdallah, H. Saidani-Scott, O.E. Abdellatif, Performance analysis for hybrid PV/T system using low concentration MWCNT (water-based) nanofluid. Solar Energy **181**, 108–115 (2019)
42. A. Radwan, M. Ahmed, S. Ookawara, Performance enhancement of concentrated photovoltaic systems using a microchannel heat sink with nanofluids. Energy Conver. Manage. **119**, 289–303 (2016)
43. S. Soltani, A. Kasaeian, H. Sarrafha, D. Wen, An experimental investigation of a hybrid photovoltaic/thermoelectric system with nanofluid application. Solar Energy **155**, 1033–1043 (2017)
44. M.M. Farid, A.M. Khudhair, S.A.K. Razack, S. Al-Hallaj, A review on phase change energy storage: materials and applications. Energy Conver. Manage. **45**(9–10), 1597–1615 (2004)
45. L. Colla, L. Fedele, S. Mancin, L. Danza, O. Manca, Nano-PCMs for enhanced energy storage and passive cooling applications. Appl. Therm. Eng. **110**, 584–589 (2017)
46. A.S. Abdelrazik, F.A. Al-Sulaiman, R. Saidur, R. Ben-Mansour, A review on recent development for the design and packaging of hybrid photovoltaic/thermal (PV/T) solar systems. Renew. Sustain. Energy Rev. **95**, 110–129 (2018)
47. R. Stropnik, U. Stritih, Increasing the efficiency of PV panel with the use of PCM. Renew. Energy **97**, 671–679 (2016)
48. K. Karunamurthy, K. Murugumohankumar, S. Suresh, Use of CuO nano-material for the improvement of thermal conductivity and performance of low temperature energy storage system of solar pond. Dig. J. Nanomater. Biostruct. **7**(4), 1833–1841 (2012)
49. M. Sardarabadi, M. Passandideh-Fard, M.J. Maghrebi, M. Ghazikhani, Experimental study of using both ZnO/water nanofluid and phase change material (PCM) in photovoltaic thermal systems. Solar Energy Mater. Solar Cells **161**, 62–69 (2017)
50. J.P. Benner, L. Kazmerski, Photovoltaics gaining greater visibility. IEEE Spectr. **36**(9), 34–42 (1999)
51. H.A. Kazem, H.A. Al-Badi, A.S. Al Busaidi, M.T. Chaichan, Optimum design and evaluation of hybrid solar/wind/diesel power system for Masirah Island. Environ. Develop. Sustain. **19**(5), 1761–1778 (2017)
52. E. Kymakis, S. Kalykakis, T.M. Papazoglou, Performance analysis of a grid connected photovoltaic park on the island of Crete. Energy Conver. Manage. **50**(3), 433–438 (2009)
53. A. McEvoy, T. Markvart, L. Castañer, T. Markvart, L. Castaner (eds.), *Practical Handbook of Photovoltaics: Fundamentals and Applications* (Elsevier, Waltham, 2003)
54. D.D. Milosavljević, T.M. Pavlović, D.S. Piršl, Performance analysis of a grid-connected solar PV plant in Niš, republic of Serbia. Renew. Sustain. Energy Rev. **44**, 423–435 (2015)
55. P.J. Braspenning, F. Thuijsman, *Artificial Neural Networks: An Introduction to ANN Theory and Practice*, vol 931 (Springer Science & Business Media, Heidelberg, 1995)
56. H. Kalani, M. Sardarabadi, M. Passandideh-Fard, Using artificial neural network models and particle swarm optimization for manner prediction of a photovoltaic thermal nanofluid based collector. Appl. Therm. Eng. **113**, 1170–1177 (2017)
57. J. Zupan, Introduction to artificial neural network (ANN) methods: what they are and how to use them. Acta Chimica Slovenica **41**, 327–327 (1994)
58. A.K. Jain, J. Mao, K.M. Mohiuddin, Artificial neural networks: a tutorial. Computer **29**(3), 31–44 (1996)
59. D.W. Patterson, *Artificial Neural Networks: Theory and Applications* (Prentice Hall PTR, Upper Saddle River, 1998)
60. M.A. Behrang, E. Assareh, A. Ghanbarzadeh, A.R. Noghrehabadi, The potential of different artificial neural network (ANN) techniques in daily global solar radiation modeling based on meteorological data. Solar Energy **84**(8), 1468–1480 (2010)
61. A.K. Rai, N.D. Kaushika, B. Singh, N. Agarwal, Simulation model of ANN based maximum power point tracking controller for solar PV system. Solar Energy Mater. Solar Cells **95**(2), 773–778 (2011)

62. A.H. Al-Waeli, K. Sopian, H.A. Kazem, J.H. Yousif, M.T. Chaichan, A. Ibrahim, S. Mat, M.H. Ruslan, Comparison of prediction methods of PV/T nanofluid and nano-PCM system using a measured dataset and artificial neural network. Solar Energy **162**, 378–396 (2018)
63. M.B. Ammar, M. Chaabene, Z. Chtourou, Artificial neural network based control for PV/T panel to track optimum thermal and electrical power. Energy Conver. Manage. **65**, 372–380 (2013)
64. A.H. Alwaeli, K. Sopian, A. Ibrahim, S. Mat, M.H. Ruslan, Nanofluid based photovoltaic thermal (PVT) incorporation in palm oil production process. IJOCAAS **3**, 292–294 (2017)
65. A. Fudholi, K. Sopian, M.H. Ruslan, M.A. Alghoul, M.Y. Sulaiman, Review of solar dryers for agricultural and marine products. Renew. Sustain. Energy Rev. **14**(1), 1–30 (2010)
66. E.C. Kern Jr, M.C. Russell, Combined photovoltaic and thermal hybrid collector systems (No. COO-4577-3; CONF-780619-24). Massachusetts Inst. of tech., Lexington, USA. Lincoln Lab (1978)

Chapter 4
PV/T Feasibility and Cost Assessment

4.1 Background

The sustained growth of individual well-being and society and the proper development of state resources in any country are measured by the demand for energy, and here we mean electricity. With this demand increasing, two major concerns arise: energy security and global warming. Most of the energy currently generated in most of the world is fossil and fossil fuels, as coal, oil, or natural gas. The problem of energy security first emerged prominently in 1973 when many Arab OPEC countries decided to cut and reduce their share of oil exports to a large proportion. The second case emerged in 2008 when oil prices rose to high record, as the price per barrel exceeded 140 US$ followed by a large and rapid decline and unprecedented deterioration in prices. This fluctuation in prices has a negative impact on the budget of producing and exporting countries and similarly for the importing countries [1]. These crises have had a positive impact on the rise in reliance on renewable energy sources and fuel alternatives [2]. These sources have the potential to enhance energy security and reduce greenhouse gases when compared to conventional fossil fuels. The concept of "climate change" has become widespread and deliberate after it has been the subject of debate among scientists and researchers. Climate change has become a reality, and the most important factor is human intervention in the composition of the Earth's air through the emission of billions of tons of pollutants into the atmosphere. This pollution was the first cause of the global warming phenomenon followed by problems one after the other. Working to reduce pollutants into the air has become universal, and everyone is working to mitigate its impact on life on Earth [3].

One of the most important solutions to this problem is the transformation of energy generation by using environmentally renewable green energy such as solar energy, wind energy, bioenergy, geothermal energy, etc. One of the most important renewable sources is solar energy and wind, both of which are free and do not require the cost of import or storage facilities. The renewable energy sector and its

A. H. A. Al-Waeli et al., *Photovoltaic/Thermal (PV/T) Systems*,
https://doi.org/10.1007/978-3-030-27824-3_4

various technologies have become the fastest growing sector [4]. Looking back over the past few years, the capacity of renewable power plants has doubled from 994 GW in 2007 to 2011 GW in 2016. With renewable power plants, global generating capacity increased by 161 GW in 2016 [5]. In 2016, investments in renewable energy technologies and plants have risen to $ 263 billion and have stabilized at this level despite several factors that inhibit such an escalation in use, such as the deterioration and decline in world oil prices, the traditional financial support of energy and increased legislative and political instability. Power generation investments in the renewable energy sector grew by 10% compared to 2013 [6, 7]. Figure 4.1 shows the cumulative photovoltaic systems construction worldwide by the end of 2015 [8].

The use of renewable energy, especially photovoltaic cells, in remote areas of developing countries has become a target for investors and has become a major market segment. There are great opportunities for using renewable energy as it provides inexpensive, flexible power supplies for construction both in desert, plains, and mountains, and even in water. Global statistics have shown that around 17% of the world's population, or nearly 1.2 billion people, still live without electricity to date. The most intense of these are concentrated in Asia (the Pacific States) and Africa (sub-Saharan Africa).

Photovoltaic cells are the appropriate treatment to improve the local energy supply in Russia, especially in remote areas, which suffer from harsh climatic conditions and the need to improve the quality of life of citizens in these areas, who cannot access central sources of energy and heat. The world's greatest challenge is to provide electricity to isolated areas in the far north of the globe, such as the Arctic and the Far East.

Many researchers, such as the References [9–13], have suggested that dependence on a single energy supplier may cause problems in supply. For example, solar photovoltaic energy has the challenge of relying on the fluctuation of photovoltaic

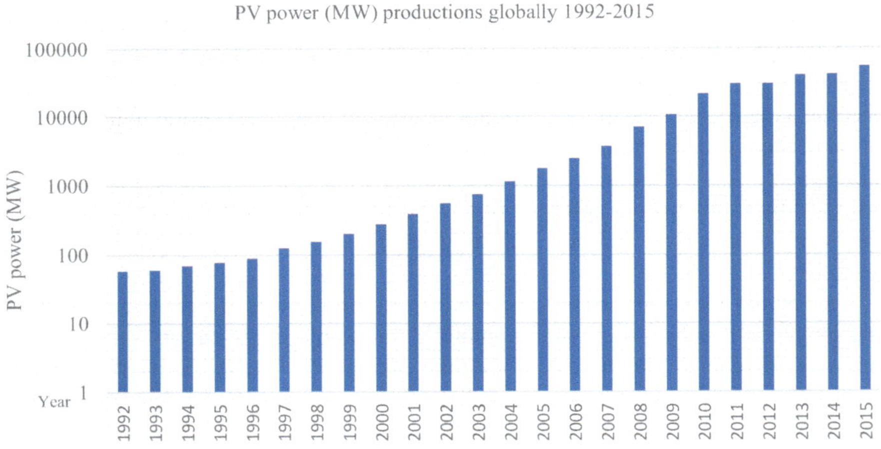

Fig. 4.1 Cumulative global solar PV capacity installations by end of 2015

output and the inability to predict the resulting efficiency because the technology is directly affected by weather conditions. This means that an independent PV system will be able to meet the demand only at times when solar radiation is available with appropriate weather conditions such as temperature, relative humidity, and wind speed. It is recommended to use a backup power supply (wind turbines or diesel generators) when using photovoltaic systems as well as the need for a storage system. Standby energy resources are concerned with the provision of electricity in the absence of solar radiation on cloudy or dusty days.

The idea of using a hybrid system (where the main part of this system will be a photovoltaic system) represents an acceptable solution especially for the use in remote areas. However, the main obstacle to its use is the high investment costs compared to the cost of fossil fuels, which remain low. Therefore, when thinking of such, stations should not only look from the environmental point of view but also focus on the economic and profitability (in the sense that the system pays its costs and starts to give profits) [14]. The use of two or more sources of energy (preferably all renewable) reduces the system size used, improves system performance, and increases the economic value of the project. Such economic systems can overcome the problems of transmission and distribution of electricity and losses in long-distance power lines [15]. Due to the specific climatic and topographical conditions of each region, the development of hybrid renewable energy systems has become a difficult task due to the different aspects of system design and technical and economic dimensions. Companies in this field tend to reduce the size of hybrid systems consisting of two or more renewable energies available at a site using the good method of bartering between economic aspects and production reliability. This process is very complex [16].

In practice, a single model cannot be provided for all situations but can be relied upon to develop and improve existing models in practice. The development and improvement of the change in hybrid renewable energy systems depends on economic criteria (e.g., net current cost) and reliability standards (ensuring that power supplies are not lost under any circumstances). The adoption of economic standards in changing the size of hybrid systems depends on a modern technology based on the algorithms of evolutionary and metaheuristic, which are expensive algorithms interested in the existence of convergence and the rate, and some of these algorithms are the algorithm of differential development [17], algorithm for the search Cuckoo [18], evolutionary metaheuristic algorithm [19], and others such as the genetic algorithm (GA) and the improved particle squadron (PSO) [20]. As for the adoption of reliability standards for scaling hybrid power plants, existing software is used. Reference [21] has reviewed 19 softwares such as HOMER, TRYNS, iHOGA, and others. The HOMER program can be considered one of the most widely used technologies for scaling hybrid renewable systems [22]. This program is used to determine the optimal hybrid system consisting of solar cells, wind turbines, and diesel generators to supply electricity to the island of Masirah (Sultanate of Oman) located deep in the Sea of Oman. Reference [23] has evaluated a hybrid system working with four types of energy sources (photovoltaic cells, fuel cell with proton exchange membrane, wind turbines, and battery) in terms of its technical and economic performance using HOMER software. Reference [24] studied the adoption of a hybrid system consisting of photovoltaic cells and cellular battery system using HOMER to verify the acceptability of the system. Reference [25] used HOMER program to optimize design for the use of a photovoltaic cell system to operate a water pump to irrigate crops in remote areas of Oman. The researchers relied in the system design on the environmental conditions of the Omani city of Sohar.

The above shows the great effect of the environmental conditions in determining the economics of any hybrid system and certainly in determining the performance, productivity, and economical PV systems. Studies that dictate the literature of photovoltaic cells have confirmed the extreme impact of weather conditions such as solar radiation, relative humidity, atmospheric temperature, wind speed, and dust density on the performance and productivity of photovoltaic systems [26]. The most important environmental conditions affecting the productivity of solar cells are the intensity of solar radiation. On one hand, the higher the solar radiation intensity, the higher the electrical efficiency of photovoltaic cells in theory, as it is practically increasing its temperature, causing a sharp decline in its productivity [27]. The adoption of photovoltaic/thermal systems has become popular among researchers in this field because of a clear improvement in the temperature of the cells, which increases the productivity and efficiency, in addition to providing thermal energy could not be provided in ordinary modules [28–33]. The problem now is that these systems do not consist of the structures of the solar cells only, the cells themselves, and the control and guidance systems, but rather the solar collector and its accessories (whether it is filled with air, water, nanofluids, nano-PCM, and other systems). These additions mean additional costs that increase the cost of setting up PV/T stations and should study the economic operation of such stations and the material payback period after paying the cost of construction and maintenance. There is a lot of research in literature on this topic, but few for the PV/T systems. In this regard, the researchers are interested in concepts such as life cycle cost analysis (LCCA), levelized cost of heat (LCOH), levelized cost of energy (LCOE), and life cycle assessment (LCA). In the coming paragraphs, we will explain these economic concepts and how they entered the world of PV/T systems and the results derived from the research published in this field.

4.2 Life Cycle Cost Analysis (LCCA)

This concept is used to analyze the total cost of an enterprise taking into account the costs of the system and the land on which the system is built. This is an important economic analysis through which alternatives can be used to reduce costs, both in current days and future. In this analysis, primary investment options are compared with the possibility of using lower cost alternatives for the next 20 years. Through this important analysis, the alternatives are offered to achieve the most cost-effective, whether in the case of purchase, ownership, operation, and maintenance and finally in the case of the disposal of a part or process, and the suitability of implementation in terms of technical purposes. This economic analysis includes all costs of construction, operation, and maintenance of the plant over a specified period of time. Through this analysis, improvements can be made in the life cycle cost of the project or reduced construction and equipment costs. This analysis is essential in system decision-making. This economic concept will focus on the path of financial gains and losses that the studied system is supposed to achieve during its estimated life span, which is expressed here in the life cycle. The cost of the life cycle is calculated by the following equation:

$$\mathrm{LCC} = C_{\mathrm{capital}} + \sum_{1}^{n} C_{\mathrm{O\&M}} . R_{\mathrm{PW}} + \sum_{1}^{n} C_{\mathrm{replacement}} . R_{\mathrm{PW}} - C_{\mathrm{salvage}} . R_{\mathrm{PV}} \tag{4.1}$$

C_{capital} represents the primary initial expenditure to be used in the design, purchase, and installation of equipment for the system, which is necessary when starting work initially. $C_{\mathrm{O\&M}}$ represents the operating and maintenance costs of the station scheduled on an annual basis and usually includes the salaries of operators and other fees (e.g., inspection fees, property tax, and insurance). It also includes the cost of replacing and repairing parts of the system during the expected operational life. The cost of replacing some system's parts is necessary to account for their end of life, for instance, replacing the inverter (which is supposed to be replaced once during the lifetime of the system). Finally, C_{salvage} represents the net value of the system in the last year of its life and is usually assessed at 20% of the cost of mechanical equipment that can be transported and is affected by a number of factors such as the condition and age of the equipment. R_{PW} represents the present value of each of the abovementioned, calculated by equation:

$$R_{\mathrm{PW}} = F / \left(1 + i\right)^{N} \tag{4.2}$$

where F is the value of the amount in the future and (i) is the interest rate. The initial value of any system depends on the parts of the system and the costs incurred on the design, installation, and civil works. The following equation is used to express the initial capital cost:

$$C_{\mathrm{capital}} = CA_i \times UC_i + ICI \tag{4.3}$$

where CA_i represents the capacity of the system, UCI the cost per unit of the system ($ per component), and ICI the total fixed cost, including installation cost and civil works (US $). Table 4.1 represents a model for calculating the present value of each system submitted by the Ref. [31].

Reference [34] studied practically and theoretically the cost of many systems that can be used, such as a single stand-alone PV system (SAPV), a water-cooled PV/T system (SAPVT-w), a nanofluid-cooled PV/T system (SAPVT-nf), and a nano-PCM nanofluid-cooled system (SAPVT nanofluid/nano-PCM). The researchers used the cost of the life cycle taking into consideration the cost of capital, operation, maintenance, and investment as well as the inclusion of inflation rates.

Table 4.1 PV systems' item costs

No.	Item	Unit price (US$)	Quantity	Price (US$)	Lifetime years
1	PV module	2.0/Wp	120	240	25
2	Support structure	–	1	–	25
3	Inventor	–	160	75	15
4	Circuit breakers	–	1	2	15
5	Civil and installation work	–	–	–	25
6	Pump	40	–	40	15
7	Heat exchanger	80	–	80	25
8	PCM tank	30	–	30	25
9	Nanofluid	74.44/liter	0.38 L	28.66	25
10	Nanofluid tank	20	–	20	25
11	Nano-PCM	0.99/kg	22 kg	21.8	1
12	Pipe	1/m	25 m	25	25
13	Insulation	5	1 m^2	5	
Total salvage value		13%		73.89	25

Table 4.2 LCCA and annual productivity of the tested system

	SAPV	SAPVT-w	SAPVT-nf	SAPVT-n-pcm-nf
Life cycle cost	568.46	774.14	1011.89	1288.37
Capital cost	482.5	733.4	762.35	808.77
Maintenance cost	32.65	29.34	34.36	35.62
Replacement cost	65.74	98.61	282.77	528.31
Annual energy productivity	141.9	164.97	209.97	230.73

Table 4.2 shows the differences in the calculation of the life cycle cost of the parts of the system.

The table results are also shown in the pie chart of Fig. 4.2. The figure represents the percentages of LCCA for SAPV, SAPVT-w, SAPVT-nf, and SAPVT nanofluid/nano-PCM components studied. The results of the study showed that the SAPV has the highest LCCA (72.0%) when compared with the SAPVT-w and SAPVT-nf systems, which were 48.0% and 44.0%, respectively. Also, the results show that photovoltaic cell is the most expensive part of the measured SAPVT nanofluid/nano-PCM system, accounting for 42.0% of the total cost, followed by the cost of the reflector at 13.0%.

Reference [35] analyzed life cycle cost analysis (LCCA) for the construction of integrated photovoltaic systems (BIPV) in India's air conditions. In this system, the researchers used air ducts to cool the system. The system was evaluated at different air flow rates and in three directions for the PV units when the cell was installed at the ideal angle of inclination or by horizontal and vertical directions taking into account the effect of the shade. The results of the study showed that the cost of energy produced by the studied system ranged between 1.61 and 3.61 US$/kWh depending on the climatic conditions of the site. The results showed that the

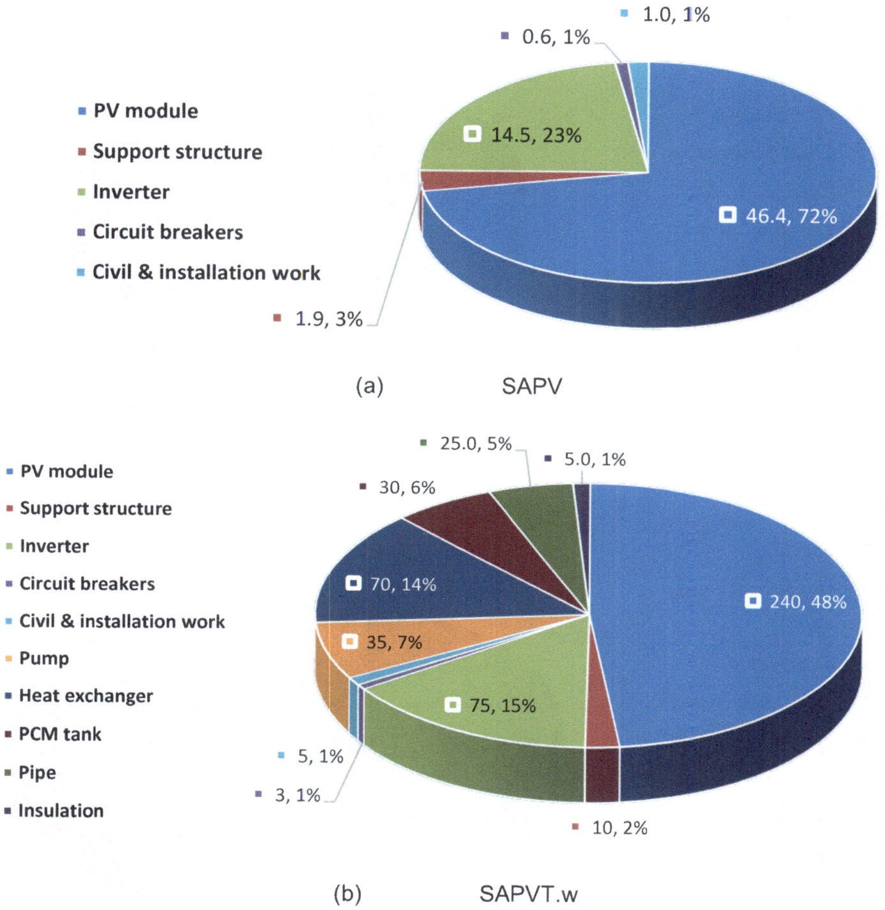

Fig. 4.2 LCCA of the tested systems by Ref. [35]

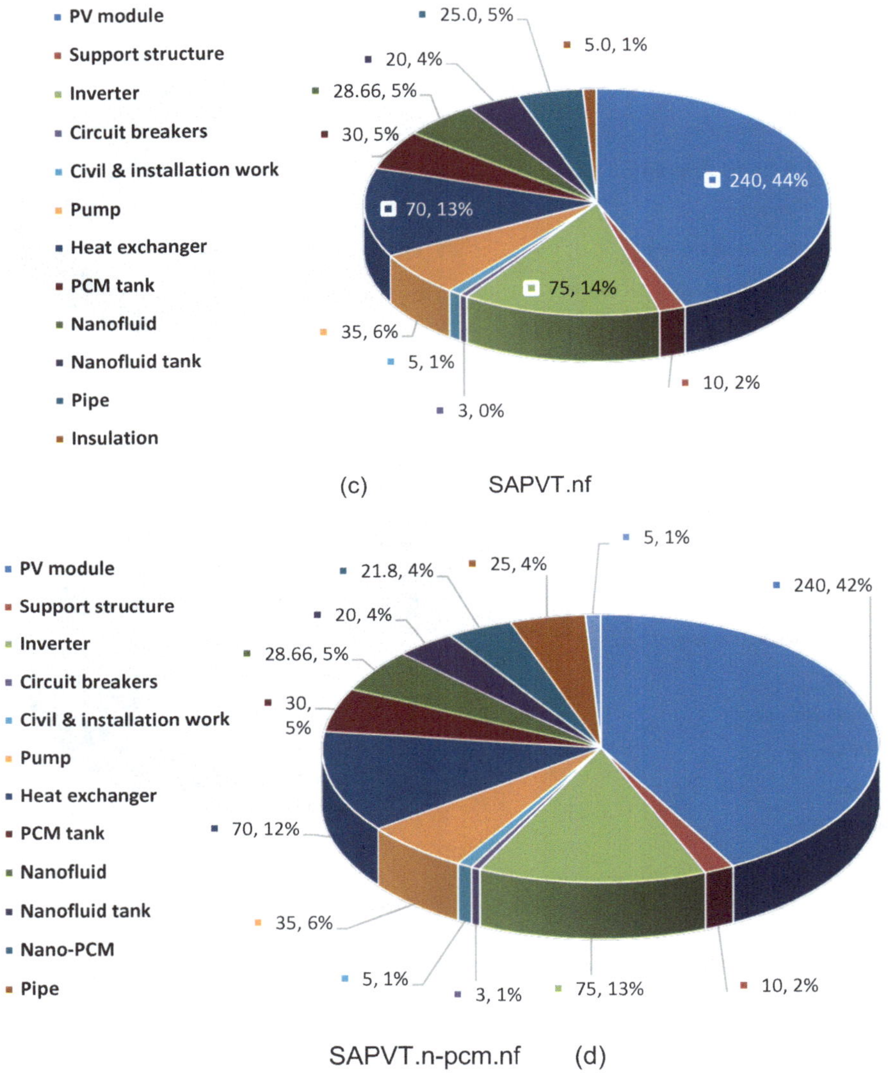

Fig. 4.2 (continued)

age of the system ranged between 7.30 years and 16.9 years, which is less than the supposed service life of such systems (30 years). The researchers considered that the reason for the decrease is due to the flow of air through the channel below the photovoltaic cell. The study concluded that the system is acceptable and suitable for application in India financially and environmentally and is ready to be a major source of renewable energy in this country.

4.3 Levelized Cost of Heat (LCOH)

It expresses the cost of heat generation for a system and more accurately reflects the working fluid temperature. This concept is an economic assessment of the cost of the heat generation system over the life of the system and includes initial investment, operation, maintenance, fuel cost, and cost of capital. Standardized cost of energy is a systematic method commonly used in the case of economic power plant assessments over a long period of operation. In this concept, the fixed cost is charged on a time-period basis, and the average cost of energy production is used over the lifetime of the power plant. When using the LCOH price model, the real fixed cost is combined with the variable cost without having to include any other fixed components. This LCOH method is characterized by its flexibility and transparency. The biggest challenge in calculating LCOH is how to estimate total heat production during the lifetime of the system. Usually when calculating the prices approved in the LCOH account, it is done on an hourly basis [36]. LCOH can be calculated using the following formula:

$$\text{LOCH} = \frac{\text{OCC.CRF} + \text{OMC}}{8.76.\text{CF}} + \text{fc.hr} \tag{4.4}$$

where OCC – overnight capital cost, [$/MW].

CRF is known by:

$$\text{CFR} = \frac{i.(1+i)^n}{(1+i)^n - 1} \tag{4.5}$$

OMC – fixed O&M cost, ($/MW)
CR– yearly capacity factor
Fc – fuel cost, ($/kWh)
hr– heat rate

LCOH represents the minimum price of the power unit generated by the heat generating system which must be sold. The maximum price determines the working fluid temperature up to fractional limit. Typically, LCOH is calculated over the age of 20 years in units of currency per kilowatt-hour (e.g., $/kWh or €/kWh or per MWh). In case of comparison between several systems to be acceptable from view point of the calculation of LCOH for a number of systems, the frameworks and costs of these systems should be determined. Additional costs, such as the cost of research and development, tax, impact costs on public health, and environmental damage, can be added and cannot be added. Any supporting policies or government support may be added to the LCOH account if available [37].

Reference [36] suggested an effective way to reform central heating pricing models and improves their efficiency. The proposed model is based on the stable thermal cost (LCOH) and the expected demand for hourly heat. Sweden's central heating prices have been studied as a reference case. In the study, the researchers adopted three methods to determine the variable fuel cost with heat production requirements. The results clearly showed that the LCOH pricing models can achieve high reliability in calculating the cost of heat production clearly. The proposed model (dynamic price model) has been more successful than other market price models in predicting changes in production cost in the case of different heat production methods.

The cost and thermal efficiency of the solar collector are important factors that must be improved to reduce the cost of heat calculated by LCOH. Reference [38] compared the LCOH for many solar thermal collectors in the weather conditions of California, USA. The results of the study showed that natural gas is the cheapest source of heat production as the LOCH has a limit of 2.9 cents/kWh, which supports the results of Ref. [39].

Reference [40] developed the LCOH model for economic and technical evaluation of hybrid PV/T systems in general. In this model, the researcher lists the technical and financial variables related to the LCOH account as in Table 4.3. Through the numerous options in the proposed design of the PV/T system and the utilization of thermal energy generated in addition to the improved electricity efficiency of the system, the researchers managed to find the LCOH capable of competing with natural gas in six states within the USA. This cost was about 41.4 ¢/kWh in Hawaii. In California, the lowest LCOH ratio for the proposed PV/T system was 1.23 ¢/kWh.

4.4 Levelized Cost of Energy

It is also called the levelized cost of electricity (LCOE); it is the net present cost of the electricity unit over the lifetime of one of the generated assets. It is often considered as a concept of the average price at which the energy generated must be in the market until its end. The use of different methods to generate electricity means there are different costs from one station to another significantly. It is natural for the

Table 4.3 List of the parameters used in LOCH calculations

Technical	Financial	Installation site
Collector size (m^2)	Thermal cost ($\$/m^2$)	CST nameplate (MW_{th})
Optical efficiency (%)	PV cost ($\$/m^2$)	Average DNI (kWh/m^2. day)
Cell area coverage (%)	Indirect mark up (%)	Cost of electricity ($/kWh)
Optical properties (abs./trans./refl.)	Debt to equity (%)	Operation lifetime (years)
Module wiring losses (%)	Interest rate (%)	–
Power conditioning losses (%)	Loan term (years)	–
Full-spectrum cell eff. (%)	Effective tax rate (%)	–
Cell bandgap (eV)	Nominal discount rate (%)	–
Thermal eff. (%)	CPV O&M ($\$/kW_{elect.}$)	–
GST degradation (5/year)	O&M escalation (5/year)	–

designer to make calculations for these costs, preferably focusing on carrying the electricity grid to the lowest level of electricity with units of kilowatt-hours or megawatt-hours. This cost includes initial capital, continuous operating costs, fuel, maintenance, and discount rate. This concept works to assist researchers and to help decision-makers and industry men during discussions and decision-making [41]. Hence, the levelized cost of energy (LCOE) is a measure of the power source of electricity generation on a constant basis and compared in various other ways. This measure is an economic assessment of the average total cost of building and operating power plants operating for its lifetime, divided by the total energy output of the asset over that lifetime. It represents the minimum average electricity selling price to enable this measure to be achieved over the lifetime of the project. The levelized cost of electricity (LCOE) can be calculated by Equation [42]:

$$\text{LCOE} = \frac{\text{Sum of costs over lifetime}}{\text{Sum of electrical energy produced over life time}} = \frac{\sum_{t=1}^{n} \frac{I_t + M_t + F_t}{(1+r)^t}}{\sum_{t=1}^{n} \frac{E_t}{(1+r)^t}} \quad (4.6)$$

I_t investment expenditures in the year t
M_t operations and maintenance expenditures in the year t
F_t fuel expenditures in the year t
E_t electrical energy generated in the year t
R discount rate
N expected lifetime of system or power station

Here, some caution should be taken into account when calculating the flat cost of energy. There are often some invisible assumptions and some factors, such as taxes with negligible effects or terms of project financing and technological analysis. LCOE is calculated as the default life span of the plant, usually 20–40 years [42]. The presumed power factor is considered to have a significant effect when calculating the LCOE. Hence, the most important requirement of the analysis is to trust that the assumptions made are fully justified.

Wind and solar energy are high-tech technologies and are intermittent, depending on the availability of solar radiation or wind movement, so they are economically disadvantaged only when they are generated with maximum availability [43]. This is because LCOE means that capital investment is available at low cost. These intermittent energy sources (wind and solar) always need to add costs associated with the provision of energy storage [44]. These sources can also be competitive at certain times such as electricity production at peak hours. As an example, solar energy peaks in the middle of the day and is the same as the peak demand for electricity [45].

Reference [46] provided a systematic study of the standard cost of electricity (LCOE) for wind and solar energy in China. The results of the study showed that taxes system should be improved for renewable energy in China as the current one does not provide adequate coverage even after the increase of the discount by 5%. There is still a need to increase subsidies for renewable electricity generation and to be dynamically modified based on LCOE to better support renewable energy development. The study concluded that the best government support for renewable energy plants to be able to compete with fossil fuel plants is to finance renewable energy projects because the invested capital has a significant impact on LCOE; and reform the electricity price over the medium and long term to make renewable energy more accepted.

Reference [47] showed that although LCOE is widely used as a benchmark for the comparison of power generation techniques, when evaluating intermittent renewable sources of energy such as photovoltaic power and wind, the costs of variation and integration are ignored. The researchers proposed a new LCOE system in which costs are calculated by developing a new mathematical definition of integration costs linked to economic theory. This new concept allows the LCOE to compare economically to power generation techniques and to obtain optimum quantities of intermittent renewable sources. The new LCOE system helps decision-makers and researchers to calculate the cost of electricity generated from renewable sources more efficiently and accurately.

Reference [48] created a mathematical model to analyze the current value of LCOE for photovoltaic and concentrated solar power stations (CPS). The researchers built such a model on annual electricity production information for past years, proven cumulative capacity, and annual costs for PV units and CSP systems. The researchers used a computer simulation program to illustrate the curves of sensitivity analysis and to chart the evolution of LCOE over the period from 2010 to 2050. The study concluded that the future LCOE designation has an important benefit in facilitating decisions to be made in energy planning policies (e.g., tariff exemptions, customs, and tax exemptions). It is also useful in making investment decisions by comparing the costs of photovoltaic power generation with other power generation technologies from 2010 to 2050. The researchers concluded that photovoltaic power plants are more suitable for use in the mid-to-high latitudes of the Earth, concentrated solar cells are preferred to be used after thermal storage has been combined with them in arid areas located on relatively low latitudes (Fig. 4.3).

4.5 Life Cycle Assessment

This concept is used by designers to assess the environmental impacts of the product during its life stages from the stage of extracting the raw material to the stages of processing of the material; the stage of manufacturing, distribution, use, repair, and maintenance; and the final stage of disposal of the final product after the end of its life. This assessment enables designers to avoid many environmental concerns

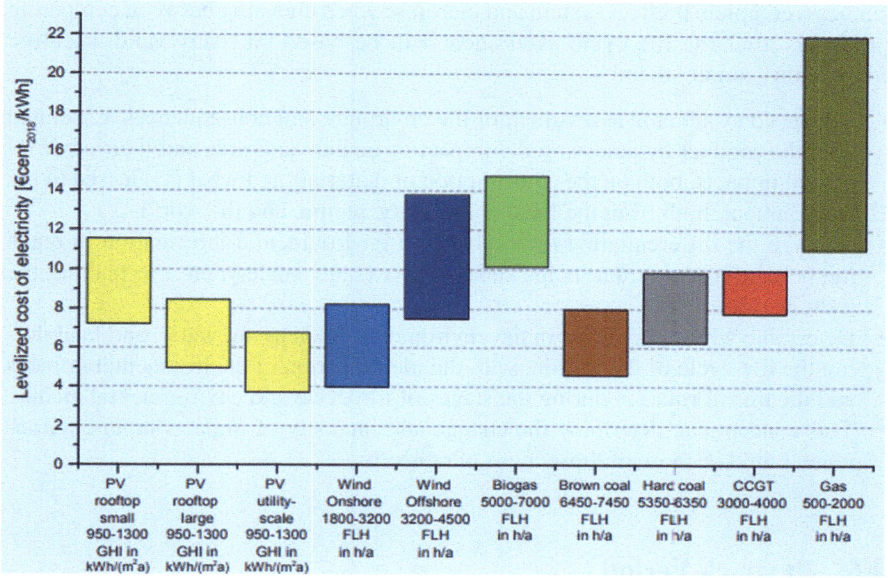

Fig. 4.3 Comparison of LCOE for renewable power plants and conventional power plants at different locations in Germany in 2018 [49]

by assessing potential impacts associated with specific inputs and releases. This concept helps decision-makers to make more objective decisions [50]. These data are therefore used to promote and support the energy policy of any government while providing a solid basis to support these decisions.

Reference [51] reviewed the latest developments in life cycle assessment methods in some areas where there has been extensive methodological development over the past years. The researchers conducted an analysis based on discussion of system boundaries and data collection and allocation. They also studied the latest developments in databases, input, and production of hybrid LCA systems. The researchers also discussed recent developments in this concept and the strengths and weaknesses of the LCA. Reference [52] explained that the decision-makers should use the LCA in general, but they are not supposed to adopt it when the difference between its results is relatively small. It should also be used because it is the most common and applied when there is no clear vision within the reach of decision-makers.

Reference [53] suggested that different marginal impacts should be included in the life cycle assessment and that their short-term effects should be incorporated into the current production power of photovoltaic power plants. The changes in productive energy appear with long-term effects, as explained by the Ref. [54]. Energy systems are usually focused on short-term effects.

Reference [55] explained that working with the LCA concept helps to avoid the transfer of environmental problems from one location to another, since this concept studies a complete product system and therefore when choosing between competing producers after the life cycle assessment will be based on many valid scientific foundations, among them:

- Advanced systematic assessment of the environmental consequences associated with the product by assessing the impact of human activities and their environmental impacts, both on the consumption of materials and what is released to the environment, both from the local community, region, and the world.
- Analyze the differentiation on the basis of environmental information to reach the best information that helps stakeholders (state, society, etc.) in making the right decision.
- Determine what is going on in the environment, such as air, water, and land during the life cycle of the station, with the identification of environmental impacts and the transformation during the stages of life cycle and environmental media.
- This concept can determine the change and intensity of impacts in an environmental field or more of those areas of concern.

4.6 Payback Period

The recovery period represents the time taken to recover the cost of the investment, the period in which the investment reaches the break-even point. Any investor in any field mostly cares about the period of recovery of capital. Investments with short-term redemption periods are always the most attractive investments. If the repayment period is long, the project will be less favorable, and investors will distance themselves from it. So, this concept gives investors and designers the ability to make quick judgments about their investments. This concept is used in the financial and capital budget. In the energy sector, the "payback period" is used to determine the cost savings and energy efficiency. The redemption period represents the relationship between the investment's costs divided by the annual cash flow. For example, if you use a photovoltaic solar system worth 5000 US$, and this system provided 100 US$ a month, meaning that the recovery period will be 4.2 years. The disadvantage of this concept is to ignore the inflation of the value of money over time, while other concepts (such as net present value (NPV), internal rate of return (IRR), and discounted cash flow) take this into account. The use of the payback period was previously used in the financial budget and capital, but it began to apply in other areas in the calculation of return on energy-saving technologies such as solar panels, insulation, maintenance, and promotions [56].

The concept of "payback period" suffers from certain dilemmas such as [57–60]:

1. Age of assets: In case the useful life of the project expires after the initial investment has been paid, additional cash flows cannot be generated, since the method of the payback period does not include the use of asset age assumptions.

2. Timely value of money: The value of money during future periods is less than the value of the money in the project capital currently due to multiple inflation factors. This is not counted in the "payback period" method. There is an advanced version of this method known as the "discounted payback period." In this concept, the depreciation of money with time is taken into consideration.
3. The complexity of cash flow: The simplicity of this concept prevents it from calculating the large number of cash flows that arise with the development of capital investment that may become after the recovery period stages. For example, but not limited to the cash expenses required for periodic promotions. These cash flows may change over time as customer demand changes and competition escalates.
4. Additional cash flows: This concept does not take into account the need for additional cash flows during the periods following full recovery arising from the investment.
5. Profitability: The "payback period" method focuses on what is required to pay off the initial investment; it is not included in the final profitability accounts of the project. Hence, this method has shown that a project has a short-term yield but does not make a gross profit and considers it a better investment when compared to a project that requires long-term recovery but at the same time makes a large gross profit over a long period of time.

Reference [24] used the following equation to calculate the payback period for the system they studied:

$$\mathrm{PBP} = C_{\mathrm{capital}}\left(\mathrm{USD}\right) / [E_{\mathrm{PV}_{\mathrm{annual}}}\left(\frac{\mathrm{kWh}}{\mathrm{year}}\right) \times \mathrm{CoE}\left(\frac{\mathrm{USD}}{\mathrm{kWH}}\right) \times R_{\mathrm{PW}} \tag{4.7}$$

where

C_{capital} The capital cost
E_{PVannual} The PV system's annual energy production
CoE The cost of energy
R_{PW} The present worth of each factor and was calculated using the formula:

$$R_{\mathrm{PW}} = F_m / \left(1 + I\right)^N \tag{4.8}$$

where

F_m The future sum of money
I The particular discount rate
N The years number

4.7 Conclusions and Recommendations

This chapter presents financial assessment of PV, solar thermal, and PV/T systems. The concepts discussed as payback period (PBP), life cycle cost (LCC), levelized cost of heat (LCOH), levelized cost of energy (LCOE), and life cycle assessment (LCA).

Conclusions

1. Increasing the complexity of PV/T system leads to increased operation and maintenance costs and stretches the payback period.
2. Payback period parameter is limited in accuracy due to several issues like additional cash flow, time value of money, etc.
3. PV/T systems have massive potential replacing standard PV and solar thermal systems due to increased energy yield which can lead to drop in cost of energy and payback period.

Recommendations

1. To investigate the costs associated with manufacturing, preparing, mixing, and delivering nanofluids for nanofluid-based PV/T systems.
2. To create correlation between nanofluids stability and added costs due to replacement of nanofluids.
3. To create a standard method for effectively assessing the economic aspect of PV/T systems.

References

1. H.M.S. Al-Maamary, H.A. Kazem, M.T. Chaichan, The impact of the oil price fluctuations on common renewable energies in GCC countries. Renew. Sustain. Energy Rev. **75**, 989–1007 (2017)
2. H.M.S. Al-Maamary, H.A. Kazem, M.T. Chaichan, Renewable energy and GCC states energy challenges in the 21st century: a review. Int. J. Comput. Appl. Sci. IJOCAAS **2**(1), 11–18 (2017)
3. H.M.S. Al-Maamary, H.A. Kazem, M.T. Chaichan, Climate change: the game changer in the GCC region. Renew. Sustain. Energy Rev. **76**, 555–576 (2017)
4. N. Chimres, S. Wongwises, Critical review of the current status of solar energy in Thailand. Renew. Sustain. Energy Rev. **58**, 198–207 (2016)
5. IRENA, Renewable Energy Statistics 2017, The International Renewable Energy Agency, Abu Dhabi. (2017). See also URL http://www.irena.org/publications/2017
6. IRENA and CPI, Global Landscape of Renewable Energy Finance, International Renewable Energy Agency, Abu Dhabi. (2018). See also URL http://www.irena.org/publications/2018
7. IRENA, Renewable Power Generation Costs in 2017, International Renewable Energy Agency, Abu Dhabi. (2018). See also URL http://www.irena.org/publications/2018
8. S. Comello, S. Reichelstein, A. Sahoo, The road ahead for solar PV power. Renew. Sustain. Energy Rev. **92**, 744–756 (2018)
9. P.E. Campana, L. Wasthage, W. Nookuea, Y. Tan, J. Yan, Optimization and assessment of floating and floating-tracking PV systems integrated in on- and off-grid hybrid energy systems. Solar Energy **177**, 782–795 (2019)

10. M.Z. Farahmand, M.E. Nazari, S. Shamlou, Optimal sizing of an autonomous hybrid PV-wind system considering battery and diesel generator. Iranian Conf. Electr. Eng. (ICEE), pp. 1048–1053 (2017)
11. M. Honarmand, A. Zakariazadeh, S. Jadid, Optimal scheduling of electric vehicles in an intelligent parking lot considering vehicle-to-grid concept and battery condition. Energy **65**, 572–579 (2014)
12. A.S.O. Ogunjuyigbe, T.R. Ayodele, O.A. Akinola, Optimal allocation and sizing of PV/Wind/split-diesel/battery hybrid energy system for minimizing life cycle cost, carbon emission and dump energy of remote residential building. Appl. Energy **171**, 153–171 (2016)
13. F. Safdarian, M.E. Nazari, Optimal sizing of a solar-thermal collector for residential applications using gravitational search algorithm. Int. J. Mech. Eng. Auto **2**, 497–505 (2015a)
14. Energy Policy and Planning Office. Energy Statistics; 2016.
15. S. Kumar, Assessment of renewables for energy security and carbon mitigation in Southeast Asia: the case of Indonesia and Thailand. Appl. Energy **163**, 63–70 (2016)
16. A. Modarresi Ghazvini, J. Olamaei, Optimal sizing of autonomous hybrid PV system with considerations for V2G parking lot as controllable load based on a heuristic optimization algorithm. Solar Energy **184**, 30–39 (2019)
17. M. Das, M.A.K. Singh, A. Biswas, Techno-economic optimization of an off-grid hybrid renewable energy system using metaheuristic optimization approaches – case of a radio transmitter station in India. Energ. Conver. Manage. **185**, 339–352 (2019)
18. M.A. Ramli, H.R.E.H. Bouchekara, A.S. Alghamdi, Optimal sizing of PV/wind/diesel hybrid microgrid system using multi-objective self-adaptive differential evolution algorithm. Renew. Energy **121**, 400–411 (2018)
19. S. Sanajaoba, E. Fernandez, Maiden application of Cuckoo Search algorithm for optimal sizing of a remote hybrid renewable energy system. Renew. Energy **96**, 1–10 (2016)
20. J.L. Bernal-Agustín, R. Dufo-López, Efficient design of hybrid renewable energy systems using evolutionary algorithms. Energ. Conver. Manage. **50**(3), 479–489 (2009)
21. Y. Sawle, S.C. Gupta, A.K. Bohre, Socio-techno-economic design of hybrid renewable energy system using optimization techniques. Renew. Energy **119**, 459–472 (2018)
22. S. Sinha, S.S. Chandel, Review of software tools for hybrid renewable energy systems. Renew. Sustain. Energy Rev. **32**, 192–205 (2014)
23. H.A. Kazem, H.A.S. Al-Badi, A.S. Al Busaidi, M.T. Chaichan, Optimum design and evaluation of hybrid solar/wind/diesel power system for Masirah Island. Environ. Develop. Sustain. **19**(5), 1761–1778 (2017)
24. S. Gangwar, D. Bhanja, A. Biswas, Cost, reliability, and sensitivity of a stand-alone hybrid renewable energy system – a case study on a lecture building with low load factor. J Renew Sustain Energy **7**(1), 013109 (2015)
25. A. Singh, P. Baredar, B. Gupta, Techno-economic feasibility analysis of hydrogen fuel cell and solar photovoltaic hybrid renewable energy system for academic research building. Energ. Conver. Manage. **145**, 398–414 (2017)
26. H.A. Kazem, A.H.A. Al-Waeli, M.T. Chaichan, A.S. Al-Mamari, A.H. Al-Kabi, Design, measurement and evaluation of photovoltaic pumping system for rural areas in Oman. Environ. Develop. Sustain. **19**(3), 1041–1053 (2017). https://doi.org/10.1007/s10668-016-9773-z. 2016
27. M.T. Chaichan, H.A. Kazem, *Generating Electricity Using Photovoltaic Solar Plants in Iraq* (Springer). ISBN: 978-3-319-75030-9. https://doi.org/10.1007/978-3-319-75031-6
28. A.H.A. Al-Waeli, M.T. Chaichan, K. Sopian, H.A. Kazem, H.B. Mahood, A.A. Khadom, Modeling and experimental validation of a PV/T system using nanofluid coolant and nano-PCM. Solar Energy **177**, 178–191 (2019)
29. A.H.A. Al-Waeli, K. Sopian, H.A. Kazem, M.T. Chaichan, Nanofluid based grid connected PV/T systems in Malaysia: A techno-economical assessment. Sustain. Energy Technol. Assessm. **28**, 81–95 (2018)
30. A.H.A. Al-Waeli, M.T. Chaichan, H.A. Kazem, K. Sopian, Evaluation and analysis of nanofluid and surfactant impact on photovoltaic-thermal systems. Case Study Thermal Eng. **13**, 100392 (2019)

31. A.H.A. Al-Waeli, M.T. Chaichan, K. Sopian, H.A. Kazem, Influence of the base fluid on the thermo-physical properties of nanofluids with surfactant. Case Study of Thermal Engineering, Accepted **13**, 100340 (2019)
32. A.H. Al-Waeli, M.T. Chaichan, H.A. Kazem, K. Sopian, A. Ibrahim, S. Mat, M.H. Ruslan, Numerical study on the effect of operating nanofluids of photovoltaic thermal system (PVT) on the convective heat transfer. Case Study Thermal Eng. **12**, 405–413 (2018)
33. A.H.A. Al-Waeli, H.A. Kazem, K. Sopian, M.T. Chaichan, Techno-economical assessment of grid connected PV/T using nanoparticles and water as base-fluid systems in Malaysia. Int. J. Sustain. Energy **37**(6), 558–578 (2018)
34. A.H.A. Al-Waeli, M.T. Chaichan, K. Sopian, H.A. Kazem, Comparison study of indoor/outdoor experiments of SiC nanofluid as a base-fluid for a photovoltaic thermal PV/T system enhancement. Energy **151**, 33–44 (2018)
35. M. Tripathy, H. Joshi, S.K. Panda, Energy payback time and life-cycle cost analysis of building integrated photovoltaic thermal system influenced by adverse effect of shadow. Appl. Energy **208**, 376–389 (2017)
36. H. Li, J. Song, Q. Sun, F. Wallin, Q. Zhang, A dynamic price model based on levelized cost for district heating. Energy Ecol. Environ. **4**(1), 15–25 (2019)
37. H. Li, Q. Sun, Q. Zhang, F. Wallin, A review of the pricing mechanisms for district heating systems. Renew. Sustain. Energy Rev. **43**, 56–65 (2015)
38. A. Hassanzadeh, L. Jiang, R. Winston, Coupled optical-thermal modeling, design and experimental testing of a novel medium-temperature solar thermal collector with pentagon absorber. Solar Energy **173**, 1248–1261 (2018)
39. P. Kurup, C. Turchi, Potential for solar industrial process heat in the United States: a look at California. In AIP Conference Proceedings 2016 May 31 (Vol. 1734, No. 1, p. 110001). AIP Publishing
40. B.C. Riggs, R. Biedenharn, C. Dougher, Y.V. Ji, Q. Xu, V. Romanin, D.S. Codd, J.M. Zahler, M.D. Escarra, Techno-economic analysis of hybrid PV/T systems for process heat using electricity to subsidize the cost of heat. Appl. Energy **208**, 1370–1378 (2017)
41. Donald J. Johnston, Projected Costs of Generating Electricity: 2005 Update (Nuclear Energy Agency/International Energy Agency/Organization for Economic Cooperation and Development, Château de la Muette, Paris, 2005)
42. K. Branker, M.J.M. Pathak, J.M. Pearce, A review of solar photovoltaic levelized cost of electricity. Renew. Sustain. Energy Rev. **15**, 4470–4482 (2011). https://doi.org/10.1016/j.rser.2011.07.104
43. V. Fthenakis, J.E. Mason, K. Zweibel, The technical, geographical, and economic feasibility for solar energy to supply the energy needs of the US. Energy Policy **37**, 387–399 (2009)
44. M. Bruck, P. Sandborn, N. Goudarzi, A Levelized Cost of Energy (LCOE) model for wind farms that include Power Purchase Agreements (PPAs). Renew. Energy **122**, 131–139 (2018)
45. Lindsay Miller, , Scott Harper, Someshwar Singh, Evaluating the link between LCOE and PPA elements and structure for wind energy, Energ. Strat. Rev., vol. 16, pp. 33–42, 2017.
46. X. Ouyang, B. Lin, Levelized cost of electricity (LCOE) of renewable energies and required subsidies in China. Energy Policy **70**, 64–73 (2014)
47. F. Ueckerdt, L. Hirth, G. Luderer, O. Edenhofer, System LCOE: what are the costs of variable renewables? Energy., vol. **63**, 61–75 (2013)
48. J. Hernández-Moro, J.M. Martínez-Duart, Analytical model for solar PV and CSP electricity costs: present LCOE values and their future evolution. Renew. Sustain. Energy Rev. **20**, 119–132 (2013)
49. C. Kost, S. Shammugam, V. Jülch, H.-T. Nguyen, T. Schlegl, Levelized Cost of Electricity Renewable Energy Technologies, Report for Fraunhofer Institute for Solar Energy Systems Ise, March 2018
50. Life Cycle Assessment (LCA) Overview. sftool.gov. Retrieved 1 July 2014
51. GHG Product Life Cycle Assessments. Ecometrica. Retrieved on: 25 April 2013

52. G. Finnveden, M.Z. Hauschild, T. Ekvall, J. Guined, R. Heijungs, S. Hellweg, A. Koehler, D. Pennington, S. Suh, Recent developments in life cycle assessment. J. Environ. Manage. **91**, 1–21 (2009)
53. S. Lundie, A. Ciroth, G. Huppes, Inventory methods in LCA: towards consistency and improvement – Final report. UNEP-SETAC Life Cycle Initiative (2007)
54. O. Eriksson, G. Finnveden, T. Ekvall, A. Björklund, Life cycle assessment of fuels for district heating: a comparison of waste incineration, biomass- and natural gas combustion. Energy Policy **35**, 1346–1362 (2007)
55. Mattsson, N., Unger, T., Ekvall, T., 2003. Effects of perturbations in a dynamic system – the case of nordic power production. In: Unger, T. (Ed.), Common Energy and Climate Strategies for the Nordic Countries – A Model Analysis. PhD thesis, Chalmers University of Technology, Göteborg, Sweden.
56. M.A. Curran, Life cycle assessment: principles and practice, Scientific Applications International Corporation (SAIC), report no. EPA/600/R-06/060, 2006
57. W.H. Cunningham, B. Joseph, Energy conservation, price increases and payback periods, in *NA – Advances in Consumer Research*, ed. by K. Hunt, vol. Volume 05, (Association for Consumer Research, Ann Abor, MI, 1978), pp. 201–205
58. N. Sengar, P. Dashora, M. Gupta, S. Mahavar, Experimental studies, energy savings and payback periods of a cylindrical building-material-housing solar cooker. Int. J. Energy Inform. Commun. **2**(3), 75–84 (2011)
59. H.A. Kazem, Renewable energy in Oman: status and future prospects. Renew. Sustain. Energy Rev. **15**, 3465–3469 (2011)
60. M. AlamImteaz, A. Ahsan, Solar panels: real efficiencies, potential productions and payback periods for major Australian cities. Sustain. Energy Technol. Assessm. **25**, 119–125 (2018)

Chapter 5
The Impact of Climatic Conditions on PV/PVT Outcomes

5.1 Background

Photovoltaic cells are installed outdoors as well as for PVT systems, making them vulnerable to external weather conditions, which in turn vary from place to place. The weather conditions have an effect on the system performance, which makes its outcomes amazing or not. From here, we find that for the same type of photovoltaic cells, the photovoltaic system outcomes (efficiency and productivity) vary from location to location. The standard conditions used by scientists, researchers, and manufacturers to measure and compare the performance of photovoltaic cells are the solar irradiance of 1000 W/m^2, the temperature of 25 °C, and the air mass of 1.5. These standard conditions are limited to the existence of one or two of them, but all of them are impossible in external weather conditions. Each location in the world has special features and circumstances that are near or far away from these standard conditions. In the next paragraphs, the impact of climate variables on the photovoltaic cell performance and the advantages of PVT systems, which make it suitable for approximating the work of photovoltaic cells from the standard conditions, will be discussed.

5.2 The Effect of Solar Radiation on PV/PVT Modules Performance

Solar energy is the origin of all other energies, and it is free, clean, and available most days of the year and at most places. The energy that reaches the Earth's surface is a small fraction of the total energy falling on the Earth from the Sun. Solar energy is a necessity of life. Today, this energy is used in many applications, such as heating homes, heating water, ventilation, and electricity production as in the case of the PV cell. As the costs and prices of fossil fuels rise and these fuels are nearing their

A. H. A. Al-Waeli et al., *Photovoltaic/Thermal (PV/T) Systems*,
https://doi.org/10.1007/978-3-030-27824-3_5

depletion according to many confirmed reports, in addition to environmental pollution resulting from the use of these types of fuels, the efficient use of solar energy is very important [1–2].

The direct solar radiation intensity from the Sun varies by about 7% during the year (from 1412.0 W/m² in January to 1321.0 W/m² in July). This change is caused by the variation in the distance between the Earth and the Sun [3–6]. Because of the fluctuations of solar radiation, the solar constant is unstable over time. The average solar constant is varied with about 1 W/m² during a typical solar cycle of 11 years (Fig. 5.1).

Solar radiation loses part of its energy as it passes through the Earth's atmosphere. This part is lost by absorption and dispersion by the air molecules. The direct part of radiation from the Sun reaching the Earth is called direct radiation, while the scattered one can be called diffused radiation. Direct radiation reaches the Earth's surface without any collision or degradation due to particulate matter in the atmosphere (the particles causing the sunrays scattering are particles of gases and aerosols such as dust particles, sulfate particles, soot, salt particles, pollen, etc.). When direct solar radiation reaches the Earth's surface, part of it is absorbed by the Earth and another part is dissolved [3]. As it passes through the atmosphere, it undergoes a complex process, its spectral distribution changes, and the ultraviolet spectrum is lost in the upper part of the atmosphere. Molecules scatter a portion of the solar radiation and disperse it, and this part is known as diffuse radiation. Also, there is a section of direct radiation reflected to the atmosphere after colliding with the Earth's surface called albedo [4]. The three parts of the solar radiation (direct, diffused, and reflected) cause all the movement of the air, such as the temperature of

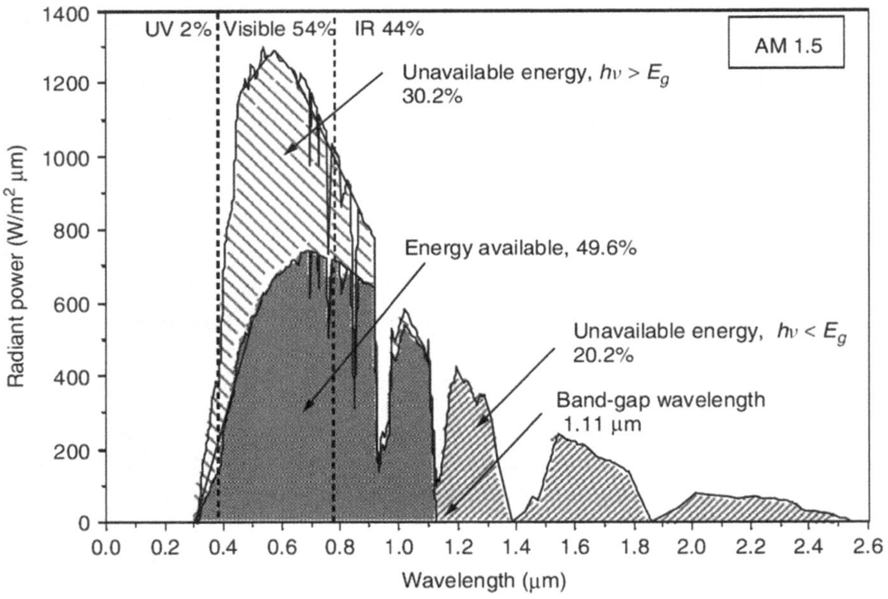

Fig. 5.1 Spectrum of solar radiation reached the Earth [6]

the wind and the variations in air and surface temperatures. 99.8% of all the energy on Earth is actually from the Sun [5]. Solar radiation loses 5% of its energy through its Earth's atmosphere (spectral change and ultraviolet absorption) and loses 15% when it collides with air particles (diffuse radiation resulting from clouds, dust, and air pollutants), so the Sun that reaches the Earth's surface is approximately 1020 W/m^2 [6, 7]. The weather conditions caused a clear reduction in the intensity of solar radiation reaching the Earth's surface, as well as the variation in color and quality of light from one region to another [8].

Solar radiation reaching the Earth is either short or long-wave radiation. The clouds and particles of air absorb the short wavelength radiation and return it as long wavelength radiation. Solar radiation with short wavelengths approaching the Earth's surface consists of direct, diffuse, and reflected radiation. Short wavelength radiation can be considered the most important part of these three types of radiation, because it contains the bulk of the incoming energy, while the energy of the other two types is directly or indirectly dependent on it [9].

Direct solar radiation depends on the shape and slope of the receiving surface of the radiation, while the diffused radiation differs somewhat in terms of inclination and gradation [10, 11]. In clear-sky conditions, the shortwave solar radiation is different depending on the height, slope, and direction of the receiver. Hence, the distribution of the receiving solar radiation is uneven and varied from one region to another on Earth [12, 13] (Fig. 5.2).

The Earth completes its orbit around the Sun in a full year and with an oval path. The line between the centers of the Sun and the Earth and its projection at the equatorial level forms an angle called the angle of solar drift. In two specific days of the year (20–21 March and 22/23 September), this angle is equal to zero [14].

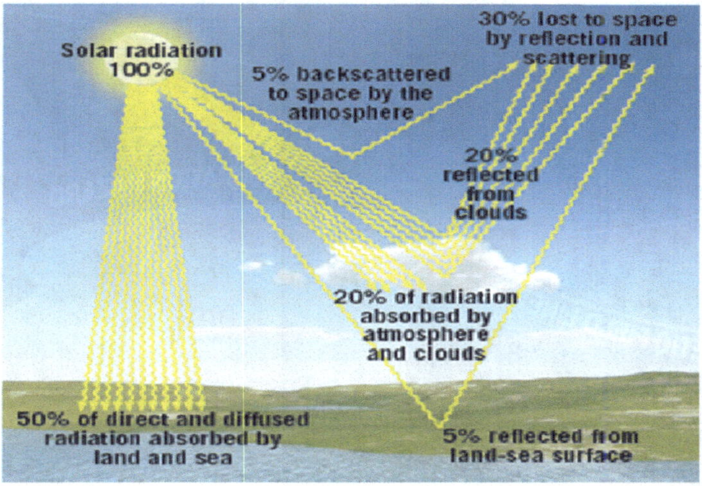

Fig. 5.2 The average distribution of solar radiation. (https://www.chegg.com/homework-help/figure-shows-30-percent-sun-s-energy-reflected-scattered-bac-chapter-16-problem-4ee-solution-9780321949752-exc)

Solar energy can be used to generate heat or produce electricity. In both options, energy is converted from solar radiation either to heat or to electricity. In solar thermal applications such as heating water for domestic and industrial purposes, heating the air for comfort purposes, or ventilating houses using Trombe walls, heat energy in solar radiation is used to provide heat to the thermal system. In the second case (solar electricity production), there are two options: first, by converting the heat in the solar radiation into mechanical energy, which rotates and powers turbines, as in the concentrated power plants (CSP) and the solar chimneys, or by converting part of this energy into direct electricity using photovoltaic cells.

Reference [3] suggested that if the world covers only a small fraction of 0.16% of the Earth's surface with photovoltaic cell systems with an efficiency of not less than 10%, this action will provide electricity capacity twice the global fossil fuel consumption rate. For photovoltaic systems (which are the focus of this book), their worldwide installed systems began to increase. In 1985, the combined PV systems were only 21 megawatts around the world. During 2008, installed systems around the world reached 6080 gigawatts, rising in 2009 to 7.3 gigawatts [15]. The stations installed in 2015 were estimated at the 54 GW line. The output of electricity in photovoltaic cells changes with the intensity of daytime irradiance, which reaches its rate in the middle of the day, almost the same time as the demand for electricity peaks. In PV systems, excess electricity can be stored in batteries [16–17] (Figs. 5.3 and 5.4).

The productivity and efficiency of photovoltaic cells vary according to the type of materials used in the manufacture of the unit and its thermal properties. It also depends largely on the intensity of the solar irradiance received, because a small part of the solar irradiance spectrum is converted into electricity, while the bulk is con-

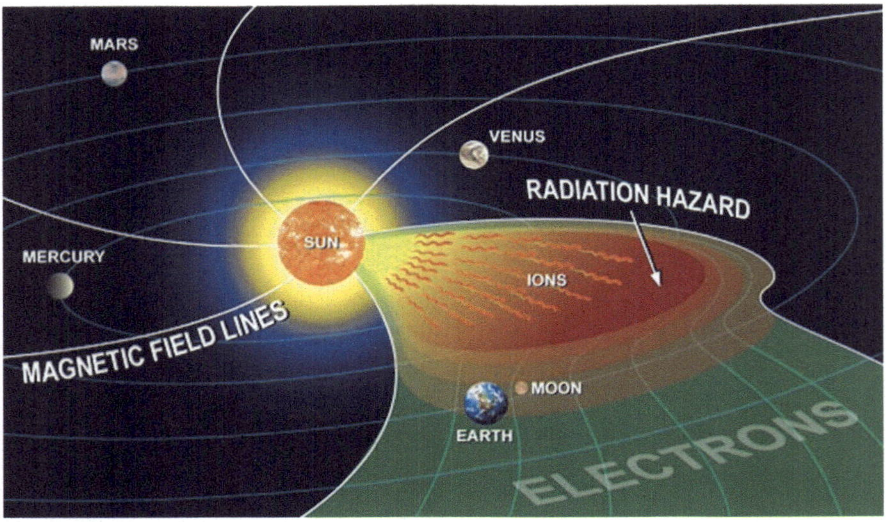

Fig. 5.3 Earth's rotation around the Sun and the effect of solar radiation on it. (http://arcturan.com/wp-content/uploads/2014/11/SolarParticleEvent.png)

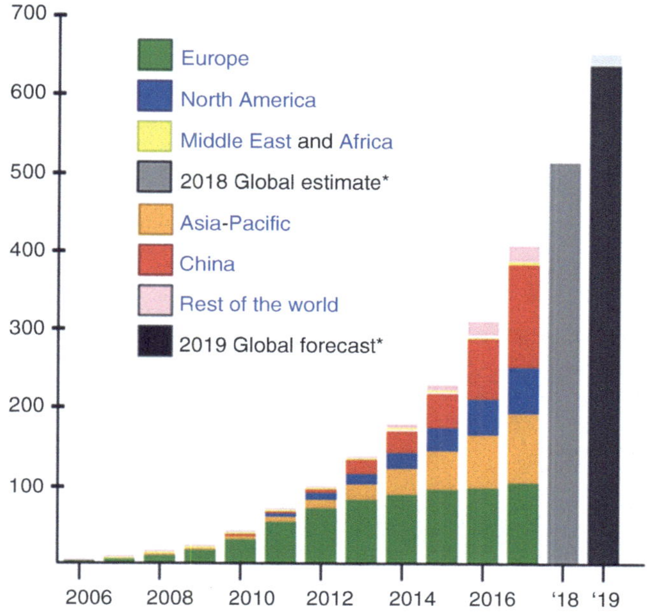

Fig. 5.4 Global growth of cumulative PV capacity in gigawatts (GWp) with regional shares (IEA estimates) (European Photovoltaic Industry Association (EPIA), "Global Market Outlook for Solar Power 2015–2019". Retrieved 14 April 2019)

verted to heat which is the direct cause of high temperature of the photovoltaic cell; as an inevitable result, it will reduce the efficiency of the electrical cell and reduce the output electricity [18]. Thousands of scientific studies have been conducted with experiments in the external air to ascertain the effect of external weather variables on the performance of many types of PV panels. Thongpao [19] measured the production of two types of photovoltaic (amorphous and polycrystalline) cells in Thailand in varying conditions of temperature and solar radiation. The study concluded that the efficiency of polycrystalline cells in temperature conditions was higher than that of amorphous thin-film cells during the hot summer months (Fig. 5.5).

The climatic weather conditions of the photovoltaic installation site affect the amount of the solar irradiance received by the cells [20]. Meteorological stations in most countries provide detailed data on solar radiation; however, many countries are clearly flawed in providing such data. The collection of solar radiation data during collector installation is a useful method [21], because the collector is characterized by its azimuth angle and tilt angle and location of the installation. Several factors (such as geographical latitude, time utilization period, climate condition, etc.) influence the optimal tilt angle [22]. The weather forecasters in developed countries record and analyze solar radiation data and produce simulated models to predict future conditions based on recorded meteorological data (e.g., humidity, ambient temperature, wind speed, etc.) (Fig. 5.6).

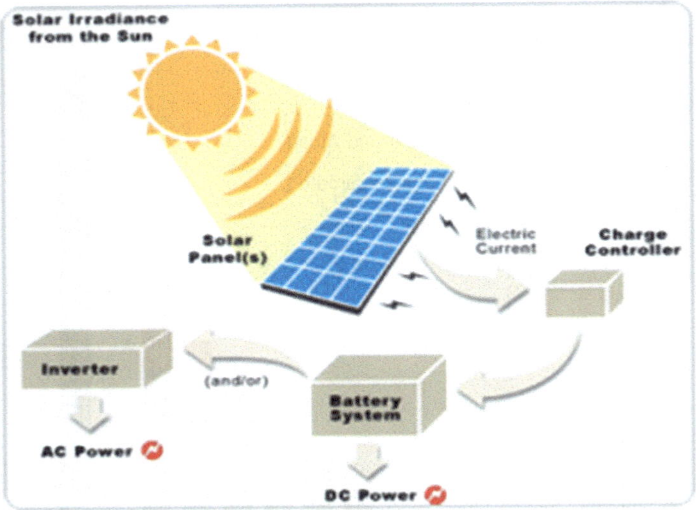

Fig. 5.5 Solar cells working process. (http://www.solarpowerenergyindia.com/home/solar-working-process/)

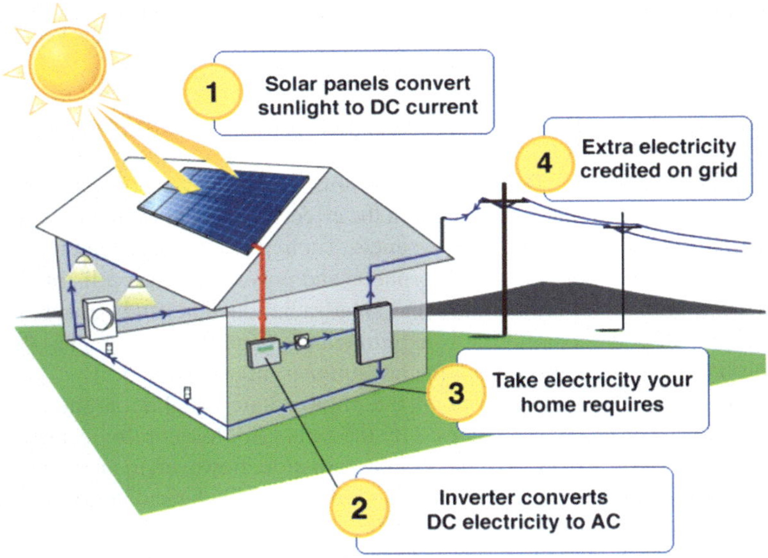

Fig. 5.6 The PV grid-connected system working process

Before designing and installing any PV system in a location, solar radiation data should be provided for this site. This data will provide information on solar energy available during the year. These data may not be readily available because of the difficulty and cost of solar radiation measurement stations. Alternative methods of addressing this point should be provided [23]. The solar radiation changes during the seasons as well as it changes during the day due to change in the location of the Sun [24]. The use of measured long-term solar radiation data enables a reliable solar radiation database. Such data have been used by many European countries to develop effective mathematical models to predict solar irradiance on tilted surfaces [25, 26]. Such models for solar radiation conversion were used to obtain data for sites where no actual solar radiation was measured, and this method yielded satisfactory results to solve this problem [27]. Many models of solar radiation prediction have been developed, and the models presented by Schaab [28] could be considered as an example. The reference to solar radiation modeling began in Jordan since the late 1970s. Salima [29] completed the work, introducing a computer model to predict the average daily, monthly, and global solar radiation intensity on horizontal surfaces, in addition to the possibility of this program to predict the rate of monthly brightness of the city of Amman, the capital.

A pyrometer is used to measure the radiation of the direct solar radiation, whereas diffused radiation is measured by other methods as thermometers, solar, or radiation. These devices must be installed at multiple locations in the study area to measure their radiation intensity. Sometimes, due to the high cost of these measurement devices, it is difficult to install a number of them in the required sites, and there are types of these devices that have a high degree of uncertainty, which means access to data is inaccurate and incomplete [30]. Mathematical relationships must be established that link the intensity of the solar irradiance to the ambient temperature, relative humidity, and wind speed by which the values of solar radiation can be defined as data that can be used in system design. These mathematical models are sufficient and can be used to determine the direct and diffuse solar radiation of sites that do not possess measured data [31].

At present, solar radiation measurements are available in acceptable quantities and qualities. Most of the current radiation measurement methods are still using older technology with some very small improvements. There have been significant problems in the uncertainty and accuracy of measurement of many of these devices in many researches. References [32, 33] explained a nonuniform temperature response in the measuring apparatus. There was also a high uncertainty in the measurements of solar irradiance carried out by References [34, 35]. Reference [36] found that there was an unobservable effect of wind velocity and ambient temperature in the measured data by pyranometers. To date, meteorologists are working to improve the quality and sensitivity of solar radiation measuring devices with the development of calibration and correction methods [37, 38, 39]. Reference [40] found that reducing uncertainty using calibration and correction did not have a significant impact on improved measured readings.

The uncertainty affects the accuracy of the measured data, so Gueymard investigated the influence of several uncertainties that caused inaccuracy of the resulting data. A high degree of uncertainty directly affects measurement data. The researchers also studied other measurement accuracy problems such as indirect determination of direct or diffuse radiation and shadow effect [41].

In India (Boni, Maharashtra), Reference [42] studied the intensity of direct and diffuse solar radiation by measuring them practically. Reference [43] studied the local meteorology conditions of the direct and diffused solar radiation. Mathematical models can also be used to estimate solar radiation in areas lacking measured data. The prediction of photovoltaic cell efficiency depends entirely on climatic variables such as air mass (AM), solar radiation, and ambient temperature. The closer the conditions are to STC, the higher the electricity output of these cells. The existing standard conditions do not meet together at the same time in nature, only rarely. Because the intensity of solar radiation changes over time, the photovoltaic cells' performance alters with time [44].

Al-Bashir's winter experiments in the external air showed the two most important variables affecting the productivity of any typical photovoltaic module: solar radiation and module temperature. The results of the study showed a decrease in the efficiency and performance ratio of the unit with increased solar radiation, which in turn causes an increase in the temperature of the back surface of the photovoltaic cell [45].

The effects of atmospheric variable overlap were studied by Cano who investigated the effect of dust accumulation on several sets of photovoltaic modules with varying tilt angles of 0, 5, 10, 15, 20, 23, 30, 33, and 40°. The results of this study indicate that the received radiation in each tilt angle varies according to the angle and the loss of energy due to the pollution increases with the approach of the angle of inclination from the horizontal position [46].

The effect of photovoltaic cells at an angle of inclination was also studied by Salih [47] using simulation of the effect of the tilt angle (azimuth and elevation) on the output of the photovoltaic cell. The researcher studied multiple angles (0°, 15°, 30°, 45°, and 60°) and calculated the solar radiation intensity falling on tilted surfaces in different directions (x and y). The results of the study showed that the solar radiation reaching the cell at the azimuth angle and the height difference was greatly affected, as the variation of the solar spectrum during daylight hours and the change of the visible spectrum by changing the angle of inclination should be measured in situ to determine the best angle of the system (Fig. 5.7).

Mahleri [48] built a mathematical model to determine the optimal angle of photovoltaic cells, which receive the greatest amount of solar radiation during a specified period. The researcher used data for this purpose to measure the accuracy of the submitted model further. During the study, the researchers built a database that enables access to the average global solar radiation of tilted surfaces, tilt angles, and different orientation during the study period. In this study, several mathematical models were proposed for the relationship between the angle of inclination, the average of the global radiation, and the effect of the slanted surfaces on the intensity of this radiation. The proposed models also enabled the acquisition of angular and optimum direction values.

In a study by Emmanuel [49], he used an algorithm based on a set of tilt angles for all latitudes to simulate. The aim of this study is to maximize the global solar radiation to an angular panel by calculating the ideal angle of PV panels in buildings and large photovoltaic power plants in Southern Italy. In this study, researchers took

Fig. 5.7 Divination of PV tilt angle. (https://pvpmc.sandia.gov/modeling-steps/1-weather-design-inputs/array-orientation/fixed-tilt/)

the diffused solar radiation to determine the tendency of the PV panels to maximize the solar radiation falling on the cell. For this purpose, they used global horizontal solar radiation data.

It can be summarized that the efficiency of photovoltaic (PV) cells decreases by increasing their temperature resulting from the high intensity of solar radiation. Kern and Rissell [50] was the first who proposed photovoltaic/thermal systems (PV/T). In such systems, electricity generation is integrated with solar heat extraction and used in two ways: the first is cooling the photovoltaic cell and the second is to utilize the heat extracted in other solar thermal applications. Thus, the efficiency of photovoltaic cells can be improved, as can the utilization of heat from cells in heating applications such as heating air for comfort purposes, heating water for domestic industrial purposes, and drying of agricultural products and other applications [51–53] (Fig. 5.8).

PV/T systems can be divided according to the cooling method for air cooling, water cooling, nanofluid cooling, or heat pipes cooling [54]. When comparing different cooling methods, air cooling systems are widely used because changes in the structure of the cell are very limited. The system is lightweight, easy to install, and low cost [55]. The disadvantage of this type of coolant is that the specific heat of the air is low, causing limited cooling, and that the surrounding air may be hot in the hot areas, thus reducing its cooling efficiency. The use of water as a coolant in PV/T systems is better than using air. Water has four times the specific heat of the air, which means it will remove more heat. If the air was used in PV/T in heating rooms in winter, it will be certainly dispensed in the hot summer while hot water can be used in summer and winter [56]. The main disadvantages of using water as a cooling fluid is its low thermal conductivity, which reduces the amount of heat transported during the time unit, thereby reducing the cooling efficiency and thermal efficiency of the system, followed by low electricity produced [57] (Fig. 5.9).

Adding nanoparticles with high thermal conductivity to water or base fluid will increase its thermal conductivity and thus increase the overall efficiency of the PV/T system by increasing its cooling, which means increasing electrical efficiency in

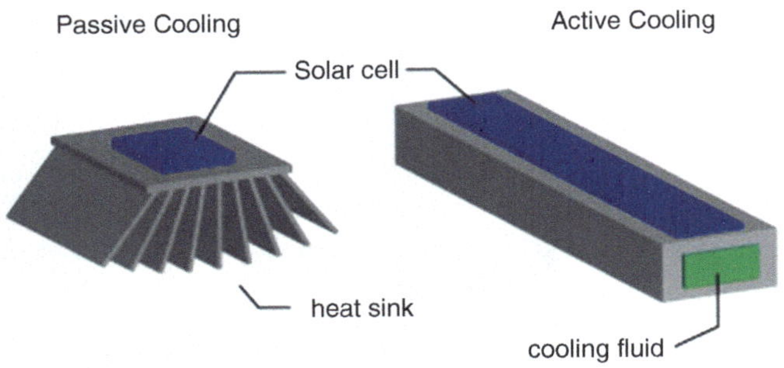

Fig. 5.8 Passive and active cooling for a hybrid PV/T. (http://www.greenrhinoenergy.com/solar/technologies/pv_concentration.php)

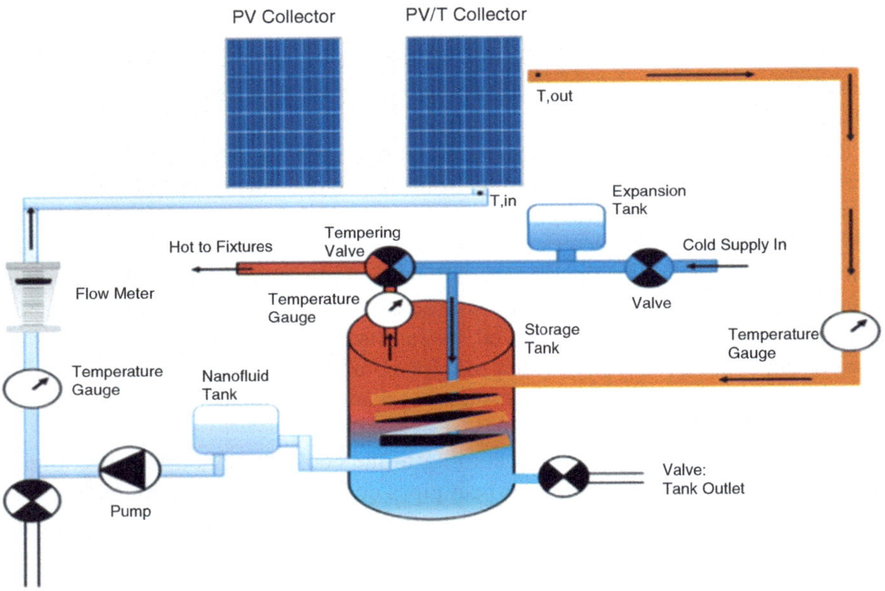

Fig. 5.9 Schematic diagram of water or nanofluid cooling PV/T system [54]

addition to increasing thermal efficiency. The use of these fluids has proven its usefulness, but it also has problems associated with the stability of nanoparticles suspension in the fluid and the deposition of nanoparticles in the system. The use of nanoparticles also increases the cost of the system [58–61]. In the latest development on PV/T systems, phase change materials (PCM) have been used in tanks adjacent to the back of the solar cell to withdraw as much heat as possible. To improve the thermal conductivity of these materials (which have low thermal conductivity), they have been mixed with high thermal conductivity nanoparticles. The researchers also circulate water or nanofluids in the PCM tank to allow them to draw

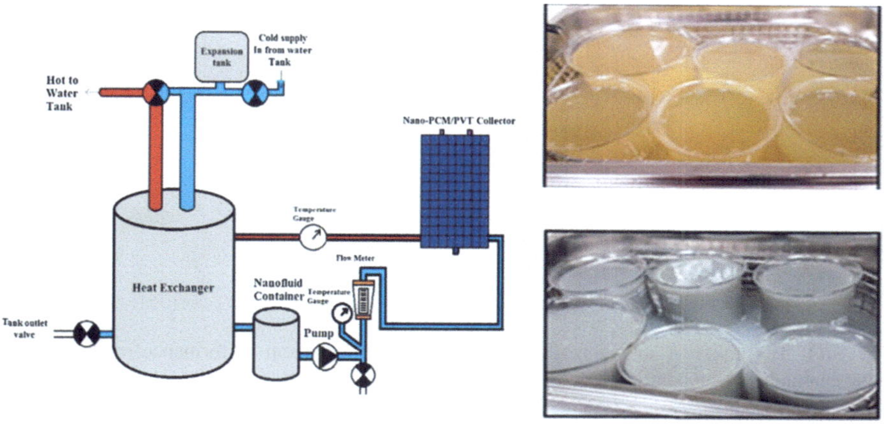

Fig. 5.10 Schematic diagram of nano-PCM nanofluid cooling PV/T system [62]

higher heat. Studies in this field are still ongoing to reach the best type of PCM for use and the best nanomaterials added to it. These systems now provide the highest overall efficiency achieved, which makes comparison with the additional cost acceptable [62–65] (Fig. 5.10).

The success of PVT systems in the cooling of the photovoltaic cell and the increase in its electrical efficiency did not make the researchers stop at this point, but they developed their ambition to take full advantage of the heat collected in the PV cell in other applications even if they had to increase this temperature using solar concentrators. The impressive results obtained by researches in the field of PV/T systems encouraged some of them to use solar concentrators to focus solar radiation on the PVT system. Previously, Reference [66, 67] used reflective mirrors to increase the solar radiation falling on the photovoltaic cell, while others used Fresnel lenses [68, 69], parabolic troughs [70, 71], composite equivalent complexes [72, 73], and parabolic dish [74, 75]. In CPVT systems, photovoltaic cells are used to withstand high temperature resulting from concentrates. In previous work, researchers with concentrates used phase change materials [76] and nanomaterials for their high cooling efficiency.

5.3 Temperature Effect on PV/PVT

The performance of the photovoltaic system is influenced by weather factors such as solar radiation, temperature, air mass, and wind speed [77]. Under standard test conditions (STC), PV module conditions consist of 1000 W/m^2, 1.5 (air mass), and the module temperature at 25 °C. Another method, called nominal cell temperature, is used as standard test conditions consisting of 0.8 kW/m^2 of solar radiation, 20 °C of ambient temperature, and 1 m/s of wind speed [78]. The electrical PV cell

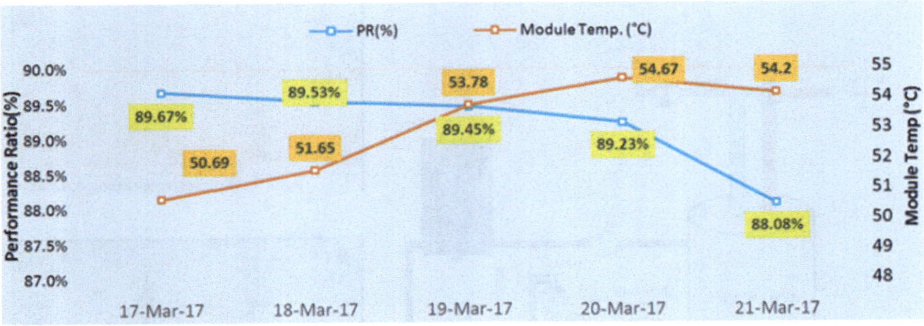

Fig. 5.11 The effect of solar module's temperature on the system performance ratio. (https://www.mahindrasusten.com/impact-temperature-pv-module-performance.php)

efficiency ranges from 6% to 18%. The existence of these conditions simultaneously is extremely difficult in real practical circumstances. Thus, the photovoltaic cell outcomes vary because of the divergence of cell conditions such as temperature and air mass (AM) from standard conditions, and this is due to many parameters such as geographical location variation and change of seasons [79]. Predicting the performance of a photovoltaic unit can be difficult [80]. Several studies have shown that the electrical efficiency of the crystalline silicon unit decreases by 0.4–0.5% with the photovoltaic temperature rising by 1 K [81, 82] (Fig. 5.11).

As shown in the previous section, the electric energy produced by the photovoltaic unit is strongly influenced by solar radiation. Photovoltaic cells produce electricity depending on solar radiation. As solar radiation increases, the photovoltaic system increases. This concept is simple, naïve, and far from reality. In practice, only 15% of the radiation reaching the solar panel causes electricity production, while the remaining 85% heat the photovoltaic module's body. The temperature of photovoltaic cells plays an important role in the PV-generated electricity. The high temperature of the cell causes a clear reduction in both the output voltage and power, which is reflected in the performance of the photovoltaic modules, causing their decrease. The climatic conditions affect PV materials (their operating temperatures) and their physical properties. Also, it is influenced by the topography of the site [83]. The results of scientific literature have supported the negative impact of high temperature PV cells on their performance.

For example, Charabi [84] investigated the effect of the geographic location of the Sultanate of Oman (this country was selected for its location at the heart of the solar belt with high solar radiation potential) on the performance of solar cells [85, 86]. In this study, it was decided to choose the suitable location for the large PV construction. The study identified only 9% of the entire area of Oman suitable for the construction of large photovoltaic power plants because of the high intensity of solar radiation and low dust rise. Among the many PV technologies tested by the researchers, CPV showed the best performance results, which means using this technology to build large PV power stations is the best for this country.

Hughes [87] studied a number of computational fluid dynamics (CFD) models to understand the methods of heat transfer and the rate of heat dissipation across the PV cell during the operating period. The climate conditions of the UAE have been used in these models (specifically the conditions of Dubai). Because of the high temperatures of photovoltaic cells in such studied conditions, the researchers suggested adding thermal tube fins to dissipate part of the high heat of the cell body. The results of the study confirmed the success of the use of the thermal fin tubes (the proposed method) in the study area (high temperature), where the temperature of the photovoltaic cells measured reached up to 70 °C during the practical experiments. CFD analyses in this study defined the best PV temperature in the UAE environment conditions at 30 °C, which can be equipped with the highest electrical efficiency (Fig. 5.12).

In Cyprus, Makridas [88] investigated the temperature impact on the productivity and stability of variable photovoltaic cell systems. The study period lasted for more than 4 years, during which the acquired and lost heat was evaluated for a number of photovoltaic systems. The results of the study showed that monocrystalline and polycrystalline systems suffer the largest annual thermal losses of electrical energy produced. The lowest thermal loss technique was thin-film technologies. To determine the seasonal losses of photovoltaic systems, the study confirmed that the thermodynamics use is very important in this field.

Many researchers used theoretical simulation papers and photovoltaic temperature models to predict cell temperature depending on the solar radiation intensity, wind speed, and surrounding air temperatures [89–94]. These studies showed that the temperature of the photovoltaic cell undergoes instantaneous changes that are not considered when calculating the temperature of the total PV cells. On the other hand, some students have adopted a traditional reference state (such as the cell temperature) to predict the typical temperature of a photovoltaic cell [95]. The hypothesis of a uniform temperature for PV cells constituting an array is unacceptable. Farr [96] demonstrated a practical comparison between two similar PV systems installed in several climates. The authors confirmed that cell temperatures vary across the array by several degrees.

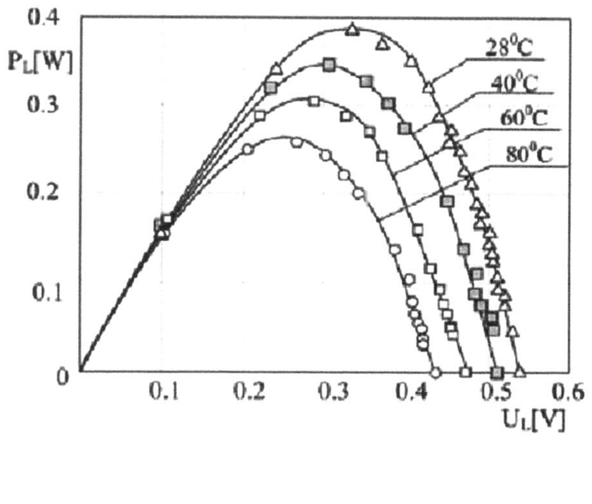

Fig. 5.12 The relation between output power and voltage of PV module for various temperatures [87]

The above studies have all used stable state conditions which, in practical conditions, are unable to predict the temperature fluctuations of PV cells, especially in periods of rapid radiation fluctuation. Such periods (rapid changes in radiation intensity) can occur within seconds as demonstrated by practical measurements. Jones [97] concluded that the photovoltaic cell had a slow thermal response to sudden changes in the intensity of solar radiation. This slow response is due to the effect of the cell's thermal mass. Tsai [98] explained that rapid changes in the intensity of the radiation can appear as an increase of a percentage point from the temperature of the measured unit and, in contrast, can increase the difference in temperature from 20 °C. This high-value error will cause significant errors in predicting the expected generated electrical potential.

Tyagia [99] investigated theoretically and practically the response of thermal PV cells to the fluctuations in the solar radiation intensity. The researchers proposed a thermodynamic model based on the thermal mass of the photovoltaic units. The researchers took into consideration all weather conditions and modes of heat transfer. The researchers adopted the proposed model to equate the unstable state and used practical measurements for 3 days in summer, winter, and spring. The results of the study gave uncertainty not more than 2 °C for 80% of study period.

Reference [100] used high quality equipment to measure the fluctuations in ambient temperature for 1 year in Sohar, Oman. The temperature profile resulted indicates high changes in temperature during the day, which causes thermal impact on the PV modules (Fig. 5.13).

Some decision-makers in many countries have taken advantage of the issue of high temperature in their countries and the effects of this rise on the productivity of PV and called for the neglect of this technology, despite the success in most of the countries that adopted this technology. The photovoltaic system's produced electricity and the efficiency could be enhanced by using air or water to cool the cell. This hybrid PVT technology combines the production of electricity and heat simultaneously.

Teo [100] has set up a series of parallel channels on the back of the PV cell for uniform distribution of airflow across it. The tests were conducted without cooling and with air cooling, and a linear relationship was found to correlate the electrical

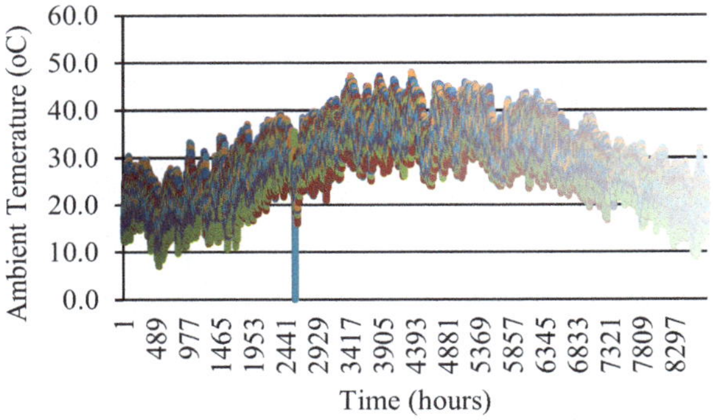

Fig. 5.13 Ambient temperature profiles for Sohar, Oman

efficiency of the photovoltaic cell and the temperature. In the case of tests without cooling, the electrical efficiency of the cell decreased by 8.6%, and the cell temperature used reached 68 °C. When using air cooling, researchers were able to stabilize the cell temperature at 38 °C, and the electrical efficiency of at least 12.5% was maintained.

Sarhaddi [101] analyzed a numerical model of a thermoelectric PV/T system cooled by air. The thermal model was enhanced by some corrections to heat loss coefficients, and an acceptable consensus was reached for numerical simulation results with the results of external experiments.

Water can be used as a coolant instead of air. This type of PV/T system has a variety of properties and can be used in many applications [102]. Krauter [103] studied the effect of using a nozzle in the PV surface of the photovoltaic cell to inject water onto its surface for cooling. The injected water reduced the reflected radiation from the PV cell surface by 2–3% and also helped to keep the surface clean from the accumulation of dust and contaminants. Injection of water in this way reduces the temperature of the cell to 22 °C and increases the electrical efficiency by 8–9% for the use of the cell without cooling. Another study conducted by Ref. [104] used percolating water on the surface of the cell from above. This process achieved a significant reduction in the temperature of the photovoltaic panel (up to 26 °C) and increased electrical capacity by 4 to 10%.

Reference [105] suggested that to remove excess heat from the PV modules surfaces, it may immerse in a liquid. For this purpose, the use of dimethyl silicon oil as a dielectric was suggested. A degree of uncertainty of 3 °C was achieved between theoretical simulation and experimental measurements. Yang [106] uses water pipes with functional gradients (FGM) to reduce the temperature of the photovoltaic cell to achieve a reduction of about 20 °C. Kirszmann [107] studied a numerical simulation using a channel for a cooling fluid made of highly efficient multi-junction cells linked to concentrated PV systems. The study demonstrated that this system succeeded in removing the excess heat and obtaining higher electrical efficiency. The PV cell's absorbed heat was stored and used in heating.

Chandrasekar [108] built a PV/T system using a cotton wick for cooling. It was placed on the back surface of the photovoltaic cell against the ground. These wicks were wet to take advantage of the capillary action of these wicks. This cooling technique caused the temperature of the photovoltaic panel to be reduced by about 20 °C. The researchers used the cotton wick and water system, but in the second set of experiments, they replaced water with nanofluids (such as nano-Al2O3-water and nano-CuO-water). The decrease in the photovoltaic cell temperature in this case increased by 11% and 17% compared with a reduction of about 30% with water. The reason for this limited decrease in cooling efficiency by nanofluids can be attributed to the association of nanoparticles with wick fibers, which causes obstruction of the capillary movement throw the wick (Fig. 5.14).

Kearn [109] used an innovative method by using heat pump system to cool solar cells. The heat pump was connected to simple PV/T collector via direct evaporating coils with photovoltaic modules that evaporate the cooling fluid when passing

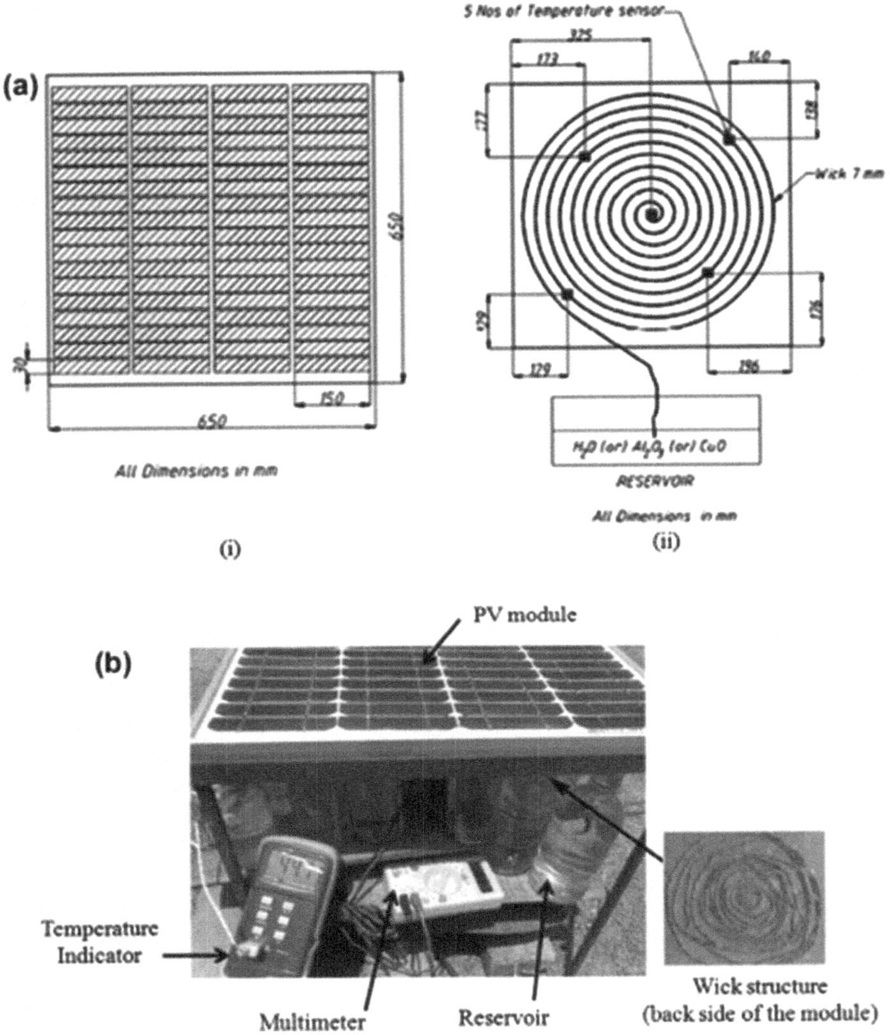

Fig. 5.14 (**a**) Schematic of (i) front side of the PV module and (ii) back side of the PV module with wick structure. (**b**) Photograph of the experimental PV module [108]

through photovoltaic cells. These coils act as evaporator of the heat pump. This design is easily vaporized at low temperatures of 0 °C to 20 °C, causing large cooling to photovoltaic cells. The results showed a high rise in the PV cell's electrical efficiency. The heat pump also provides heating through the high steam pressure produced when it reaches the condenser. The compressor can be recycled using a portion of the photovoltaic electricity generated. Research in the field of PV/T cooling systems using heat pipes is still very limited. Reference [110] studied the use of flat-plate and rotary tubes to cool the photovoltaic cells. The researchers concluded that heat tubes can be used to overcome solar panel heat and cooling problems in the PV/T system. Researchers have benefited from increased thermal conductivity in the transfer of greater heat through heat tubes. Zhao [111] suggested using a heat pipe set in PVT systems to generate electricity and hot air/water together. When comparing the proposed system with a single-cell operation, there is an increase of 15–30% in the electrical efficiency.

In Iraq, Amauri [112] analyzed the electrical and thermal performance of two specific sites using mathematical modeling for the hybrid photovoltaic/thermal system. The two studied sites are the cities of Baghdad and Falluja, and the study was conducted for a winter and a summer day. The studied variables (photovoltaic current, voltage, ambient air temperature, cell temperature, and cell's thermal gain) and their effect on electrical, thermal, and total efficiencies were investigated numerically. The authors claimed that the use of hybrid photovoltaic/thermal systems will work with acceptable efficiency in Iraq's harsh desert climate.

References [54, 60] have conducted a comprehensive review of research published in the field of PV/T systems and focused on their future eligibility to work in different weather conditions. The studies showed that many researchers have worked hard to develop these systems and enhance the use of their applications. The articles compared the outputs of PV/T systems according to the cooling fluid used (e.g., air, water, and nanofluid cooling systems). The study showed a significant improvement in the efficiency of the PV/T systems when using nanofluids for cooling. Both studies concluded that research and development work in this field is still required to achieve optimal efficiency of such systems, reduce their costs, and improve their technical integrity

Reference [61] added nano-SiC to water (3% of nanoparticles as a weight added to water) and used it to cool a PV/T system. The nanofluid used was denser and more viscous than water, and its thermal conductivity was clearly higher than that of water. The used nanofluid stability had good stability enough to be used for more than 6 months, according to researchers. The system studied had a higher electrical efficiency than the use of PV panels alone by 24.1%, while this system thermal efficiency was increased by 100.19% compared to the case of water cooling (Fig. 5.15).

Reference [59] compared the cooling effect of three types of nanofluids (made up of nanomaterials added to water) when used in PV/T systems. The tested nanoparticles were Al2O3, CuO, and SiC. The addition of nanoparticles to water increased its thermal conductivity accompanied with very slight increase in density and viscosity compared to water. Nano-SiC-water had the highest thermal conductivity compared to other nanoparticles and that the stability of the nanofluid lasted longer than the rest of the tested nanoparticles. Nano-CuO-water has higher thermal conductivity than nano-alumina water; however, the stability of this nanofluid lasts for a shorter period.

Al-Waeli et al. [62] proposed a PVT system with PCM tank connected to the back of the photovoltaic cell to control the thermal system capacity. The researchers mixed the used PCM (paraffin wax) with nano-SiC particles to enhance the thermal

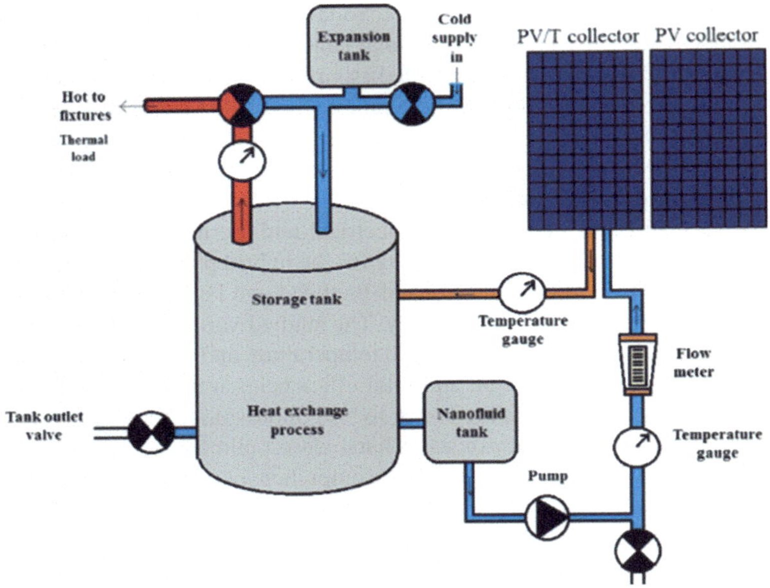

Fig. 5.15 Schematic diagram of the nano-SiC cooling PV/T system used in Ref. [61]

conductivity of paraffin as they did by adding nano-SiC to water to improve the thermal conductivity of cooling fluid. Field trials were made in Malaysia. The new PV/T caused a significant reduction in the photovoltaic cell's temperature which reached 30 °C during the peak period of solar radiation. The system's thermal and electrical efficiencies were also increased to 72% and 21%, respectively.

5.4 Humidity Effect on PV/PVT

Water causes serious harmful effects on photovoltaic cells, as it causes degradation of the polymeric constituents of these cells, resulting in decreased efficiency generated. It also accelerates corrosion processes both for glass, metal joints, and metal grids [113]. Water vapor in the air determines air moisture, which is defined as the actual water vapor pressure in the air to the saturated water vapor pressure at the same temperature; the researchers prefer to release the term relative humidity of the air (RH) on this definition. Water vapor in the air (air moisture) spreads inside the solar cell through cracks or breathing vents [114]. Air relative humidity depends entirely on the temperature of the air, and this factor variation is the cause of the differences in relative humidity rate [62]. When photovoltaic systems work in humid and hot environments, the high content of moisture in the air causes a clear decline in the performance of photovoltaic cells [115]. The changes that occur within photovoltaic cells depend entirely on the way in which the moisture propagates, which is a slow and long process [116]. When the humidity reaches the polymer, the humidity is then reached to the interbonding links of the solar cells and causes its effects based on oxidation and weakening, resulting in many effects [117]. Examples of these effects include erosion of welded joints [118], photoelectric discharge [84], and increased number of input paths [118]. Hence, studying the effect of relative humidity on photovoltaic units is necessary and binding for those who deal with photovoltaic cells to reduce their harm.

For Rajasthan (India), Bhattacharya [119] investigated the effect of scarcity of rain, dust, relative humidity, and ambient temperature on the efficiency of photovoltaic cells in this state. The relative humidity average in Rajasthan is about 42%, with a maximum of 70% in August and a minimum of less than 25% in March. The researchers concluded that the main effect of high relative humidity is corrosion, especially if high relative humidity meets high temperature. The availability of air temperature above 40 °C and relative humidity of not less than 60% facilitates corrosion and degradation of long-term cell efficiency. Also, this atmosphere is suitable for the growth of fungus which is growing in conditions of high relative humidity (between 75% and 95%). One important moisture hazard on photovoltaic cells forms a viscous surface that works on adhesion with dirt and dust particles and increases its accumulation. Most importantly, it works to bond these molecules with the surface of the cell, forming a difficult layer of cleaning. References [120, 121] reviewed the performance of photovoltaic modules in a humid atmosphere and their degradation because of corrosion due to high relative humidity. The researchers concluded that the high humidity of the air caused the erosion of the network resulting in a significant reduction in the conductivity between the emitter and the network, which is the main cause of the deterioration of the performance of photovoltaic cells that increasing temperature with the penetration of moisture and low strength of adhesion all combine to promote this deterioration in performance.

Tuati [122] studied the effects of relative humidity, temperature, and dust on the PV system performance. The researchers used crystallized and noncrystalline photovoltaic panels in the State of Qatar. The researchers concluded that the accumulated dust in the studied environment reduces the performance of the photovoltaic cells to a higher degree than the effect of both ambient air temperature and relative humidity. Also, higher air temperatures to rates higher than 40 °C caused low performance of amorphous PV cells.

Reference [54] investigated the impact of relative humidity on the photovoltaic cells' efficiency. The study showed that wind speed is directly related to relative humidity. As this speed increases, a decrease in the relative humidity in the air results in an improvement in the PV units' efficiency. High wind speeds, on the other hand, cause dust particles to rise in the air and then spread and then deposits to the surface of the photovoltaic cells, causing a clear deterioration in cell performance. The researchers concluded that effects of humidity, air velocity, and dust when combined together caused a sharp deterioration in the performance of photovoltaic cells, which calls for studying the mutual effects between these variables together to assess their effects on the solar modules' efficiency.

Gwandu [123] investigated the impact of high humidity in air on the solar radiation intensity, as sunlight is subjected to reflection or diffraction as a result of its collation with water vapor molecules. The rise in relative humidity in the air causes a clear reduction of the solar radiation intensity reaching photovoltaic cells resulting in a nonlinear variation in the open circuit voltage coupled with large linear changes in the short circuit current.

Reference [124] investigated the influence of humidity on cell performance; the authors concluded that the high relative humidity of the air could result (if the moisture penetrated inside the crystalline silicon cell through the cracks in the cell) a malfunction of the cell. Fragility in photovoltaic cells can also result from corrosion caused by relative air humidity. The researchers considered that changing the color of photovoltaic cells is evidence of low service life and near failure. The penetration of air moisture inside PV cells manufactured from crystalline silicon causes most of the failure of the electrical connection of the cell due to the erosion of thin films and sensitive units.

Prakash [125] used the physical properties of glass as a reference for photovoltaic cells considering it as a glass cell. In this study, the researchers used an uncompressed cooling system to cool the solar cells while providing limited quantities of cooling water for PV surface cooling. The researchers concluded that photovoltaic cells can be cleaned and cooled at the same time using the proposed water cooling system in the study in all hot and dusty areas, but it is necessary to start cooling the cells after they reach a temperature of 42 °C. It should be noted here that researchers in this study neglected the long-term effects such as corrosion caused by cooling water.

Darwish [126] put a mathematical relationship between the solar radiation intensity, relative humidity, and their effect together on the short circuit. The researchers concluded that wind speed reverses the effect of relative humidity on the received solar radiation by the photovoltaic cell. The exposure of photovoltaic modules to high humidity for a long time causes the penetration of water inside the solar cell, causing degradation of its performance.

The influence of RH on the performance of solar cells has been studied by Omobo [127]. The study showed a proportional mathematical relationship between relative humidity, PV cell efficiency, solar radiation, and alternating current, since relative humidity reduces the solar radiation intensity causing a decrease in the electrical voltage produced by photovoltaic cells. Kattkar [128] examined the possibility of evaluating the efficiency of solar modules in different weather conditions (such as the effect of cell body temperature and relative humidity), as researchers have shown, because the environmental conditions surrounding photovoltaic cells directly affect the transformation of solar radiation into electrical energy. The researchers conducted their experiments in harsh environments (ambient air temperature of 58 °C and relative humidity not less than 60%). The results of the practical measurements showed that the studied cells had decreased their electrical efficiency by up to 32.42% due to work in such harsh conditions.

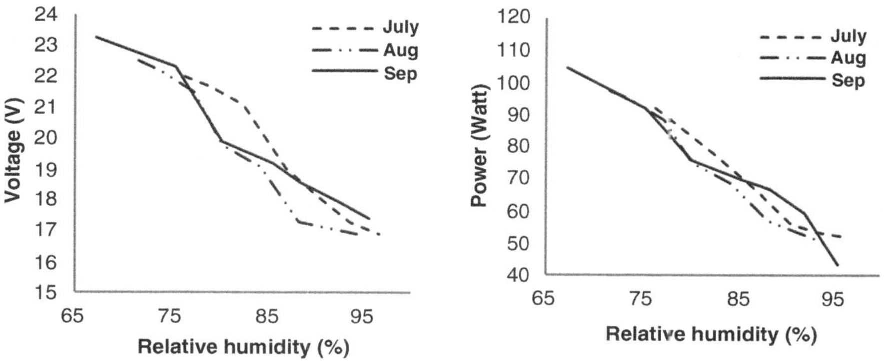

Fig. 5.16 The impact of relative humidity on PV module's voltage and power [129]

In Oman (a country with high relative humidity), Kazem studied [129] the performance of three types of solar cells. In low relative humidity conditions, solar cell performance improved as the study results showed. The electrical voltage and electrical power generated increased. Low relative humidity results in greater electrostatic efficiency for monochromatic modules than crystalline and amorphous silicon modules (Fig. 5.16).

In the United Arab Emirates (which enjoys a high humidity environment most of the year), Al-Hinai [130] studied the effect of the varied weather conditions of this country (dry and cold in the winter, offset by hot and humid weather in the summer). The researchers found that air humidity affects the performance of the photovoltaic cell with an inverse relation to the temperature. Therefore, the correlation between these two variables should be determined to assess their real effect on cell performance (Fig. 5.17).

In Kalabar, Nigeria, Ettah [131] explained in his study that low relative humidity (between 69% and 75%, as the researchers claimed) caused an increase in electricity generated by photovoltaic cells, as these rates caused an increase in electrical voltage. These results were also confirmed by reference [132], who concluded that low relative humidity results in enhanced PV performance.

Klampaftis [133] found that n-type silicon chips resist harsh weather conditions such as 75 °C and relative humidity of 100% for varying periods of time depending on the extent of moisture penetration within the solar cell body. The exposure of the solar cell to air humidity for a long time causes the cell's surface recombination and degradation of its output through three mechanisms: first, the diffusion of moisture within SiO2 and the interaction of hydrogen atoms with Si-SiO2 components. In the second mechanism, the permeable moisture is reacted with SiO2 via the formation of silica acid, which increases the stresses on the film, causing an increase in the density of this acid in the Si-SiO2 interface. In the third mechanism, the surface is charged with precipitation after exposure to sunlight.

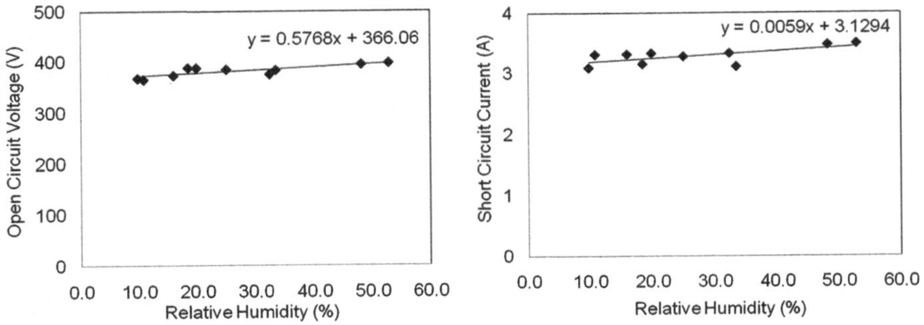

Fig. 5.17 Relative humidity impact on open circuit voltage and short circuit current [130]

Panjwani [134] found that the use of photovoltaic cells in a climate of relative humidity ranging from 40% to 78% causes the loss of 15–30% of the generated electrical energy. High relative humidity of about 70% causes deterioration in electricity production of up to 55–60%. The researchers explained this deterioration because the water vapor layer in photovoltaic cells facing the Sun directly suffers from its low absorption capability of solar radiation.

Omubo [135] studied the performance of photovoltaic systems established in areas far from the seashore (characterized by low relative humidity). The study found that the efficiency of the solar modules was higher than that of the beach. The decrease in relative humidity causes increased solar flow, which improves the current generated and the electrical efficiency of photovoltaic cells.

In his study, Rachman [136] focused on the effect of the use of a cooling system at different external weather conditions on the efficiency of the PV system. The high relative humidity of the air results in increased dehumidification by the cooling system. The best performance of such systems can be achieved when low temperature and relative humidity were available at the entrance.

In Miami, Florida, and the United States (considered to be a geographic area with high RH for its beachfront surroundings), Kemp [137] showed that the relative humidity penetration inside a photovoltaic cell body causes corrosion to the unit hull, causing corrosion of metals and cell welding joints, resulting in complete degradation of the system performance. The results of the study correlated the rate of decomposition within the solar cell and the permeable moisture. The best way to prevent moisture penetration into the PV cell hull is by closing it with low-density foams consisting of dried materials.

Wolgimuth [138] studied experimentally using a test called 85/85 and means cell work at $T = 85$ °C and RH = 85% to speed up the effects of both relative humidity and external temperatures on cells depending on Ref. [139] results. After 1000 hours of solar cell exposure to these accelerated weather conditions, corrosion began in the parts of the studied cell.

Fig. 5.18 PV module with severe delamination [142]

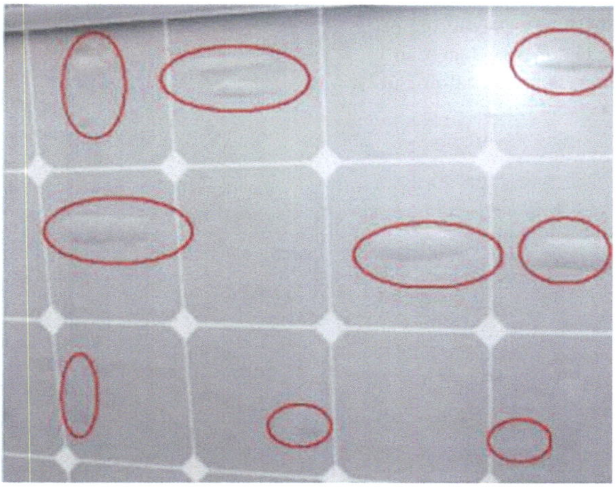

Fig. 5.19 Bubbles in the back side of a PV module [142]

The researchers [140–143] studied the effect of photovoltaic cell color change due to corrosion and degradation on its performance. The researchers explained that the case of delamination (a tape in which the adhesion between PV cells and the windshield is lost, or the loss of adhesion between the cells itself and the polymer membrane) causes an increase in light reflection rather than absorption and facilitates the penetration of moisture inside the cell hull (Figs. 5.18, 5.19, and 5.20).

Skoczek [144] investigated the impact of pollution (commonly expressed in hot and humid climates) to evaluate the degradation of solar cell performance. The appearance of sheets on the edges of the solar panel causes a high risk of leakage of electricity from the photovoltaic cell. Pollution usually causes the penetration of moisture into the structure of the photovoltaic cell, resulting in negative effects of the metal connections within these cells, whether chemical or physical. Also, the penetrated moisture transfers salts dissolved in it as well as dissolved pollutants into the PV cell's body (Fig. 5.21).

Fig. 5.20 Bubbles in the front side of a PV module [142]

Fig. 5.21 Decoloring of cells due to a change in antireflective coating [142]

Reference [145] demonstrated that the moisture's penetration into the structure of the photovoltaic cell causes the formation of hydrofluoric acid made up of fluoride and tin oxide found in the solar cell. This acid digs and necrosis in the interconnections between TCO and glass, hence, changing the color of the solar modules to yellow and brown, and causing glass adhesives and photovoltaic cells fail; as a final result, the packaging unit fails to function, and moisture penetration increases. The color change of photovoltaic cells causes a reduction in the light exerted by the cell resulting in reduced electrical power generation [146] (Fig. 5.22).

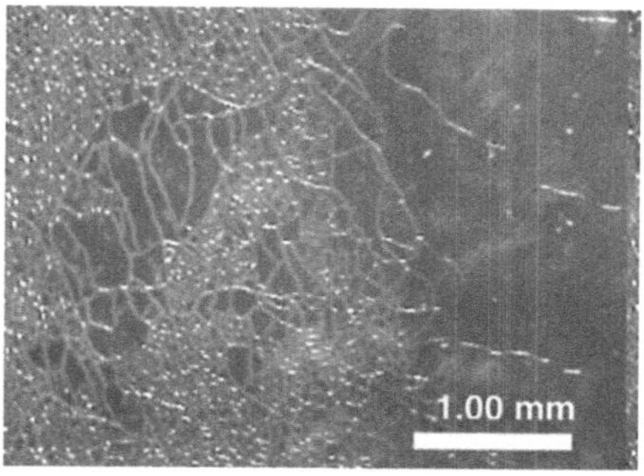

Fig. 5.22 Micrograph of a ZnO film after treatment showing extensive cracking of the transparent conductive oxide (TCO) layer [145]

Oreski [147] explained that the meeting of high temperatures (above 50 °C), water (from the permeable moisture inside the hull), and ultraviolent radiation (UV) causes color change in several non-contiguous regions of the solar system. The shape and extensions of these spots depend on the properties of the polymers used in the cell industry. These different color spots are a major cause of solar system degradation.

Kojima [148] achieved a change in the color of the solar cell to yellow by exposing the studied cells to artificial radiation. The researchers concluded that solar panels become more sensitive to light after exposure of more than 400 hours, causing production degradation, and the more areas yellowing the deterioration of electricity production becomes more. The slow degradation of the solar cell takes place over a long period of time with the continuous exposure of these cells to ultraviolet light [149].

In most cases, photovoltaic glass is broken or cracked during transport, installation, and maintenance [150]. Cells with broken or cracked glass may continue to function properly for a long time, but due to moisture leakages, the risk of electric shock can increase, as they cause corrosion, discoloration, and delamination. The cracks in the new solar cell's glass discovery with the naked eye are very difficult, but this is possible using visual inspection equipment [151]. The production of low glass thickness cells has increased in recent years, as producers tend to reduce the cost of producing solar panels, increasing the risk of fracturing and cracking of glass due to its fragility and high risk of fracture during transport and installation [152].

5.5 Wind Effect on PV/PVT

Solar energy is free and available most of the year. It is a safe source that is not controlled by other countries and thus allows stable economic development and growth. The clear growth in PV technology and the increase in its efficiency have made it universally accepted as a promising future for the production of sustainable, environmentally friendly electricity instead of fossil fuels. These panels are fixed outdoor exposed to the outside changing weather conditions. These ritual variables affect the output of photovoltaic cells. As shown in the preceding paragraphs, solar radiation, ambient temperature, and relative humidity have significant effects on cell productivity [153]. The optimum conditions for photovoltaic work are not available due to constant change in weather conditions daily, monthly, or seasonally. So, the basic characteristics of solar panels are limited in the performance of the PV system due to these changes in weather conditions [154].

Wind is an environmental variable that has a direct effect on the performance of PV photovoltaic cells. Wind movement around the Earth is caused by the Sun (albeit indirectly), and wind energy is used as the electrical power generation technology, one of the best sustainable and green technologies in producing electricity around the world [155].

Because of the installation of solar cell systems in the external atmosphere, they are definitely exposed to wind. The effect of wind speed on solar cells varies. It is positive, as it works to remove part of the heat of the cell hull, which means normal cooling of photovoltaic cells to enhance their performance and productivity. Winds opposite the surface of the solar panel can cause natural cleaning and remove some of the accumulated dust on the surface of the photovoltaic panel, helping to improve the productivity of these panels. The negative impact of the wind is that it puts external power on the solar module, which causes the need for additional costs to support the structure of the panels and, without it, causes damage in this structure. On the other hand, the rapid wind movement causes dust and dirt to irritate and raise high in the air and move it to vast areas away from its source. This dust will eventually settle and accumulate on the surface of the solar panels, causing a sharp decrease in the performance of these plates [156].

The abovementioned wind effects depend on the intensity of the wind (moving air speed) and its direction, as well as the problem of the structure of the solar system and the conditions of exposure [157]. Since solar cells are either fixed directly on the ground (on a low altitude) or on the roofs of buildings, therefore they are fixed at the lower boarder of the air layer. The prediction of wind speeds and its direction changes can be considered very complicated because of severe disturbances and different topographical features. Ground terrain and its severity have a clear effect on average wind speed in an area, which differs from one area to another for this reason as well as for the roughness of the surface caused by the variation in size, shape, density, and distribution of terrain, as well as obstacles such as buildings, trees, etc. [158]. Since the solar cells are stabilized in a tilt angle with the horizon, they vary according to the position of these cells in the latitudes in order to obtain optimal electrical energy generated and to capture the best and greatest solar radiation possible. Before offering solar cells for sale, an important series of qualification tests are carried out to ensure that the structure can withstand long-term durability. These tests include mechanical, thermal, and electrical tests [159].

The influence of wind forces on photovoltaic cells varies according to the location topography, the rows' shape, and their elevation from the surface of the Earth, as well as their tilt angle [160, 161]. Cells on the edges of a PV row system are affected by forces that are different from those in the middle of the rows. Similarly, the cells facing the wind movement in the first row bear high wind loads and protect the rest of the rows. In a study of the impact of wind on photovoltaic cells,

Ref. [162] found that wind movement in a parallel direction to PV cells did not cause much load on these cells. For photovoltaic systems installed in the north of the equator, usually facing the south, the northern wind movement carries high loads on the back row of cells while the southern wind will cause high loads on the first row of solar panels. Reference [163] explained that setting the rows of PV cells apart can cause rapid airflow between the resulting channels between the rows, which means an increase in the load of wind movement on the internal rows of the system.

To simulate wind load on a photovoltaic system, mechanical and dynamic loads are distributed uniformly on the back and front surfaces of the solar cell. Test results depend on 10,000 cycles and no more than 20 cycles per minute. This test is intended to simulate environmental conditions to ascertain the reliability of the mechanical design of the structure. In fact, no test can accomplish a realistic simulation of wind conditions. It is impossible to determine the speed and angle of motion of the wind, and it does not often shed the same load on the whole cell space evenly. The variation of the type and shape of the site causes a significant difference in the size of the row of photovoltaic cells and tilt angle. Most manufacturers rely on multiple tests to ensure the validity of their designs before they are released to markets [164]. For systems installed on the roofs of buildings, these systems are subjected to many calculations of the minimum pressure or wind load. Many manufacturers and researchers have carried out studies on the effect of wind on cells within the rows of solar systems using wind tunnels [165]. The results of the tests in the wind tunnel showed the possibility of reducing the impact of wind on the rows of solar cells by putting obstacles to intercept the movement of wind and reduce the speed resulting in lower effects on solar cells.

Reference [166] conducted a study to determine to which extent the photovoltaic cells were affected by winds using wind tunnel. The study was carried out on three different types of flat cell classes. The front row panels were always subjected to high and large loads, which are inflated by the airflow and hold a large part of this load and prevent them from reaching the next rows. To find out the effect of the wind's lifting force on the structure of the photovoltaic cell, Ref. [167] carried out a practical study which concluded that wind loads on solar cells could be reduced using appropriate building materials and methods. Reference [168] used solar cells fixed in parallel inside wind tunnel. Experimental tests were performed using 15 m/s wind speed to determine the maximum torque position applied to the solar cells. In this study, the researchers also proposed a mathematical model that allows the prediction of torque in photovoltaic at work now. In India, Ref. [169] studied the effect of photovoltaic cell efficiency at wind speed and ambient temperature using statistical analysis. The results of the study showed that the wind speed has a somewhat positive relationship with the efficiency of the photovoltaic cell. Reference [170] investigated the cooling effect of wind and its usefulness to photovoltaic cells. The researchers recognized the importance of taking advantage of wind distribution and real velocity in different locations. The results showed that the wind cooling effect can improve the efficiency of converting solar energy into electricity when the solar cells are distributed on the roof. The real benefits of thermal energy can be up to 17% on a sunny day if the studied proposed method was employed.

The negative impact of wind on air-cooled PV/T systems was studied by the Ref. [171] who concluded that they caused a decrease in system efficiency. The researchers interpreted this condition to the heat transferred from the cell's surface to the air because of its high speed, resulting in a decrease in the mass of air flowing on the back surface of the cell. BIPV/T systems are affected by their electrical and thermal efficiency toward wind movement directions. While another study showed that the performance of solar cells was improved at high wind speeds because of their direct effect on reducing the temperature of the body of the cell, the study showed an increase in electrical efficiency by 0.7% with an increase in wind speed of 1 km/h [172]. Such an increase in efficiency can be considered a big gain, because the maximum efficiency of solar cells is usually between 8% and 16%. Reference [173] studied the effect of direct and indirect air movement on the performance of a photovoltaic system mounted on the ceiling in an environment characterized by high ambient temperature. The researchers concluded that an increase in wind speed to 5 m/sec caused an increase in the efficiency of the solar panels higher than the state of its work at speeds of winds exceeding 5 m/sec.

Some researchers have used airflow control on the front surface of a photovoltaic module to measure the response time of solar irradiance oscillation during the day. Reference [174] conducted his experiments in the dark to avoid the effect of radiation in heat loss. The researchers explained that the external operating factor conditions' impact (such as ambient temperatures, change the air speed and change in its direction) on the PV cell's temperature are not controlled in the outdoor tests, while in the laboratory experiments, the control of these factors is available. Reference [175] studied the thermal behavior of a solar cell exposed to wind blowing on its face, increasing the speed of the wind gradient from least to the largest. The researchers proposed a new thermal model that takes into account the external weather conditions and their impact on the performance of photovoltaic cells and even on the way they are fixed to the ground or the surface of the building. Reference [176] studied the effect of wind speed on the performance of solar cells in the city of Bolzano (Italy). The researchers used wind speed data with solar cell temperature data as models for weather forecasting instead of practical measurements of spot wind speed. The method used by the study resulted in improved predictability of benchmarking comparisons (Fig. 5.23).

We have previously shown that the adoption of standard test conditions (STC) for measuring the efficiency of PV panels (the temperature of PV cells from 25 °C, 1000 W/m^2, and the air mass of 1.25) is difficult to provide together in external weather conditions. In such circumstances, the temperature of the solar cell varies dynamically but remains associated with the remaining environmental variables [177]. Changing cell hull temperature can be considered the main effect on cell output. In summer, the body of the cell can reach a temperature of 60 °C in Europe, while it can be more in the deserts of the Middle East, for example. The increased temperature of the cell hull results in a deterioration in the generated electricity. Of course, there are other effects such as solar radiation and wind, which can be considered the most important weather variables that positively affect the temperature of the cell. At present, the standard approach used to predict cell temperature

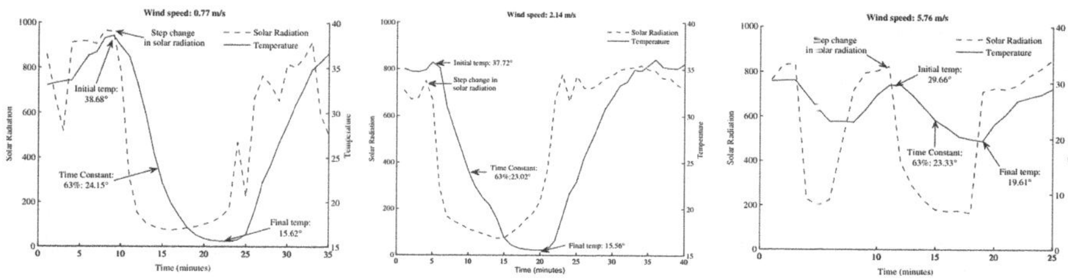

Fig. 5.23 Effect of wind speed variation on PV panel's temperature [172]

depends only on the data of the radiation intensity and the temperature of the surrounding air and does not enter the wind speed and its effects on cell temperature [173]. Refs [175–178]. used a new method that included the effect of wind in estimating the temperature of a solar cell. Reference [174] demonstrated that the wind can cause a drop in solar cell temperature by about 15–20 °C when the wind speed is about 10 m/s and the conditions of solar radiation intensity is about 1000 W/m^2. Refs [177–180]. used wind data for multiple sites with the other weather conditions to test different models and validate the results. In these locations, unfortunately, the wind measured data were rare, making the researchers rely on numerical prediction instead of measured data.

Reference [180] used passive cooling based on natural convection to cool solar cells by adding fins to the photovoltaic cell to improve natural convection. In this study, some effects are the cells' tilt angle, the solar radiation intensity, the surrounding air temperature, and the wind speed in addition to the measurement of the fin. The efficiency of solar panels has decreased linearly by increasing the temperature of the surrounding air, while by increasing wind speed, this efficiency has increased. The measurement of fin size had limited effect on the efficiency of the solar cell. However, the study showed that the efficiency of finned panels was higher than that without fins with about 0.27–1.14% according to different study conditions.

Reference [181] studied the overlapping effects of dust, relative humidity, and wind velocity on the efficiency of photovoltaic cells. The deposition of dust and pollutants on the solar cell surface causes a significant reduction in its electrical efficiency. The increase in wind speed caused a reduction in the temperature of the PV cell and the subsequent improvement of its efficiency. Also, high wind speeds cause an in situ reduction of atmosphere relative humidity, resulting in improved photovoltaic efficiency. That high wind speeds to high limits cause dust and sand to volatilize in the atmosphere that reduce the solar radiation intensity, resulting in a decrease in the solar panels' performance. The study concluded that the resulted efficiency of the solar cell is the outcome of the overlap effects of dust, relative humidity, and wind speed. Therefore, it is preferable to study these variables together and not to study each individually in the evaluation of the performance of photovoltaic cells because the effects of these positive and negative factors overlap. Reference [182] investigated the installation of a solar cell system on the surface of the building and its effects on thermal gain of construction. The temperature of photovoltaic cells increases when the dimensions between the solar cell system and the surface of the building are not suitable, causing a clear reduction in the system's performance and increasing the walls and surfaces heat gain. The study suggested a design that could change the tilt angle of the solar panel and practical experiments to verify the accuracy of the simulation model. The study found variations in wall surface temperatures and predicted solar cell temperatures. The study concluded that the appropriate air gap caused a reduction in heat transfer across walls and ceilings up to 1.85 kW/m^2/year. The researchers suggested that the air gap would be limited to determine the increase in the area of unshaded areas of the construction that would cause increased thermal gains in the building.

Reference [183] tested experimentally the outside wind forces on the solar cells installed on the roofs of buildings. Pressure gauges were placed above and below the solar cell, and the wind speed and direction were measured to assess wind's lift forces. The maximum lifting force applied to a photovoltaic cell depends mainly on the direction of the wind. Reference [184] used a solar cell model installed in the wind tunnel to measure the drag and lift power on these models using power transformers. Power coefficients on solar panels are increased by increasing the tilt angle of the solar panel from 0° to 90°.

Reference [185] conducted a practical study to measure the loads of a solar cell from top and back. The experiments were carried out using a wind tunnel, and the air movement was directed to meet the four directions of the studied cells. The wind pressure was distributed on the surface of the unit similar to what is exposed to this surface facing front winds at angles of 0° and 180°. For other wind direction angles, the pressure was distributed asymmetrically. The large distance between panels was considered, by the study, to be necessary because the pressure of the cell surface was clearly affected. Increasing the tilt angle increases the pressure exerted on the cell surface when facing the wind (Fig. 5.24).

Reference [186] studied experimentally the effect of airborne dust and its concentrations on the performance of photovoltaic cells in the wind tunnel. The accumulation of dust on the surface of solar panels depends mainly on wind speed. In this study, four wind speeds were used with four concentrations of dust. As was known, the accumulation of dust, especially the fine elliptical dust on the surface of

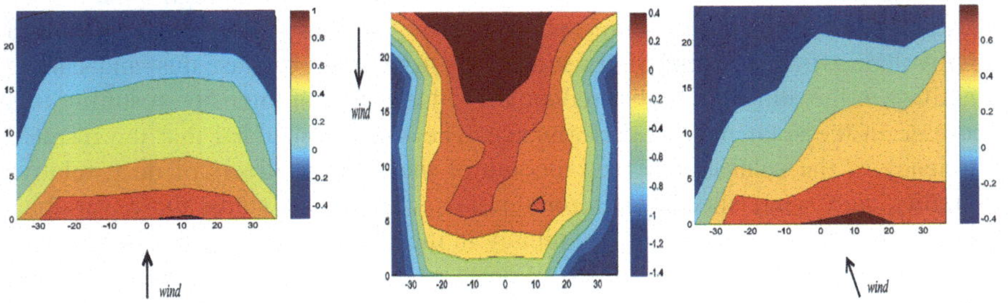

Fig. 5.24 The cooling effect variation of wind direction on the PV module [185]

the solar panels, negatively affected the efficiency of electricity generation. A notable reason which causes the low performance of the cells is the wind and the effects of dust in the air; although the low efficiency of the cell increases by increasing the speed of the wind, which reduces the permeability of light to form similar to the shadow covering the cell [187]. The increase in wind speed causes a high increase in dust concentrations rising in the air, resulting in dust that will accumulate on the surface of the solar panels. The study concluded that high concentrations of atmospheric dust have limited impact when wind direction is in the back of the cell.

Wind loads represent the biggest reason for increasing the cost of solar cell systems, whether directly installed on the ground or on the roofs of buildings. Completion of wind loads on solar cells is practically impossible, but by setting barriers, their effects can be reduced, thus reducing costs. Reference [188] studied the possibility of operating a photovoltaic cell system without wind effects and found that structural costs could be reduced by up to 75%. To address the high costs added to solar cell structures, the researchers proposed several solutions: to install the cells inside rows separated from one another to allow the passage of wind without symptoms or using a fence of the symptoms on the perimeter of the system. Reducing the cost of solar cell structures requires significant work to reduce wind power. Reference [189] showed that a 20% lower wind speed would cause a reduction in the cost of cell structures by 20%. It should be emphasized here that reducing wind loads will have a clear impact on the performance of the system, and must be checked before it takes action, as the low flow of air on the solar cells cause a reduction in the cells' cooling process, which will lead to lower solar cell performance.

5.6 Dust Effect on PV/PVT

The net energy reaches the Earth from the Sun within an hour equals what the Earth energy needs for a year. This energy is the cause of life on Earth and is the source of heat, wind, and rain. Scientists estimated the temperature of the surface of the Sun at 5800 Kelvin, and what reaches up to the Earth's atmosphere is estimated at 1367 W/m^2 [190]. Direct radiation and the most important energy scattering within the atmosphere contain the visible portion of the wavelengths, while at low temperatures the portion of the infrared spectrum is located, whereas the ultraviolet radiation, in particular, is prevented from entry into the atmosphere because of the ozone layer [191].

In the atmosphere, the radiation is divided into three types: a beam absorbed by the ozone layer in the upper part of the atmosphere (UV), a beam that is accumulated as the particles of carbon dioxide and water vapor absorb a large part of the infrared radiation and its heat is accumulated in the atmosphere, and a transient beam, which is the visible light that the photovoltaic cells take advantage from [192, 193].

The solar cells are established in areas completely exposed to sunlight to do its work by converting the solar radiation that reaches it into electricity. Since these cells are located outdoors, they are exposed to all other weather conditions as well as sunlight. In the previous parts of this chapter, we discussed the effect of solar radiation, temperature, humidity, and wind on photovoltaic cells, and in this part we will focus on the effect of accumulation of dust and pollutants on the output of photovoltaic cells [194].

Dust is defined as all solid, airborne particles with a diameter less than 500 microns. So, pollen, bacteria, and fungus are dust. The minutes of tissue, clothing, carpets, and fabrics are dusty; salts, motor engines, and large vehicles pollution can be considered as dust. Dust is deposited depending on several environmental factors and weather conditions. The movement of wheels in the roads as well as pedestrians' movement causes dust to be stirred by air as well as the movement of cars which causes consumption of tires and turning small parts consumed into dust particles. Volcanic eruptions cause millions of tons of dust to pour out, and quick winds stir up dust and move it from areas that are generated to distant places, known as sand- and dust storms [195, 196] (Fig. 5.25).

Dust deposition (its quantity and sediments shape) depends mainly on the type of dust and its physical and chemical properties and depends also on external weather conditions and location of solar cells. The mass of dust and its distribution on the surface of the solar panel are affected by several factors such as cell surface roughness, angle of inclination, relative humidity, and wind speed [197].

All field experiments confirmed that the accumulation of sand and dust on the surface of solar cells causes their electrical efficiency and their performance to decrease [198]. The suspended particles of dust in the air dispersing the solar radiation and scattering it reduce the direct part of them reaching the cell, which inhibits the performance of the cell. On the other hand, the accumulation of dust on the surface of the solar cell is a barrier between the incoming radiation and this surface, which causes a clear reduction of the electricity generated and the performance of the cell. Photovoltaic systems in dry and hot areas (such as the Middle East and North Africa) suffer from this problem clearly and explicitly. These areas are characterized by large basins of deserts (the Great Desert and the Al-Robbie Al-Khali). In the presence of rapid winds, particles of sand and dust are stirred and raised to hundreds of meters in the air and pushed away from the place they generate for hundreds of kilometers [199].

Fig. 5.25 Model of the system after exposure to external conditions for 1 month

Deserts are characterized by large flat areas that are free from urbanization, making them very suitable for installing photovoltaic plants. Also, these areas are characterized by high levels of solar radiation, almost all of these locations either inside or adjacent to the solar belt. The importance of sustainability in the countries of these regions such as Saudi Arabia, Kuwait, Qatar, and the UAE has directed the governments of these countries to series of projects of PV stations, which its first obstacle is the dust. To this day, it is impossible to claim the possibility of eliminating the problem of dust accumulation and its harmful effects on solar cells, and the world has not yet been provided with definitive solutions to this dilemma [200].

The most influential parameters in the deposition of dust on the surfaces of solar cells are dust (chemical and physical properties) and the local environment. The local environment is represented by vegetation and natural obstacles that cause shade or prevent wind movement, as well as the nature of human activities in this area (agriculture, construction, drilling, etc.). The method of establishing solar cells and their final finishes, such as their elevation from the ground and tilt angle, can be considered as a key factor in the quantity and quality of accumulated dust. Weather conditions play an important role in determining the properties of dust (physical, chemical, biological, and static electricity, as well as the size, shape, and weight of dust particles). The PV surfaces' final finishes are necessary because the surface is viscous (furry, coarse, with adhesive residue or electrostatic accumulation) making it more susceptible to dust accumulation than surfaces that have less viscosity. The presence of a layer of dust on the PV cell's surface promotes the accumulation of other layers of dust on this surface (dust after settling on the cell surface tends to attract more dust) [201, 202].

Due to gravitational forces, dust tends to accumulate on horizontal or slanted surfaces in close proximity to the horizon, and the amount of accumulated dust depends on the speed of the wind prevailing (high-speed winds resist the stability of the dust and raise it, while in the case of a slight wind speed, it encourages the accumulation of dust). When installing photovoltaic systems, the direction of the prevailing wind movement must be taken into consideration because of the important role it plays in increasing the accumulation of dust or reducing it (natural cleaning of the solar cell) [203].

Despite the growing interest in photovoltaic systems to generate electricity as a result of their environmental (reduced air pollution) and economic (by converting free solar radiation into electricity) benefits, the studies have agreed on the importance of reducing the factors that reduce the efficiency and performance of photovoltaic cells, especially the accumulation of dust and pollutants [204] (Fig. 5.26).

Hottel and Woertz had sudied the dusts' influence on photovoltaic cells. In 1940, the researchers measured the degradation of electricity generation efficiency for three sets of solar cells located near the US thermal power plant. The rate of deterioration that researchers found from the accumulation of dust on a slanted glass panel at a 30° tilt angle was about 1%, while the maximum reduction recorded was 4.7% after exposure to external conditions for 2 months [205].

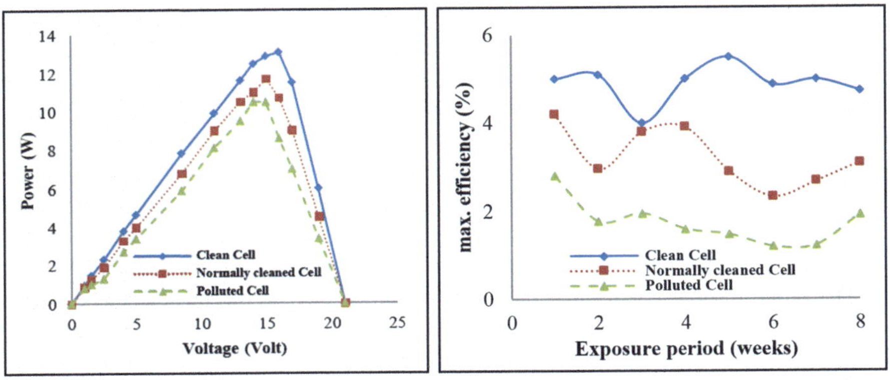

Fig. 5.26 The impact of cleaning method on the PV module characteristics [204]

In Saudi Arabia (most of which are desert areas), Nemo and Seid [206] investigated the deterioration of the efficiency of photovoltaic panels after exposure to external conditions for more than 6 months. The measured results showed a significant degradation up to 40% in the performance of the studied system. This study failed to determine levels of increase and decrease of dust accumulation during the study period. Also in Saudi Arabia (in the capital city of Riyadh), Salem et al. [207] studied the effect of dust accumulation on two sets of photovoltaic cells for a long time (8 months). The PV systems' tilt angle was used according to the geographical location from the equator (24.6° from horizon). The researchers cleaned the first group of cells every day and left the second group of cells without cleaning. The accumulation of dust on the dirty group caused a degradation in performance of 32% more than the clean group.

Maghrabi et al. [208] studied weather variables such as infrared temperature, the optical depth of aerosol, and the intensity of solar irradiance before, during, and after the storm that struck the capital of Saudi Arabia, Riyadh, on 10 March 2009. This dusty storm, which lasted for only a few hours, was considered one of the worst dust storms in Saudi Arabia in two decades. The heavy dust carried by this storm caused the lack of vision. The study showed significant changes in the factors studied. The theoretical simulations showed a significant change in the components of the solar spectral irradiance and the intensity of the incoming irradiance. A high density of dust caused a 44% increase in the scattered radiation intensity compared to a clear day. A temperature rises of 24 °C was observed 2 hours after the storm.

In the State of Kuwait (which is a desert country), Wakim [209] studied the effect of the dust and sand accumulation on the efficiency of solar cells and found that it causes deterioration in electricity generated by up to 17% after 6 days of exposure to external environments. The study indicated that the accumulation of dust increased during the summer and spring and decreased during winter and autumn. Sayigh [210] studied the effect of tilt angle on the amount of dust accumulated on the surfaces of PV panels. The researchers used seven sets of flat panels arranged each in pairs (two) at angles of 0°, 30°, and 60°, while the seventh panel was at angle of 90°. In each pair of cells, one was cleaned daily, and the other cell was left without cleaning to compare the resulting performance. Dust accumulated on photovoltaic cells during the months of April and June (most dusty months of the year) at a rate of 2.5 $g/m^2/day$. The accumulation of dust after 38 days of exposure to external weather conditions on slanted solar cells ranges from 64% to 17% for the studied inclination angles (0° to 60°, respectively). As for the horizontal tilt angle cell, its performance decreased by 30% after 3 days of exposure to external atmospheric conditions [211].

In Qatar (also a desert country), Touati et al. [212] studied the sensitivity of solar cells to the local environment such as temperature, dust, and relative humidity. The accumulation of dust on noncrystalline and crystalline silicon cells caused a clear reduction in electrical efficiency. When comparing the degradation caused by the accumulation of dust by the effect of temperature or humidity, dust showed the greatest degradation compared to other parameters. In a second study, Touati et al. [213] found that solar cells made of crystallized silicon are more suitable for the desert climate (as a climate of Qatar) than monocrystalline silicon cells. Measurements showed that exposure of these cells to local weather conditions for 100 days caused a 10% reduction in electricity generation efficiency for monocrystalline cells only, which means the use of solar cells as suitable source of reliable electrical power generation (Figs. 5.27 and 5.28).

Darwish et al. [214] studied the existing interference between the atmospheric variables on the properties of accumulated dust and on the outcomes of the solar cell. The study concluded that this overlap is so complex that it requires many other studies and further research. The local climate of a geographic location determines the properties of dust (dust mass, size of dust particles, particle shape, and electrostatic deposition behavior). Geographic location is the only determinant of optimal

Fig. 5.27 Dusty day in Qatar [213]

Fig. 5.28 Dust deposition on rooftop PV panels [213]

cell tilt angle from the horizon. In order to achieve the best reception for the falling solar radiation, attention must be paid to the cell's tilt angles. Also, the height from the surface of the Earth must be considered, as we have already shown. In the United Arab Emirates (Desert State), Darwish et al. [215] investigated the effect of a number of environmental factors on the performance of solar cells, including dust. The accumulation of dust is one of the factors causing the deterioration of the electrical productivity of solar cells, whether cells within large or small systems.

The physical and chemical properties of the dust content represent a heterogeneous mix of different pollutants, which are produced due to human activities in a geographical location, in addition to the soil type in this site and the surrounding areas. Reference [216] studied the effect of dust and pollutants inside and outside the laboratory. This article reviewed the different effects of multiple pollutants on the performance of many types of solar cells experimentally.

In the Sultanate of Oman (adjacent to the Empty Quarter), Kazem et al. [217] studied the effect of the accumulation of desert dust on the performance of solar cells. Experimental tests were conducted using specific types of dust (red soils, sand, ash, calcium carbonate, and silica) on the performance of a photovoltaic cell. The type of dust deposited in addition to the amount of deposition quantities clearly affect the electricity generated. The results of the study showed a great effect of ash dust on the performance of photovoltaic units compared to other types of dust. The degradation of the system performance reached 25%.

Kazem et al. [218] investigated experimentally the effect of the physical properties of dust from multiple regions in the northern region of the Sultanate of Oman on the performance of solar cells. The researchers found that the gathered dust from two specific areas (Sohar and Saham) because of their moisture content (52.21% in Saham dust and 45% in Sohar dust) caused the highest decrease in solar cell perfor-

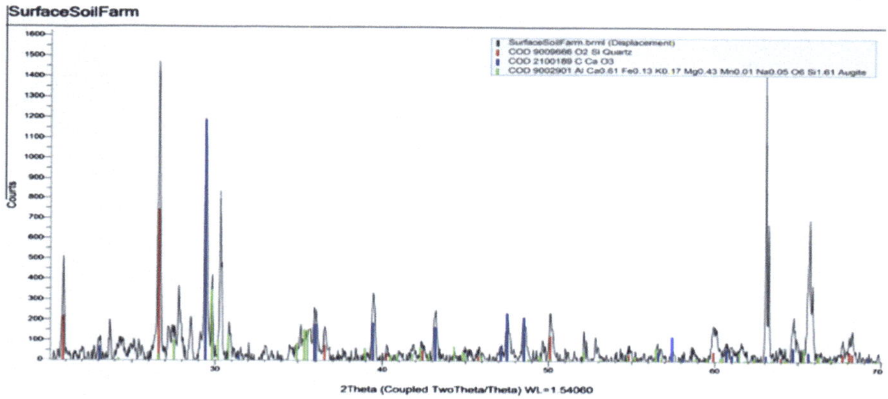

Fig. 5.29 XRD patterns of dust collected from Sohar, Oman [218]

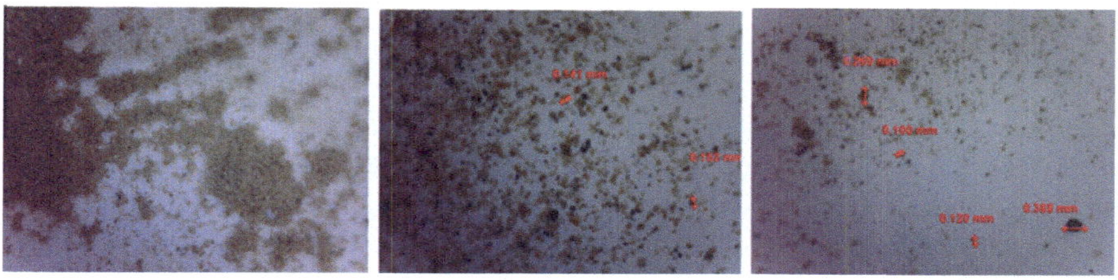

Fig. 5.30 Microscopic image of Sohar city (Oman) dust [218]

mance compared to other regions. After a large number of experiments, the study confirmed that the studied area (north of the Sultanate of Oman) has dust characteristics that make the impact on solar cells less than those of the neighboring Arabian Gulf States (Figs. 5.29 and 5.30).

From Iraq, Kazem et al. [219] demonstrated that the Fertile Crescent has turned into a source of dust to neighboring countries of Iraq. The exposure to two consecutive decades of drought, in addition to overgrazing and low levels of water in the Tigris and Euphrates rivers due to the dams that Turkey, Iran, and Syria are building on both rivers, has caused other factors in the birth of this new dust basin. For the period from 1980 to today, this country, which is rich in natural resources such as oil and natural gas, has suffered from long wars and unjust siege, and we add to it some of the activities carried out by successive governments such as drying the marshes and deteriorating irrigation infrastructure, which reduced the area of irrigated land and increased salinity of most lands. All this has led to an increase in the frequency of dust storms, which today is a worrying concern for both decision-makers and citizens. Dust storms are living proof of the extent of desertification on agricultural areas and soil erosion due to wind erosion. In this study, the researchers examined the different conditions of dust in Iraq and the extent of the impact on the performance of solar systems. The study concluded that the great interest in

Fig. 5.31 Space image of the dust storms strikes Iraq (5/7/2009) [219]

periodic cleaning as well as the forced cleaning after each dust storms will reduce the effect of this factor on the performance of photovoltaic cells and make the possibilities of using them not only acceptable but high (Fig. 5.31).

The establishment of solar cell plants in the desert is considered the most appropriate solutions, as these areas enjoy abundant solar radiation throughout the year, as Sukhatme described [220]. Several studies have discussed a set of ideas for installing and equipping photovoltaic power plants in deserts of some countries and exporting exports the surplus electricity to other countries. Because these countries are located in the desert, which means that they are exposed to dust storms in addition to their usual dusty atmosphere, cleaning labor and equipment staff will be added to the daily and weekly cleaning of the solar cells.

In the Thar Desert of India, Nahar and Gupta [221] studied the effect of solar cell tilt angle and the accumulation of dust on the permeability of solar radiation from a set of glass panels. The deposition of the deposited dust increased with the cells' tilt angle from the horizon. For tilt angles of 0°, 45°, and 90°, the rate of degradation of sunlight through glass was 19.17%, 13.81%, and 5.67%, respectively.

To provide solar electricity around the world, Sayah et al. [222] claimed that vast areas of deserts should be utilized with the use of various solar technologies (photovoltaic cells, concentrated solar systems (CSPs), photovoltaic cells (CPVs), and wind power). In this study, it was found that the use of this proposal in the Mojave Desert (in the United States) alone can meet the need for the entire US states of electricity. The free solar energy available throughout the year in these arid zones can provide electricity generation that may exceed the need of the market by stages. But, it is very important that the solar cells remain clean from accumulated dust to keep the system working at its maximum value. During the study, the researchers measured the mass of dust accumulated after a sand storm on the surface of a solar

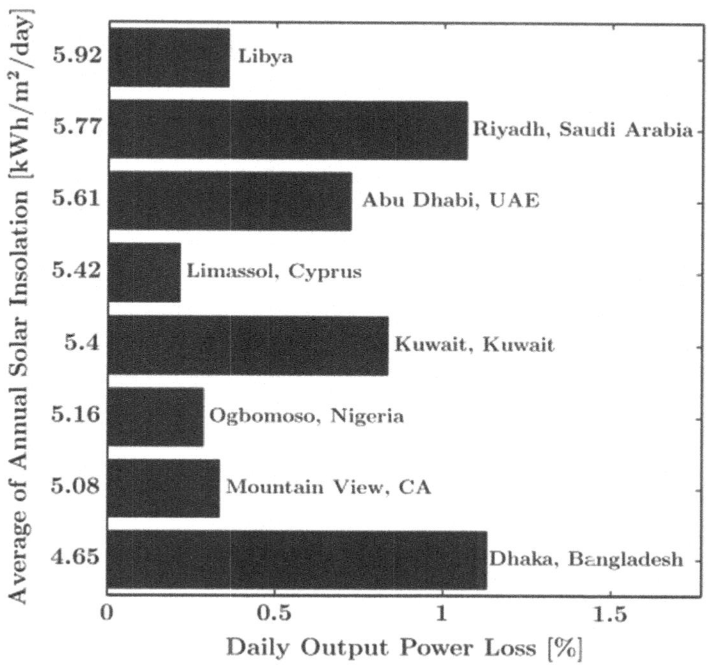

Fig. 5.32 Power loss due to soiling in different areas of the world [22]

cell and found that the permeability of solar radiation has decreased at rates ranging from 5% to 30% per year; this variation in the accumulated amounts of dust is due to change in the number and time of dust storms in the study area From month to another (Fig. 5.32).

5.7 Conclusions and Recommendations

This chapter investigates the influence environment holds on the hybrid PV/T collector and systems in terms of electrical and thermal production. The effects are studied for factors such as solar irradiance, ambient temperature, relative humidity, dust accumulation, and wind speed. Understanding the behavior of PV/T corresponding to variations of these factors will help in determining the optimum conditions and location for installing PV/T systems and assessing their feasibility more accurately.

1. Thermal stress causes negative effects on the lifetime and power production of PV cells.
2. Increased shading and dust accumulation on surface of PV/T causes degradation in the overall efficiency, particularly for the PV cell.

3. Increase of solar irradiance and wind speed correlates to the increase of PV efficiency. However, the later may cause increase in thermal losses and hence drop in thermal efficiency of PV/T. Moreover, further costs are associated with areas of increase wind speed due to need for a structure that can protect the PV/T system.
4. Ambient temperature and relative humidity are found to cause decrease in the electrical efficiency of PV systems. Moreover, these lead to optical and thermal losses as well and hence reduction in the overall efficiency of PV/T.

Recommendations

1. To consider shading, cloudy weather, and dust accumulation for sizing and performance prediction of PV/T systems
2. To consider methods to determine the optimum wind and ambient temperature for PV/T systems by balancing between the benefit of cooling PV cell through natural convection of wind and limiting the thermal losses due to increase wind speed simultaneously, which is certainly somewhat of a dilemma

References

1. J.A. Duffie, W.A. Beckman, *Solar Engineering of Thermal Processes* (Wiley, New York, 1991)
2. C. Fröhlich, History of solar radiometry and the world radiation reference. Metrologia **28**, 111–115 (1991)
3. H. Mousazadeh, A. Keyhani, A. Javadi, H. Mobli, K. Abrinia, A. Sharifi, A review of principle and sun-tracking methods for maximizing solar systems output. Renew. Sustain. Energy Rev. **13**, 1800–1818 (2009)
4. W.C. Dickinson, P.N. Cheremisinoff, *Solar Energy Technology Handbook* (Butterworths, London, 1980)
5. L. Kumar, A.K. Skidmore, E. Knoles, Modeling topographic variation in solar radiation in a GIS environment. Int. J. Geogr. Inf. Sci. **11**(5), 475–497 (1997)
6. A. Sayigh, *Practical Photovoltaic for Sustainable Electricity and Buildings*, 1st edn. (Springer, Basel, 2017)., ISBN10 3319392786, ISBN13 9783319392783
7. International energy agency (http://www.iea.org/)
8. Earth radiation Budgest earth Radeation budget (http://marine.Rutgers.edu/mrs/education/classs/yuri/erb.html) NASA langlely research center (2006-10-17) Retrieved on 2006-10-17
9. C.H. Duncan, R.C. Willson, J.M. Kendall, R.G. Harrison, J.R. Hickey, Latest rocket measurements of the solar constant. Sol. Energy **28**, 385–390 (1982)
10. K.Y. Kondratyev, *Radiative Heat Exchange in the Atmosphere* (Pergamon Press, New York, 1965)
11. L.D. Williams, R.G. Barry, J.T. Andrews, Application of computed global radiation for areas of high relief. J. Appl. Meteorol. **11**, 526–533 (1972)
12. T. Khatib, A. Mohamed, M. Mahmoud, K. Sopian, Estimating global solar energy using multilayer perception artificial neural network. Int. J. Energy **6**(1), 25 (2012)
13. R.M. El-Shazly, Feasibility of concentrated solar power under Egyptian conditions, MSc Thesis, University of Kassel, Cairo University (Egypt) (2011)
14. M.T. Chaichan, K.H. Abaas, H.A. Kazem, F. Hasoon, H.S. Aljibori, A. Ali, K. Alwaeli, F.S. Raheem, A.H. Alwaeli, Effect of design variation on saved energy of concentrating solar power prototype, Proceedings of the World Congress on Engineering, WCE III 2012, London, U.K. (2012)

15. Asia/Pacific PV Markets 2010 (http://www.solarbuzz.com/AP10.htm)
16. S. Duryea, S. Islam, W. Lawrance, A battery management system for stand-alone photovoltaic energy systems. IEEE Ind. Appl. Soc **7**(3), 67–72 (2001)
17. W. Durisch, H. Kiess, Crystalline silicon cells and technology, third generation concepts for photovoltaic conference, Osaka, Japan, May 12–16 2003
18. A. Chouder, S. Silvestre, N. Sadaoui, L. Rahmani, Simulation of a grid connected PV system based on the evaluation of main PV. Simul. Model. Pract. Theory **20**(1), 46–58 (2012)
19. K. Thongpao, P. Sripadungtham, P. Raphisak, K. Sriprapha, O. Ekkachart, Solar cells based on the influence of irradiance and module temperature, in *Electrical Engineering/Electronics Computer Telecommunications and Information Technology (ECTI-CON), International Conference*, (Chiang Mai, Thailand, 2010), pp. 153–160
20. F.W. Burari, A.S. Sambo, Model for the prediction of global solar radiation for Bauchi using meteorological data, Nigeria. J. Renew. Energy **91**, 30–33 (2001)
21. A.M. Al-Salihi, M.M. Kadum, A.G. Mohammed, Estimation of global solar radiation on horizontal surface using meteorological measurement for different cities in Iraq. Asian J. Sci. Res **3**(4), 240–248 (2010)
22. S. Kalogirou, *Solar Energy Engineering: Process and Systems* (Academic Press, London, 2009)
23. J. Jarras, *Feasibility of a Fund for Financing Solar Water Heaters and Projects Related to the Promotion of Renewable Energies in Jordan* (MEMR Press, Amman, 1987)
24. D.H.W. Li, J.C. Lam, An analysis of climatic variables and design implications. Archit. Sci. Rev. **42**(1), 15–25 (1999)
25. T. Muneer, Solar radiation model for Europe. Build. Serv. Eng. Res. Technol. **11**(4), 153–163 (1990)
26. D.H.W. Li, J.C. Lam, Predicting solar irradiance on inclined surfaces using sky radiance data. Energ. Conver. Manage. **45**(11–12), 1771–1783 (2004)
27. E. Vartiainen, A new approach to estimating the diffuse irradiance on inclined surfaces. Renew. Energy **20**(1), 45–64 (2000)
28. G. Schaab, Modeling and visualization of the spatial and temporal variability of the irradiance by means of a geo-information system, Cartographic building blocks, Dresden, p. 160. (2000)
29. A. Al-Salaymeh, Modeling of global daily solar radiation on horizontal surfaces for Amman city. Emirates J. Eng. Res **11**(1), 49–56 (2006)
30. T. Khatib, A. Mohamed, K. Sopian, A review of solar energy modeling techniques. Renew. Sustain. Energy Rev. **16**, 2864–2869 (2012)
31. J. Kreider, F. Kreith, *Solar Energy Handbook* (McGraw-Hill, New York, 1981)
32. D.I. Wardle, D.C. McKay, Recent advances in pyranometry, in *Proceedings of International Energy Agency, Task IX, Solar Radiation and Pyranometer Studies*, (Norrkoping, Sweden), p. 984
33. G.A. Zerlaut, Solar radiometry instrumentation, calibration techniques, and standards. Sol. Cells **18**, 189–203 (1986)
34. B.C. Bush, Characterization of thermal effects in pyranometers: A data correction algorithm for improved measurement of surface insolation. J. Atmos. Oceanic Tech. **17**, 165–175 (2000)
35. E.G. Dutton et al., Measurement of broadband diffuse solar irradiance using current commercial instrumentation with a correction for thermal offset errors. J. Atmos. Oceanic Tech. **18**, 297–314 (2001)
36. S. Wilcox, Improving global solar radiation measurements using zenith angle dependent calibration factors, in *Proceedings Forum 2001 Solar Energy: The Power to Choose*, (American Solar Energy Society, Washington, DC, 2001)
37. J. Michalsky, Optimal measurement of surface shortwave irradiance using current instrumentation. J. Atmos. Oceanic Tech. **16**, 55–69 (1999)
38. J.J. Michalsky, Toward the development of a diffuse horizontal shortwave irradiance working standard. J. Geophys. Res. (2005). https://doi.org/10.1029/2004JD00526

39. I. Reda, Using a blackbody to calculate net-longwave responsively of shortwave solar pyranometers to correct, for their thermal offset error during outdoor calibration using the component sum method. J. Atmos. Oceanic Tech. **22**, 1531–1540 (2005)
40. C.A. Gueymard, D.R. Myers, Solar radiation measurement: Progress in radiometry for improved modeling, in *Modeling Solar Radiation at the Earth Surface*, ed. by V. Badescu, (Springer, Berlin, Heidelberg, 2008), pp. 1–27
41. C.A. Gueymard, D.R. Myers, Evaluation of conventional and high-performance routine solar radiation measurements for improved solar resource, climatological trends, and radiative modeling. Sol. Energy **83**, 171–185 (2009)
42. A.K. Rajput, R.K. Tewari, A. Sharma, Utility base estimated solar radiation at destination Pune, Maharashtra, India. Int. J. Pure Appl. Sci. Technol **13**(1), 19–26 (2012)
43. J. Almorex, Estimating global solar radiation from common meteorological data in Aranjuez-Spain. Turk. J. Phys. **35**, 53–64 (2011)
44. B. Marion, B. Kroposki, K. Emery, J. del Cueto, D. Myers, C. Osterwald, *Validation of a Photovoltaic Module Energy Ratings Procedure at NREL* (Golden, National Renewable Energy Laboratory, 1999)
45. M.A. Bashir, H.M. Ali, S. Khalil, M. Ali, A.M. Siddiqui, Comparison of performance measurements of photovoltaic modules during winter months in Taxila, Pakistan. Int. J. Photoenergy, Hindawi Publishing Corporation **2014**, Article ID 898414,. 8 pages (2014)
46. J. Cano, Photovoltaic modules: Effect of tilt angle on soiling, MSC Thesis, Arizona State University (2011)
47. S.M. Salih, L.A. Kadim, Effect of tilt angle orientation on photovoltaic module performance. ISESCO J. Sci. Technol **10**(17), 19–25 (2014)
48. E.D. Mehleri, P.L. Zervas, H. Sarimveis, J.A. Palyvos, N.C. Markatos, Determination of the optimal tilt angle and orientation for solar photovoltaic arrays. Renew. Energy, Elsevier **35**(11), 2468–2475 (2010)
49. C. Emanuele, The disagreement between anisotropic-isotropic diffuse solar radiation models as a function of solar declination: Computing the optimum tilt angle of solar panels in the area of southern-Italy. Smart Grid Renew. Energy **3**, 253–259 (2012)
50. E.C.J. Kern, M.C. Rissell, Combined photovoltaic and thermal hybrid collector systems. 13th IEEE photovoltaic specialists conference, Washington, D.C., 1978
51. M.J.M. Jong, H.A. Zondag, System studies on combined PV/Thermal panels. 9th international conference on solar energy in high latitudes, Leiden, the Netherlands, 2001
52. P. Barnwal, G.N. Tiwari, Grape drying by using hybrid photovoltaic-thermal (PV/T) greenhouse dryer: An experimental study. Sol. Energy **82**(12), 1131–1144 (2008)
53. J. Guo, S. Lin, J.I. Bilbao, S.D. White, A.B. Sproul, A review of photovoltaic thermal (PV/T) heat utilization with low temperature desiccant cooling and dehumidification. Renew. Sustain. Energy Rev. **67**, 1–14 (2017)
54. A.H.A. Al-Waeli, K. Sopian, H.A. Kazem, M.T. Chaichan, PV/T (photovoltaic/thermal): Status and future prospects. Renew. Sustain. Energy Rev. **77**, 109–130 (2017)
55. R. Kumar, M.A. Rosen, A critical review of photovoltaic–thermal solar collectors for air heating. Appl. Energy **88**(11), 3603–3614 (2011)
56. A.H.A. Al-Waeli, M.T. Chaichan, K. Sopian, H.A. Kazem, Influence of the base fluid on the thermo-physical properties of nanofluids with surfactant. Case Stud. Therm. Eng . Accepted **13**, 100340 (2019). https://doi.org/10.1016/j.csite.2018.10.001
57. A.H.A. Al-Waeli, H.A. Kazem, K. Sopian, M.T. Chaichan, Techno-economical assessment of grid connected PV/T using nanoparticles and water as base-fluid systems in Malaysia. Int. J. Sustain. Energy **37**(6), 558–578 (2018). https://doi.org/10.1080/14786451.2017.1323900
58. A.H.A. Al-Waeli, M.T. Chaichan, K. Sopian, H.A. Kazem, Comparison study of indoor/outdoor experiments of SiC nanofluid as a base-fluid for a photovoltaic thermal PV/T system enhancement. Energy **151**, 33–44 (2018)
59. A.H.A. Al-Waeli, M.T. Chaichan, H.A. Kazem, K. Sopian, Comparative study to use nano-(Al2O3, CuO, and SiC) with water to enhance photovoltaic thermal PV/T collectors. Energ. Conver. Manage. **148**(15), 963–973 (2017). https://doi.org/10.1016/j.enconman.2017.06.072

60. A.H.A. Al-Waeli, K. Sopian, H.A. Kazem, M.T. Chaichan, Photovoltaic thermal PV/T systems: A review. Int. J. Comput. Appl. Sci **2**(2), 62–67 (2017)
61. A.H.A. Al-Waeli, K. Sopian, M.T. Chaichan, H.A. Kazem, H.A. Hasan, A.N. Al-Shamani, An experimental investigation on using of nano-SiC-water as base-fluid for photovoltaic thermal system. Energy Conserv. Manag **142**, 547–558 (2017)
62. A.H. Al-Waeli, K. Sopian, M.T. Chaichan, H.A. Kazem, A. Ibrahim, S. Mat, M.H. Ruslan, Evaluation of the nanofluid and nano-PCM based photovoltaic thermal (PVT) system: An experimental study. Energ. Conver. Manage. **151**, 693–708 (2017)
63. A.H. Al-Walei, M.T. Chaichan, K. Sopian, H.A. Kazem, Energy storage: CFD modeling of thermal energy storage for a Phase Change Materials (PCM) added to a PV/T using nanofluid as a coolant. J. Sci Eng. Res **4**(12), 193–202 (2017)
64. A.H.A. Al-Waeli, K. Sopian, H.A. Kazem, J.H. Yousif, M.T. Chaichan, A. Ibrahim, S. Mat, M.H. Ruslan, Comparison of prediction methods of PV/T nanofluid and nano-PCM system using a measured dataset and artificial neural network. Sol. Energy **162**, 378–396 (2018)
65. A.H.A. Al-Waeli, M.T. Chaichan, K. Sopian, H.A. Kazem, H.B. Mahood, A.A. Khadom, Modeling and experimental validation of a PV/T system using nanofluid coolant and nano-PCM. Sol. Energy **177**, 178–191 (2019). https://doi.org/10.1016/j.solener.2018.11.016
66. M. Mohsenzadeh, R. Hosseini, A photovoltaic / thermal system with a combination of a booster diffuse re fl ector and vacuum tube for generation of electricity and hot water production. Renew. Energy **78**, 245–252 (2015)
67. S. Calnan, Applications of oxide coatings in photovoltaic devices. Coatings **4**, 162–202 (2014)
68. J.I. Rosell, X. Vallverdu, M.A. Lechon, M. Ibanez, Design and simulation of a low concentrating photovoltaic/thermal system. Energ. Conver. Manage. **46**, 3034–3046 (2005)
69. N. Xu, J. Ji, W. Sun, W. Huang, J. Li, Z. Jin, Numerical simulation and experimental validation of a high concentration photovoltaic / thermal module based on point focus Fresnel lens. Appl. Energy **168**, 269–281 (2016)
70. J.S. Coventry, Performance of a concentrating photovoltaic / thermal solar collector. Sol. Energy **78**, 211–222 (2005)
71. L.R. Bernardo et al., Performance evaluation of low concentrating photovoltaic/thermal systems: A case study from Sweden. Sol. Energy **85**, 1499–1510 (2011)
72. D.B. Singh, G.N. Tiwari, Performance analysis of basin type solar stills integrated with N identical photovoltaic thermal (PVT) compound parabolic concentrator (CPC) collectors: A comparative study. Sol. Energy **142**, 144–158 (2017)
73. D.B. Singh, G.N. Tiwari, Exergo-economic, enviro-economic and productivity analyses of basin type solar stills by incorporating N identical PVT compound parabolic concentrator collectors: A comparative study. Energ. Conver. Manage. **135**, 129–147 (2017)
74. H. Helmers, K. Kramer, Multi-linear performance model for hybrid (C) PVT solar collectors. Sol. Energy **92**, 313–322 (2013)
75. S. Yilmaz, H. Riza, Energy supply in a green school via a photovoltaic-thermal power system. Renew. Sustain. Energy Rev. **57**, 713–720 (2016)
76. S. Sharma, A. Tahir, K.S. Reddy, T.K. Mallick, Solar energy materials & solar cells performance enhancement of a building-integrated concentrating photovoltaic system using phase change material. Sol. Energy Mater. Sol. Cells **149**, 29–39 (2016)
77. M.S. Buday, Measuring irradiance, temperature and angle of incidence effects on photovoltaic modules in Auburn Hills, Michigan, MSC, University of Michigan, (2011)
78. S. Al-Yahyai, Y. Charabi, A. Gastli, Review of the use of numerical weather prediction (NWP) models for wind energy assessment. Renew. Sustain. Energy Rev. **14**(9), 3192–3198 (2010)
79. A. Al-Tarabsheh, S. Voutetakis, A.I. Papadopoulos, B. Seferlis, I. Etiera, O. Saraereh, Investigation of temperature effects in efficiency improvement of non-uniformly cooled photovoltaic cells. Chem. Eng. Trans. **35**, 1387–1392 (2013)
80. J.A. del Cueto, *PV Module Energy Ratings Part II: Feasibility of Using the PERT in Deriving Photovoltaic Module Energy Ratings* (National Renewable Energy Laboratory, Golden, 2007)

81. D.T. Lobera, S. Valkealahti, Dynamic thermal model of solar PV systems under varying climatic conditions. Sol. Energy **93**, 183–194 (2013)
82. E. Zambolin, D. Del Col, Experimental analysis of thermal performance of flat plate and evacuated tube solar collectors in stationary standard and daily conditions. Sol. Energy **84**, 1382–1396 (2010)
83. A. Gregg, T. Parker, R. Swenson, A "real world" examination of PV system design and performance. Conf. Rec. IEEE Photovoltaic Spec. Conf. **31**, 1587–1592 (2005)
84. Y. Charabi, B.H.M. Rhouma, A. Gastli, GIS-based estimation of roof-PV capacity & energy production for the Seeb region in Oman. IEEE Int. Energy Conf. Renew. Energy **57**, 635–644 (2013)
85. A. Gastli, Y. Charabi, S. Zekri, GIS-based assessment of combined CSP electric power & seawater desalination plant for Duqum-Oman. Renew. Sustain. Energy Rev. **14**(2), 821–827 (2010)
86. S.A. Ghaznafar, M. Fisher, *Vegetation of the Arabian Peninsula* (Kluer Academin Publishers, Dordrecht, 1998), pp. 5–38
87. B.R. Hughes, N.B.S. Cherisa, O. Beg, Computational study of improving the efficiency of photovoltaic panels in the UAE. World Acad. Sci. Eng. Technol. **5**(1), 33 (2011)
88. G. George Makrides, B. Zinsser, A. Phinikarides, M. Schubert, G.E. Georghiou, Temperature and thermal annealing effects on different photovoltaic technologies. Renew. Energy **43**, 407–417 (2012)
89. F.A. Mutlak, Design and fabrication of parabolic trough solar collector for thermal energy applications, Ph. D Thesis, University of Baghdad, (2011, March)
90. M.T. Chaichan, K.I. Abaas, H.A. Kazem, The effect of variable designs of the central receiver to improve the solar tower efficiency. Int. J. Eng. Sci. **1**(7), 56–61 (2012)
91. H.A. Kazem, M.T. Chaichan, Status and future prospects of renewable energy in Iraq. Renew. Sustain. Energy Rev. **16**(8), 6007–6012 (2012)
92. Y. Franghiadakis, P. Tzanetakis, Explicit empirical relation for the monthly average cell-temperature performance ratio of photovoltaic arrays. Prog. Photovolt. Res. Appl. **14**, 541–551 (2006)
93. R. Chenni, M. Makhlouf, T. Kerbache, A. Bouzid, A detailed modeling method for photovoltaic cells. Energy **32**, 1724–1730 (2007)
94. W. Durisch, B. Bitnar, J.C. Mayor, H. Kiess, K.H. Lam, J. Close, Efficiency model for photovoltaic modules and demonstration of its application to energy yield estimation. Sol. Energy Mater. Sol. Cells **91**, 79–84 (2007)
95. E. Skoplaki, J.A. Palyvos, Operating temperature of photovoltaic modules: A survey of pertinent correlations. Renew. Energy **34**, 23–29 (2009)
96. M.G. Farr, J.S. Stein, Spatial variations in temperature across a photovoltaic array, proceeding to IEEE 40th photovoltaic specialist conference (PVSC), 2014
97. A.D. Jones, C.P. Underwood, A thermal model for photovoltaic systems. Sol. Energy **70**, 349–359 (2001)
98. H.F. Tsai, H.L. Tsai, Implementation and verification of integrated thermal and electrical models for commercial PV modules. Sol. Energy **86**, 654–665 (2012)
99. V.V. Tyagia, S.C. Kaushik, S.K. Tyagi, Advancement in solar photovoltaic/thermal (PV/T) hybrid collector technology. Renew. Sustain. Energy Rev. **16**, 1383–1398 (2012)
100. H.G. Teo, P.S. Lee, M.N.A. Hawlader, An active cooling system for photovoltaic modules. Appl. Energy **90**, 309–315 (2012)
101. F. Sarhaddi, S. Farahat, H. Ajam, A. Behzadmehr, M. Adeli, An improved thermal and electrical model for a solar photovoltaic thermal (PV/T) air collector. Appl. Energy **87**, 2328–2339 (2010)
102. M.A.M. Rosli, S. Mat, M.K. Anuar, K. Sopian, M.Y. Sulaiman, S. Ellias, Progress on flat-plate water based of photovoltaic thermal (PV/T) system: A review. Iranica J. Energy Environ **5**(4), 407–418 (2014)
103. S. Krauter, Increased electrical yield via water flow over the front of photovoltaic panels. Sol. Energy Mater. Sol. Cells **82**, 131–137 (2004)

104. S. Odehand, M. Behnia, Improving photovoltaic module efficiency using water cooling. Heat Transfer Eng **30**(6), 499–505 (2009)
105. A.S. Joshi, A. Tiwari, G.N. Tiwari, I. Dincer, B.V. Reddy, Performance evaluation of a hybrid photovoltaic thermal (PV/T) (glass-to-glass) system. Int. J. Therm. Sci. **48**, 154–164 (2009)
106. D.J. Yang, Z.F. Yuan, P.H. Lee, H.M. Yin, Simulation and experimental validation of heat transfer in a novel hybrid solar panel. Int. J. Heat Mass Transf. **55**, 1076–1082 (2012)
107. T. Kerzmann, L. Schaefer, System simulation of a linear concentrating photovoltaic system with an active cooling system. Renew. Energy **41**, 254–261 (2012)
108. M. Chandrasekar, S. Suresh, T. Senthilkumar, M.G. Karthikeyan, Passive cooling of standalone flat PV module with cotton wick structures. Energ. Conver. Manage. **71**, 43–50 (2013)
109. J.R.E.C. Kern, M.C. Russell, Combined photovoltaic and thermal hybrid collector systems, in *Proceedings of the 13th IEEE PV Specialist Conference*, (1978), pp. 1153–1157
110. X. Zhang, X. Zhao, S. Smith, J. Xub, X. Yuc, Review of R&D progress and practical application of the solar photovoltaic/thermal (PV/T) technologies. Renew. Sustain. Energy Rev. **16**, 599–617 (2012)
111. X. Zhao, X. Zhang, S.B. Riffat, X. Su, Theoretical investigation of a novel PV/T roof module for heat pump operation. Energ. Conver. Manage. **52**, 603–614 (2011)
112. K.E. Amori, H.M.T. Al-Najar, Analysis of thermal and electrical performance of a hybrid (PV/T) air based solar collector for Iraq. Appl. Energy **98**(100), 384–395 (2012)
113. M. Koehl, M. Heck, S. Wiesmeier, Modeling of conditions for accelerated life time testing of humidity impact on PV-modules based on monitoring of climatic data. Sol. Energy Mater. Sol. Cells **99**, 282–291 (2012)
114. M.D. Kempe, Modeling of rates of moisture ingress into photovoltaic modules. Sol. Energy Mater. Sol. Cells **90**(16), 2720–2738 (2006)
115. Rotronic Instrument Corp., The Rotronic Humidity Handbook, All you never wanted to know about humidity and didn't want to ask! 12/2005. www.rotronic-usa.com
116. A.H. Fanney, M.W. Davis, B.P. Dougherty, D.L. King, W.E. Boyson, J.A. Kratochvil, Comparison of photovoltaic module performance measurements. Trans. ASME **128**, 152 (2006)
117. K. Morita, T. Inoue, H. Kato, I. Tsuda, Y. Hishikawa, Degradation factor analysis of crystalline-Si PV modules through long-term field exposure test, in *Proceedings of the 3rd World Conference on Photovoltaic Energy Conversion*, (2003, May), pp. 1948–1951
118. N.G. Dhere, N.R. Raravikar, Adhesional shear strength and surface analysis of a PV module deployed in harsh coastal climate. Sol. Energy Mater. Sol. Cells **67**(1–4), 363–367 (2001)
119. B. Bhattachary, S. Dey, B. Mustaphi, Some analytical studies on the performance of grid connected solar photovoltaic system with different parameters, proceeding to 3rd international conference on material processing and materials science characterization (ICMPC-2014), vol. 6, pp. 1942–1950, 2014
120. R. Laronde, A. Charki, D. Bigaud, Lifetime estimation of a photovoltaic module subjected to corrosion due to damp heat testing. J. Sol. Energy Eng **135**(2), Article ID 021010 . 8 pages (2013)
121. C. Peike, S. Hoffmann, P. Hulsmann, Origin of dampheat induced cell degradation. Sol. Energy Mater. Sol. Cells **116**, 49–54 (2013)
122. F. Touati, A. Massoud, J. Abu Hamad, S.A. Saeed, Effects of environmental and climatic conditions on PV efficiency in Qatar, International conference on renewable energies and power quality (ICREPQ'13), Bilbao (Spain), 20th to 22th March, 2013
123. B.A.L. Gwandu, D.J. Creasey, Humidity: A factor in the appropriate positioning of a photovoltaic power station. Renew. Energy **6**(3), 313–316 (1995)
124. M.D. Kempe, Modeling of rates of moisture ingress into photovoltaic modules. Sol. Energy Mater. Sol. Cells **90**(16), 2720–2738 (2006)
125. J.K. Prakash, N. Gopinath, V. Kirubakaran, Optimization of solar PV panel output: A viable and cost effective solotion, International Journal of Advanced Technology & Engineering Research (IJATER) National Conference on "Renewable Energy Innovations for Rural Development" ISSN No: 2250–3536 20, New Delhi (2014)

126. Z.A. Darwish, H.A. Kazem, K. Sopian, M.A. Alghoul, M.T. Chaichan, Impact of some environmental variables with dust on solar photovoltaic (PV) performance: Review and research status. Int. J. Energy Environ **7**(4), 152–159 (2013)
127. V.B. Omubo-Pepple, C. Israel-Cookey, G.I. Alamunokuma, Effects of temperature, solar flux and relative humidity on the efficient conversion of solar energy to electricity. Eur. J. Sci. Res. **35**, 173–180 (2009)
128. A.A. Katkar, N.N. Shinde, G.C. Koli, S.P. Gaikwad, Evaluation of industrial solar cell w.r.t. temperature. IOSR J. Mech. Civ. Eng (IOSR-JMCE) **3**, 27–38 (2013)
129. H.A. Kazem, M.T. Chaichan, I.M. Al-Shezawi, H.S. Al-Saidi, H.S. Al-Rubkhi, J.K. Al-Sinani, A.H.A. Al-Waeli, Effect of humidity on the PV performance in Oman. Asian Trans. Eng **2**(4), 29–32 (2012)
130. T. Al-Hanai, R.B. Hashim, L. El Chaar, L.A. Lamont, Environmental effects on a grid connected 900 W photovoltaic thin-film amorphous silicon system. Renew. Energy **36**, 2615–2622 (2011)
131. E.B. Ettah, A.P. Udoimuk, J.N. Obiefuna, F.E. Opara, The effect of relative humidity on the efficiency of solar panels in Calabar, Nigeria. Univers J. Manag. Soc. Sci **2**(3), 8–11 (2012)
132. H.A. Kazem, M.T. Chaichan, Effect of humidity on photovoltaic performance based on experimental study. Int. J. Appl. Eng. Res. **10**(23), 43572–43577 (2015)
133. E. Klampaftis, K.R. McIntosh, B.S. Richards, Degradation of an undiffused SI–SIO2 interface due to humidity, 22nd European Photovoltaic Solar Energy Conference, Milan, Italy, 3–7 Sept 2007
134. M.K. Panjwani, B. NarejoG, Effect of humidity on the efficiency of solar cell (photovoltaic). Int. J. Eng. Res. General Sci **2**(4), 499–503 (2014)
135. V.B. Omubo-Pepple, I. Tamunobereton-ari, M.A. Briggs-Kamara, Influence of meteorological parameters on the efficiency of photovoltaic module in some cities in the Niger delta of Nigeria. J. Asian Sci. Res **3**(1), 107–113 (2013)
136. A. Rachman, K. Sopian, S. Mat, M. Yahya, Feasibility study and performance analysis of solar assisted desiccant cooling technology in hot and humid climate. Am. J. Environ. Sci. **7**(3), 207–211 (2011)
137. M.D. Kemp, Control of moisture ingress into photovoltaic modules, in *31st IEEE Photovoltaic Specialists Conference and Exhibition*, (Lake Buena Vista, Florida, 2005)
138. J.H. Wohlgemuth, S. Kurtz, Reliability testing beyond qualification as a key component in photovoltaics progress toward grid parity, proceeding in IEEE international reliability physics symposium monterey, California, April 10–14, 2011
139. IEC: International Electro-technical Commission, Standard IEC61215: Crystalline silicon terrestrial photovoltaic (PV) modules, design qualification and type approval IEC Central Office, Geneva, Switzerland, 1987
140. M. Vazquez, R.S. Ignacio, Photovoltaic module reliability model based on field degradation studies. Prog. Photovolt. Res. Appl. **16**, 419–433 (2008)
141. M.A. Quintana, D.L. King, T.J. McMahon, C.R. Osterwald, Commonly observed degradation in field-aged photovoltaic modules, in *Proc. 29th IEEE Photovoltaic Specialists Conference*, (2002), pp. 1436–1439
142. M.A. Munoz, M.C. Alonso-Garcia, V. Nieves, F. Chenlo, Early degradation of silicon PV modules and guaranty conditions. Sol. Energy **85**, 2264–2274 (2011)
143. A. Ndiaye, A. Charki, A. Kobi, C.M.F. Ke'be´, P.A. Ndiaye, V. Sambou, Degradations of silicon photovoltaic modules: A literature review. Sol. Energy **96**, 140–151 (2013)
144. A. Skoczek, T. Sample, E.D. Dunlop, H.A. Ossenbrink, Electrical performance results from physical stress testing of commercial PV modules to the IEC61215 test sequence. Sol. Energy Mater. Sol. Cells **92**, 1593–1604 (2008)
145. K.W. Jansen, A.E. Delahoy, A laboratory technique for the evaluation of electrochemical transparent conductive oxide delamination from glass substrates. Thin Solid Films **423**, 153–160 (2003)
146. G. Oreski, G.M. Wallner, Aging mechanisms of polymeric films for PV encapsulation. Sol. Energy **79**, 612–617 (2005)

147. G. Oreski, G.M. Wallner, Evaluation of the aging behavior of ethylene copolymer films for solar applications under accelerated weathering conditions. Sol. Energy **83**, 1040–1047 (2009)
148. T. Kojima, T. Yanagisawa, The evaluation of accelerated test for degradation a stacked a-Si solar cell and EVA films. Sol. Energy Mater. Sol. Cells **81**(1), 119–123 (2004)
149. C.R. Osterwald, A. Anderberg, S. Rummel, L. Ottoson, Degradation analysis of weathered crystalline-silicon PV modules, in *29th IEEE Photovoltaic Specialists Conference*, (New Orleans, Louisiana, 2002)
150. J.H. Wohlgemuth, D.W. Cunningham, A.M. Nguyen, G. Kelly, D. Amin, Failure modes of crystalline silicon modules, in *Proceedings of PV Module Reliability Workshop*, (2010)
151. E. Rueland, A. Herguth, A. Trummer, S. Wansleben, P. Fath, Optical l-crack detection in combination with stability testing for inline inspection of wafers and cells, in *Proceedings of 20th EUPVSEC*, (Barcelona, 2005), pp. 3242–3245
152. W. Dallas, O. Polupan, S. Ostapenko, Resonance ultrasonic vibrations for crack detection in photovoltaic silicon wafers. Meas. Sci. Technol. **18**, 852–858 (2007)
153. V.J. Fesharaki, M. Dehghani, J.J. Fesharaki, The effect of temperature on photovoltaic cell efficiency, in *Proceedings of the 1st International Conference on Emerging Trends in Energy Conservation (ETEC '11), Tehran, Iran*, (2011, November)
154. R. Siddiqui, U. Bajpai, Deviation in the performance of solar module under climatic parameter as ambient temperature and wind velocity in composite climate. Int. J. Renew. Energy Res **2**(3), 486–490 (2012)
155. E.B. Ettah, E.E. Eno, A.B. Udoimuk, The effects of solar panel temperature on the power output efficiency Calabar, Nigeria. J. Assoc. Radiograph Nigeria **23**, 16–22 (2009)
156. W.C. Hinds, *Aerosol Technology: Properties, Behavior, and Measurement of Airborne Particles* (Wiley, New York, 1999)
157. J.K. Kaldellis, M. Kapsali, K.A. Kavadias, Temperature and wind speed impact on the efficiency of PV installations. Experience obtained from outdoor measurements in Greece. Renew. Energy **66**, 612–624 (2014)
158. H.W. Tieleman, Wind tunnel simulation of the turbulence in the surface layer. J. Wind Eng. Ind. Aerodyn. **36**, 1309–1318 (1990)
159. J. Burdick et al., Qualification testing of thin-film and crystalline photovoltaic modules. Sol. Energy Mater. Sol. Cells **41/42**, 575–586 (1996)
160. B. Edwards, Collector deflections due to wind gusts and control scheme design. Sol. Energy **25**, 231–234 (1980)
161. A. Radu, E. Axinte, C. Theohari, Steady wind pressures on solar collectors on flat- roofed buildings. J. Wind Eng. Ind. Aerodyn. **23**(1–3), 249–258 (1986)
162. A. Radu, E. Axinte, Wind forces on structures supporting solar collectors. J. Wind Eng. Ind. Aerodyn. **32**, 93–100 (1989)
163. Y. Zhou, A. Kareem, Definition of wind profiles in ASCE 7. J. Struct. Eng. **128**(8), 1082–1086 (2002)
164. National Research Council, *National Building Code of Canada*, 13th edn. (Associate Committee on the National Building Code, Ottawa, 2010)
165. G.S. Wood, R.O. Denoon, K.C. Kwok, Wind loads on industrial solar panel arrays and supporting roof structure. Wind Struct. **4**(6), 481–494 (2001)
166. G.A. Kopp, D. Surry, K. Chen, Wind loads on a solar array. Wind Struct. **5**(5), 393–406 (2002)
167. T. Bhattacharya, A.K. Chakraborty, K. Pal, Effects of ambient temperature and wind speed on performance of monocrystalline solar photovoltaic module in Tripura, India, Hindawi Publishing Corporation. J. Sol. Energy, Article ID 817078, 5 pages (2014). https://doi.org/10.1155/2014/817078
168. N. Vasan, T. Stathopoulos, Wind tunnel assessment of the wind velocity distribution on vertical façades, Proceedings of eSim 2012: The Canadian conference on building simulation, Page 61 of 614, Halifax Nova Scotia, Canada, May 1–4 2012

169. M.A. Green, K. Emery, Y. Hishikawa, W. Warta, Solar cell efficiency tables (version 37). Prog. Photovolt. Res. Appl. **19**, 84–92 (2011)
170. A. Rao, M. Mani, Evaluating the nature and significance of ambient wind regimes on solar photovoltaic system performance, Building Simulation Applications BSA 2013, 1st IBPSA Italy conference, Bozen-Bolzano, 2013, pp. 395–405
171. H. Matsukawa, K. Kurokawa, Temperature fluctuation analysis of photovoltaic modules at short time interval, in *31st IEEE Photovoltaic Specialists Conference*, (2005), pp. 1816–1819
172. S. Armstrong, W.G. Hurley, A thermal model for photovoltaic panels under varying atmospheric conditions. Appl. Therm. Eng. **30**, 1488–1495 (2010)
173. C. Schwingshackla, M. Petittaa, J.E. Wagnera, G. Belluardoc, D. Moserc, M. Castellia, M. Zebischa, A. Tetzlaff, Wind effect on PV module temperature: Analysis of different techniques for an accurate estimation. Energy Procedia **40**, 77–86 (2013)
174. M. Mattei, G. Notton, G. Cristofari, M. Muselli, P. Poggi, Calculation of the polycrystalline PV module temperature using a simple method of energy balance. Renew. Energy **31**, 553–567 (2006)
175. E. Skoplaki, J.A. Palyvos, Operating temperature of photovoltaic modules: A survey of pertinent correlations. Renew. Energy **34**, 23–29 (2009)
176. E. Skoplaki, A.G. Boudouvis, J.A. Palyvos, A simple correlation for the operating temperature of photovoltaic modules of arbitrary mounting. Sol. Energy Mater. Sol. Cells **92**, 1393–1402 (2008)
177. M. Koehl, M. Heck, S. Wiesmeier, J. Wirth, Modeling of the nominal operating cell temperature based on outdoor weathering. Sol. Energy Mater. Sol. Cells **95**, 1638–1646 (2011)
178. S. Kurtz, K. Whitfield, D. Miller, J. Joyce, J. Wohlgemuth, M. Kempe, et al., Evaluation of high-temperature exposure of rackmounted photovoltaic modules, in *34th IEEE Photovoltaic Specialists Conference (PVSC)*, (2009), pp. 2399–2404
179. H. Chen, X. Chen, S. Li, H. Ding, Numerical study on the electrical performance of photovoltaic panel with passive cooling of natural ventilation. Int. J. Smart Grid Clean Energy **4**(4), 395–400 (2014)
180. S. Mekhilef, R. Saidur, M. Kamalisarvestani, Effect of dust, humidity and air velocity on efficiency of photovoltaic cells. Renew. Sustain. Energy Rev. **16**, 2920–2925 (2012)
181. P. Trinuruk, C. Sorapipatanal, D. Chenvidhya, Effects of air gap spacing between a photovoltaic panel and building envelope on electricity generation and heat gains through a building. Asian J. Energy Environ **8**(1 and 2), 73–95 (2007)
182. C.P.W. Geurts, R.D.J.M. Steenbergen, Full scale measurements of wind loads on stand-off photovoltaic systems, in *5th European & African Conference on Wind Engineering (EACWE)*, (Florence, Italy, 2009)
183. R. Velicu, G. Moldovean, I. Scaletchi, B.R. Butuc, Wind loads on an azimuthal photovoltaic platform. Experimental study, in *Proceeding of International Conference on Renewable Energies and Power Quality*, (Granada, Spain, 2010)
184. M. Shademan, R.M. Barron, R. Balachandar, H. Hangan, Numerical simulation of wind loading on ground-mounted solar panels at different flow configurations. Can. J. Civ. Eng **41**, 728–738 (2014)
185. A.A. Ogedengbe, H. Hangan, K. Siddiqui, Experimental investigation of wind effects on a standalone photovoltaic (PV) module. Renew. Energy **78**, 657–665 (2015)
186. D. Goossens, E. Van Kerschaever, Aeolian dust deposition on photovoltaic solar cells: The effects of wind velocity and airborne dust concentration on cell performance. Sol. Energy **66**(4), 277–289 (1999)
187. H.A. Kazem, M.T. Chaichan, A.H.A. Al-Waeli, K. Mani, Effect of shadows on the performance of solar photovoltaic, in *Mediterranean Green Buildings & Renewable Energy*, (2017), pp. 379–385. https://doi.org/10.1007/978-3-319-30746-6_27
188. U.S. Energy Information Administration, International energy statistics—total electricity installed capacity (2010, Aug)

189. L. Bony, S. Doig, C. Hart, E. Maurer, S. Newman, Achieving low-cost solar PV: Industry workshop recommendations for near-term balance of system cost reductions, Opportunity 1: Efficient design for wind forces, Rocky Mountain Institute, RMI.org, (2010)
190. C.A. Gueymard, The sun's total and spectral irradiance for solar energy applications and solar radiation models. Sol. Energy **76**(4), 423–453 (2004)
191. H. Li, W. Ma, X. Wang, Y. Lian, Estimating monthly average daily diffuse solar radiation with multiple predictors: A case study. Renew. Energy **36**, 1944 (2011)
192. C.A. Gueymard, Direct and indirect uncertainties in the prediction of tilted irradiance for solar engineering applications. Sol. Energy **83**(3), 432–444 (2009)
193. A.I. Kudish, E.G. Evseev, The analysis of solar UVB radiation as a function of solar global radiation, ozone layer thickness and aerosol optical density. Renew. Energy **36**, 1854 (2010)
194. M.T. Chaichan, H.A. Kazem, A.A. Kazem, I. Abaas Kh, K.A.H. Al-Asadi, The effect of environmental conditions on concentrated solar system in desertec weathers. Int. J. Sci. Eng. Res. **6**(5), 850–856 (2015)
195. M.T. Chaichan, K.I. Abass, H.A. Kazem, Dust and pollution deposition impact on a solar chimney performance. Int. Res. J. Adv. Eng. Sci **3**(1), 127–132 (2018)
196. H.A. Kazem, M.T. Chaichan, A.H.A. Alwaeli, The impact of dust's physical properties on photovoltaic modules outcomes, Solar energy conference, London UK, 2018
197. M.T. Chaichan, K.I. Abass, H.A. Kazem, Energy yield loss caused by dust and pollutants deposition on concentrated solar power plants in Iraq weathers. Int. Res. J. Adv. Eng. Sci **3**(1), 160–169 (2018)
198. H.K. Eliminir, A.E. Ghitas, R.H. Hamid, F.E. Hussainy, M.M. Beheary, K.M. Abdel-Moneim, Effect of dust on the transparent cover of solar collectors. Energ. Conver. Manage. **47**, 3192–3203 (2006)
199. H. Jiang, L. Lu, K. Sun, Experimental investigation of the impact of airborne dust deposition on the performance of solar photovoltaic (PV) modules. Atmos. Environ. **45**, 4299–4304 (2011)
200. M.C. Peel, B.L. Finlayson, T.A. McMahon, Updated world map of the Köppen- Geiger climate classification. Hydrol. Earth Syst. Sci. **11**, 1633–1644 (2007)
201. D. Goosens, E.V. Kerschaever, Aeolian dust deposition on photovoltaic solar cells: The effects of wind velocity and airborne dust concentration on cell performance. Sol. Energy **66**, 277–289 (1999)
202. R.T.A. Hamdi, S.H. Hafed, M.T. Chaichan, H.A. Kazem, Dust impact on the photovoltaic outcomes. Int. J. Comput. Appl. Sci **5**(2), 385–390 (2018)
203. M.T. Chaichan, H.A. Kazem, Effect of sand, ash and soil on photovoltaic performance: An experimental study. Int. J. Sci. Eng. Sci **1**(2), 27–32 (2017)
204. M.T. Chaichan, B.A. Mohammed, H.A. Kazem, Effect of pollution and cleaning on photovoltaic performance based on experimental study. Int. J. Sci. Eng. Res. **6**(4), 594–601 (2015)
205. H. Hottel, B. Woertz, Performance of flat-plate solar-heat collectors. Trans. Am. Soc. Mech. Eng. (USA) **64**, 91 (1942)
206. B. Nimmo, S.A.M. Said, Effects of dust on the performance of thermal and photovoltaic flat plate collectors in Saudi Arabia—preliminary results, in *Proceedings of the 2nd Miami International Conference on Alternative Energy Sources*, ed. by T. N. Vezirogluv, (1979, Dec 10–13), pp. 223–225
207. A. Salim, F. Huraib, N. Eugenio, PV power-study of system options and optimization, in *Proceedings of the 8th European PV Solar Energy Conference*, (1988)
208. A. Maghrabi, B. Alharbi, N. Tapper, Impact of the March 2009 dust event in Saudi Arabia on aerosol optical properties, meteorological parameters, sky temperature and emissivity. Atmos. Environ. **45**(13), 2164–2173 (2011)
209. F. Wakim, *Introduction of PV Power Generation to Kuwait*, Kuwait Institute for Scientific Researchers [Report No. 440] (1981)
210. A.A.M. Sayigh, Effect of dust on flat plate collectors, in *Sun: Mankind's Future Source of Energy; Proceedings of the International Solar Energy Congress, New Delhi*, ed. by F. de Winter, M. Cox, vol. 2, (Pergamon Press, New York, 1978), pp. 960–964

211. A.A.M. Sayigh, S. Al-Jandal, H. Ahmed, Dust effect on solar flat surfaces devices in Kuwait. Proceedings of the workshop on the physics of non-conventional energy sources and materials science for energy, ICTP, Triest, Italy. pp. 353–67, 1985
212. F. Touati, M. Al-Hitmi, H. Bouchech, Towards understanding the effects of climatic and environmental factors on solar PV performance in Arid Desert Regions (Qatar) for various PV technologies", World renewable energy congress, Indonesia, international conference on renewable energy and energy efficiency, Bali, Indonesia, 17–19 Oct 2011
213. F. Touati, A. Massoud, J. Abu Hamad, S.A. Saeed, Effects of environmental and climatic conditions on PV efficiency in Qatar, International conference on renewable energies and power quality (ICREPQ'13), Bilbao (Spain), 20–22 March, 2013
214. Z.A. Darwish, H.A. Kazem, K. Sopian, M.A. Al-Goul, M.T. Chaichan, Impact of some environmental variables with dust on solar photovoltaic (PV) performance: Review and research status. Int. J. Energy Environ. **7**(4), 152 (2013)
215. Z.A. Darwish, H.A. Kazem, K. Sopian, M.A. Al-Goul, H. Alawadhi, Effect of dust pollutant type on photovoltaic performance. Renew. Sustain. Energy Rev. **41**, 735–744 (2015)
216. H.A. Kazem, T. Khatib, K. Sopian, F. Buttinger, W. Elmenreich, A.S. Albusaidi, Effect of dust deposition on the performance of multi-crystalline photovoltaic modules based on experimental measurements. Int. J. Renew. Energy Res **3**(4), 850–853 (2013)
217. H.A. Kazem, M.T. Chaichan, S.A. Saif, A.A. Dawood, S.A. Salim, A.A. Rashid, A.A. Alwaeli, Experimental investigations of dust type effect on photovoltaic systems in North Region, Oman. Int. J. Sci. Eng. Res. **6**(7), 293–298 (2015)
218. H.A. Kazem, M.T. Chaichan, Experimental effect of dust physical properties on photovoltaic module in northern Oman. Sol. Energy **139**, 68–80 (2016). https://doi.org/10.1016/j.solener.2016.09.019
219. A.A. Kazem, M.T. Chaichan, H.A. Kazem, Effect of dust on photovoltaic utilization in Iraq: Review article. Renew. Sustain. Energy Rev. **37**, 734–749 (2014)
220. S.P. Sukhatme, *Solar Energy: Principles of Thermal Collection and Storage* (Tata McGraw-Hill, New Delhi, 2003)
221. N. Nahar, J. Gupta, Effect of dust on transmittance of glazing materials for solar collectors under arid zone conditions of India. Sol. Wind Technol. **7**, 237–243 (1990)
222. Al-Sayyah, J. Stark, T. Abuhamed, W. Weisinger, M. Horenstein, M.K. Mazumder, Energy yield loss caused by dust deposition in solar power plants, Proc. 2012 joint electrostatics conference, 2012

Chapter 6
Applications and PV/T Systems

6.1 Background

In this chapter, the latest data on the state of the global solar thermal industry and market today and future possibilities have been included. Previous chapters have shown the concept of PV/T and the positive reasons for the trend toward PV/T systems over the past few years. In this chapter, the criteria for evaluating the technical, economic, and environmental performance of the PV/T systems used by the researchers to determine the optimal utilization of these systems are outlined. Also, an extensive review of literature in the research, and development to grasp the gap between research and industry of this technology, is provided. After identifying key features, emphasis is placed on examining the difficulties and barriers that cause the commercial proliferation of PV/T systems to be delayed. In the end, some of the products marketed by the best PV/T systems are displayed.

6.2 Introduction

After the great development in urbanization that began with the reconstruction of the aftermath of the Second World War of destruction and with the great scientific development that significantly increased the levels of human comfort, there has been a sharp increase in energy use to reach very high levels. In 2008, the world consumed 474 exa-joules (474 × 1018 J), the main source of which was fossil fuels (coal, crude oil, and natural gas) [1]. After the economic crisis of 2008 and 2014, and the fluctuation of oil prices was high, this caused a slight decline of world energy consumption by only 1.1% in 2009. Despite the unexpected global economic recession, global energy consumption has not been affected. On the contrary, in many developing countries, this consumption has increased, especially in Asia, which has been growing. This reliance on fossil fuel consumption and combustion

A. H. A. Al-Waeli et al., *Photovoltaic/Thermal (PV/T) Systems*,
https://doi.org/10.1007/978-3-030-27824-3_6

of power generation causes a continuous rise in carbon emitted into the atmosphere causing global warming [2, 3]. Most of the world has gone to produce part of the energy using clean renewable energies such as wind, geothermal, and solar. Unfortunately, till 2007, solar energy currently accounts for only 0.5% of the world's total energy market, while the PV technology counted for 0.04% of the energy generated [4]. However, things changed in the last 10 years for the benefit of solar energy systems.

As noted earlier, global decision-making policies, especially after the Paris 2018 conference, are aimed at reducing fossil fuel consumption as a source of energy generation and replacing it with alternative renewable energies. The most important energy source and technologies are solar thermal and photovoltaic technologies, which are today ready to occupy a large part of the global energy market with very high growth potential for the near future [5]. These technologies are characterized by continuous technical developments, which will reduce the fears of depletion of oil globally and the decline of the global capacity to save energy and protect the environment. The next stage will be a global approach to harnessing solar radiation as a free fuel to run various solar thermal and electrical applications, thereby achieving a significant reduction in carbon emissions, thereby reducing the existing damage and improving part of it for the benefit of future generations [6].

Solar thermal energy is widely used today in heating water for homes and factories. It is also used for cooling purposes for comfort, air heating, and cooling fluids in industrial processes. Its marketing potential and global reach are huge. Nearly 90% of the world's solar energy is using this application. The global solar thermal market has grown since the early 1990s and continues to grow globally. Global statistics show that the solar thermal market has tripled over the years 2002–2006 and is still growing. The report of the European Solar Thermal Technology (ESTP) Vision Plan states that the solar thermal market will process about 50% of the low and medium heat by 2030 [7]. The European Union of Solar Thermal Energy (ESTIF) has forecast that the total amount of thermal energy will be between 91 and 320 GW by 2020. This amount of energy will provide the equivalent of 5600 tons of crude oil. By 2050, these systems will generate only 1200 gigawatts of solar thermal power in the European Union [8]. The use of solar thermal systems is currently a mature technology both technically and commercially and is able to provide a large amount of electricity using solar energy.

Photovoltaic cells can also be considered a promising technique in the field of electricity generation. Although the amount of technology produced by this technology globally is very small in 80 decades, it provides only 0.1% of the world's electricity generation. However, this percentage clearly improved today. Global studies show that the global PV market is growing at an annual rate of 40% [9]. Technical and scientific progress of this technology is ongoing; as prices have declined, and the volume of installation has increased, and many laws have been put in place to encourage their use. The use of photovoltaic cells is expected to continue to capture a large proportion of the world-generated energy. The International Energy Agency (IEA) predicted that photovoltaic electricity systems would take at least 5% of global energy generated by 2030 and would rise to 11% by 2050 [4].

Today, researchers and manufacturers are turning to the use of hybrid systems combining photovoltaic and solar thermal systems in one unit called PV/T [10]. This technique clearly improves the efficiency of converting solar energy falling on the system, so it improves the economic feasibility of this system. PV/T systems can generate electricity and heat at the same time, taking advantage of the specifications of solar thermal collectors and photovoltaic solar systems, so the rate of utilization of solar radiation is higher than in PV cells or solar heaters if they were used separately. Due to these great potentials, their spread and commercialization are expected to be higher and faster than photovoltaic systems or solar thermal systems. Since PV/T systems are technically emerging and modern, there are many points to be confirmed before they are widely marketed and competitive with older solar systems. For example, it is necessary to study the possibilities of the technical system and choose the most suitable ones. Also, to examine the problems that arise in these systems in relation to maintenance for working many years without interruption. It is also necessary to study the market potential to receive such systems and obstacles to practical applications.

The adoption of economic results for the use of PV/T systems will be different from one country to another and sometimes even in the same country, since the costs, tariffs, and policies vary from country to country. Accurate and wise economic assessment is achieved through the inclusion of domestic policies and the composition of the energy market in the studied country. The rate of installed new PV systems continues to increase. For example, in Sweden in 2016, 79.2 MWp was installed, representing an annual growth of 63% in the PV market compared with 48.4 MWp installed in 2015 [6]. Here we should emphasize that the interest in PV applications is little, and investing in it represents a very small share of the energy market. On the contrary, the market for photovoltaic systems connected to the network is growing rapidly. The EU tax cuts of 98% on large-scale photovoltaic systems exceeding 255 kW have encouraged growth in investment in this field and increased PV market [6].

In the previous chapters, a review study was carried out on past and current research work on PV/T technology. The research work is under way to establish a technical advantage of PV/T systems to achieve the selection, design, and installation of the best commercially available PV/T system. Further work is under way to develop associated coding standards and systems. The current PV/T research aims at developing its products to be commercially viable and competitive with what is currently available in the global market and encouraging the trend toward PV/T technologies worldwide. Finally, as with all new technologies, there is a need for further work, research, and study of multi-species PV/T systems. Today, PV/T systems that use air, water, or both as refrigerant coolers can be considered commercially ready, although their thermal efficiency is low as previously indicated. Today, however, researchers suggest the use of high-conductivity cooling fluids to maximize the impact of heat on the system. The utilization of nanoparticle in fluids such as water, oil, and glycols has significantly increased the thermal efficiency of PV/T systems. The addition of phase change materials (PCMs) in the system as a heat storage media has increased this efficiency. The addition of nanoparticles to PCMs has enhanced the conductivity of these thermal materials, thus increasing the thermal efficiency of the PV/T systems very clearly. Here, there is still some research hurdles that need to be further studied in order to reach the optimal system for global marketing. On the one hand, the best nanoparticles that should be used in the cooling nanofluids, their quantities, and the appropriate type and correct surfactant value should be determined. The type of PCM suitable for use in PV/T systems should be determined, and specific types or types of nanoparticles that must be added to the PCMs should be agreed upon to be used in this application. These systems face some technical and research challenges that require more time, research, and investigation. There is still scope for more business to reach:

1. Working on the development of advanced systems that are economically feasible and have high efficiency, whether electrically or thermally.
2. Access to the design of the optimal structure of these systems, which bear all the stresses and pressures and the lifetime of long operation.
3. Exploring different installation methods and approaches for PV/T systems on roofs of real buildings to determine their economic feasibility.
4. Studying and analyzing the future environmental impact and economic viability of these systems.

Modern technology marketers offer new opportunities for such innovations. The innovative PV/T technology combines both electricity and heat in the same space, which means less installation cost compared with the installation of thermal and PV assemblies. Low-cost versus high overall performance of PV/T systems will lead to the development of the solar thermal technology market alongside PV technologies. Here the focus on an important point is that PV/T technology reduces the price of energy per square meter.

6.3 Overview of Latest R&D on PV/T Systems for Commercial

Researchers around the world have conducted many research and development studies to develop the concept of PV/T in terms of components, shapes, potentials, and particularly performance. Today, there are many kinds of PV/T which have efficient and high electrical and thermal efficiencies as there are many shapes proposed and developments occurred over time to optimize the design in each case individually. As the development in this area depends on theoretical and practical approaches, the latest developments resulting from scientific research in these fields will be listed in the following sections.

6.4 Practical Research

There have been many practical researches and several ways of cooling PV cells, including air, water, nanoparticles, etc. The studies are extensive in each of these types; a small part of these research works will be mentioned in the coming paragraphs because the full containment needs too many volumes.

6.4.1 *PV/T Cooling by Air*

This method is cheaper and easier than other existing types, as the air is rotated in the channels under the solar panels to remove the possible heat. In this aspect, research has been carried out.

Komp and Reeser [11] designed, manufactured, and installed a composite pneumatic compound on the roof with a central pressure and finned it to enhance heat transfer from the back of the solar panels to the air. The researchers took advantage of this air to heat a winter house by using a fan. Both Moshtegh [12] and Sandbergy [13] studied the effect of airflow caused by buoyancy. In these works, heat transfer was carried out using a vertical channel mounted on the side wall of the cell. The results of the studies showed that the induced velocity affects the irregular flow of heat inside the channel and that this effect depends on the size of the channel outlet and its geometric shape.

Sopian et al. [14] analyzed the performance of both single and double air collectors. The results of the study show that the dual air collector results in higher efficiencies. The study found that the use of PV/T gave a saving in the surface area used up to 38%, resulting from compensation in the electrical efficiency produced using this system. Ji et al. [15] used the facade of residential buildings in Hong Kong to install an integrated PV/T system. The results of the study found that the thermal efficiency of the system is about 48% for cells made of thin silicon and 43% for cells made of crystalline silicon. The use of this technique in covering buildings reduced the cooling load of the building to a large extent, because the PV/T cover caused a reduction in the heat aborted by walls.

Tonui and Tripanagnostopoulos [16] built a low-cost air-cooled PV/T system and studied the possibility of improving heat transfer between the flowing air and the photovoltaic cell. The airway is fitted with fins made of metal sheets and fastened to the rear wall of the channels to increase the amount of heat extracted from photovoltaic cells. The results showed that the air mass flow decreased by increasing the temperature of the surrounding air and increasing the angle of inclination of the system to a certain level at the same temperature, which reduces thermal efficiency. Practical experiments showed that the optimum channel depth was between 0.05 and 0.1 m for the studied system. This system was acceptable in terms of cost, and the possibility of integration in the building is appropriate.

Shahsavar and Ameri [17] compared the performance of the air-cooled PV/T system when glass covers were added or not to the system in the city of Kerman, Iran. To increase the heat transfer area and to improve the heat extraction from photovoltaic cells, the researchers used thin aluminum foil suspended in the middle of the air channels. The air circulation system is operated by forced convection. The results of the empirical study showed an acceptable consensus with the calculations derived from the theoretical simulation model. The study found that there is ideal number of fans to be used in a system of this form to achieve maximum electrical efficiency. Also, the results of the study showed that the addition of glass cover in higher PV panels increased thermal efficiency and resulted in a decrease in the electrical efficiency.

6.4.2 PV/T Cooling by Water

This technique is better than its predecessor, because water has a higher heat capacity than air, which means that it can remove more heat from the PV module. It is worth to mention the several difficulties faced by researchers since the optimal design depends on the movement of water (flow), its little thermal conductivity, and its large weight, which means a clear change in the system combinations.

Huang et al. [18] used a commercial PV module and made a PV/T system; the researchers relied on the system thermal performance evaluation used to test traditional solar hot water heaters. The study results were promising as temperature difference between the water tank and the photovoltaic panel was achieved at 4 °C. Initial power of 61.3% was provided in the proposed system.

Agarwal [19] and Grag [20] designed several models of water heaters to be included in PV/T systems. Bergene and Lovvik [21] also investigated heat transfer through a water-cooled PV/T system. The results of the study showed the possibility of reaching the total competencies (electric + thermal) up to 60% up to 80%. The studied system produced hot water that could be used for household purposes.

Kalogirou [22] found that the economic viability of the PV/T system cooled much higher than the system that cooled by air.

Bazilian et al. [23] proposed a PV/T system cooled by water. This system has a solar water collector, and it is connected to a PV cell to store heat extracted from the solar panel. The proposed system also has an additional heater and an inverter (connected to the network). Figure 6.1 shows a schematic diagram of the proposed PV/T system. According to researchers, this system addressing one of the main defects of PV, which is the cell, has low electrical efficiency when its temperature is rising.

Chow et al. [24] check experimentally a PV/T system consisting of water collector that works as a water heater and PV cell. These complexes were installed vertically and studied the effect of climatic conditions for different seasons on their performance. The results of the study showed that water circulating by natural convection gave better results than forced water recycling of the proposed system. The maximum thermal efficiency of the studied system was 38.9%, which was met with

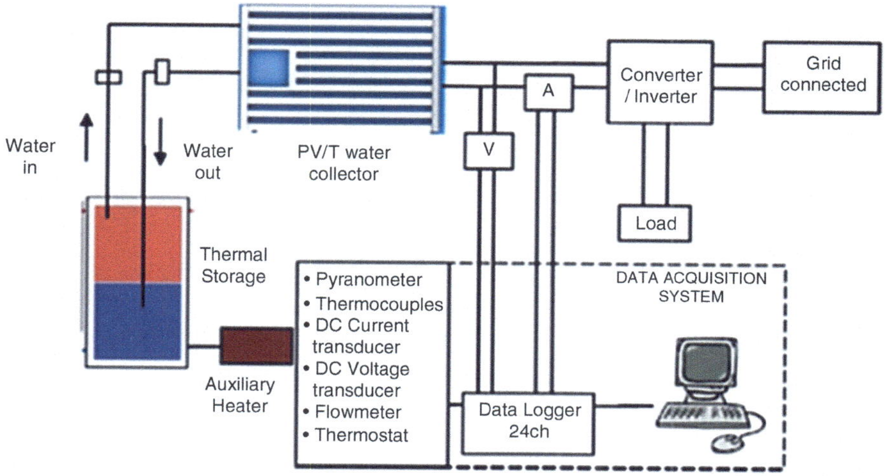

Fig. 6.1 Schematic diagram of the proposed PV/T system in reference [23]

electricity conversion efficiency of 8.56% during the summer period of the city of Hong Kong. The researchers proposed the development of channels that allow the flow of water below the photovoltaic cells to optimize the overall efficiency of the PV/T system. The study concluded that PV/T systems using single glass panels and pipes are the most suitable designs for commercial execution because they are characterized by high efficiency.

Chow et al. [25] used concentrated solar cells to heat the water in a solar collector below the cells and implemented this system (BIPV/T) as illustrated in Fig. 6.2. For the purpose of optimizing the absorbed heat, the system is placed adjacent to the outer wall of a room to increase thermal gain. The interior walls of the room were thermally insulated. The room has been designed and constructed to be heat controlled by heaters that must maintain the interior and exterior walls at a pre-designed temperature of 22 ± 0.5 °C throughout the year.

The PV system consists of a photovoltaic cell and a water collector installed on a brick wall so that the water tank is attached to the room's surface. The results of heat gains across the PV/T wall compared to the reference wall in the peak summer and winter period showed that the PV/T wall reduces cooling load of the room space and the heating loads can be greatly reduced in summer and winter. The thermal efficiency of the system was 38.9%, while electrical efficiency was 8.56% in the late summer of Hong Kong.

Dubey and Tay [26] evaluated several PV/T systems operated in tropical climatic conditions of Singapore. The monocrystalline solar cells were used with a tube- and plate-type thermal collector in the first system. The second system consisted of a photovoltaic panel with multi-crystalline solar cells with a parallel plate heat tank. Researchers used basic energy balance equations to verify the performance of the two thermal systems. The results of the study showed that the electrical

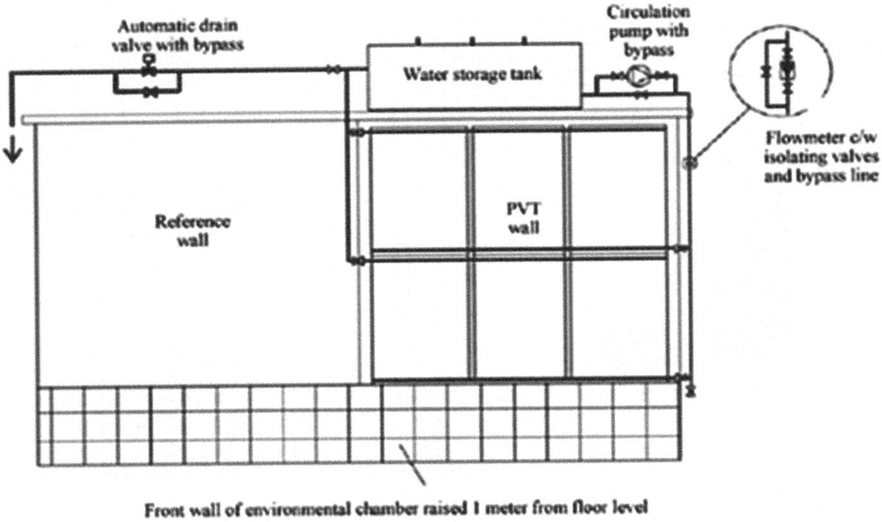

Fig. 6.2 Schematic diagram of the BiPV/T setup showing the water heating circuit proposed in reference [25]

efficiency and thermal efficiency of the first system were 11.8% and 40.7%, respectively, and for the second system were 11.5% and 39.4%, respectively.

Jouhara et al. [27] studied and analyzed the results of six PV systems, five of which were equipped with a cooling cycle; two photovoltaic cells were monocrystalline silicon modules and two other polycrystalline silicon modules. The fifth one was a solar PV system without cooling for comparison purposes. Use the recirculating water as a cooling fluid was enough to heat 350 L tank to cover the hot water demands of a single family. The results of the study showed that the heat from the PV/T systems studied is sufficient to cover about 60% of the needs of hot water under the conditions of low solar radiation levels, and in cases of high solar radiation, these systems are able to cover the needs of the family of hot water up to 100%.

6.4.3 PV/T Cooling by Heat Pipes

The heat pipe is one of the most efficient means of transmission of energy, consisting of the main structure (composed of hollow tubes of air and partially filled with a liquid and steam used in solar energy applications). This tube has very high thermal conductivity, allowing it to transfer large amounts of heat, the temperature of the pipe is determined by the temperature of the tube and the heat transferred to the upper end of the tube, but the two-way temperature pipe is called the wickless heat pipe. Cooling the PV cells within PV/T systems by heat pipes has got satisfactory results.

Nishikawa et al. [28] used a heat pump to cool the PV/T system. The cooling fluid used in the study was R22. The study showed that the system studied COP has achieved a greater value than conventional heat pump with the cooling of solar panels more effectively. Zhao et al. [29] designed a PV/T module installed on the roof to generate electricity and act as an evaporator in the associated heat pump system. During the study, the operating conditions of the system were checked, and the system performance was analyzed. The study results indicated that this system should work with 10 °C evaporation and 60 °C of condensation temperatures. The monocrystalline PV cells have higher electrical efficiency than crystallized films and thin films. The assumed system was able to produce thermal efficiency of 55% and electrical efficiency of 19%. The heat pump system had more than 70% total efficiency. On the economic side, the combination of photovoltaic cells and evaporation coils in one system caused significant savings in capital and operating costs compared to using PV cells and heat pumps for each individual bug.

Moradgholi et al. [30] proposed a PV/T system that uses a thermal tube (thermosyphon type) to absorb excess heat from solar cells. The researchers studied the system in the spring and summer, where they installed the system at an angle of 30° from the horizon with the use of methanol as a working fluid in the thermal tube. In this test, the generated electrical power increased by 5.67% compared to the individual photovoltaic generation. In the summer, researchers changed the angle of inclination to 40° and replaced methanol with acetone as a working fluid. In the summer, the generated electricity increased by 7.7% from the production of a single cell, and the system produced thermal efficiency of 45.14%. In the proposed system, the temperature of the photovoltaic panel decreased significantly to 15 °C. The experimental results supported the idea of using heat pipe technology in PV/T systems to enhance electrical efficiency generated and increase thermal energy.

6.4.4 PV/T Cooling by Nanofluids

Nanofluids have attracted a lot of interest from researchers because these fluids have high thermal conductivity, which makes them have better ability to transfer heat. Many studies have shown that using such fluids in a PV/T system will enhance the efficiencies of the systems. Commonly used fluids, such as water, oil, glycol ethylene, and glycol propylene, have a low thermal conductivity, which adversely affects heat transfer. Therefore, mixing these fluids with material of high volume, whether metal or metal oxides have a very high conductivity, can improve this property in the solution and increase the rate of heat transfer significantly. The following are some of these studies that show the evolution of these fluids and their use in the PV/T systems.

Yusufi et al. [31] studied the effect of addition of two ratios of aluminum oxide to water (2% and 4%) on the efficiency of a PV/T system. Several samples of nanofluid were prepared with and without surfactant. Experiments were performed using different flow rates. The results of the study showed clearly that the use of nanofluid in the studied system increased its stability and enhanced its performance better. For the studied system, thermal efficiency improved by 28.3% when using 2% by weight of nano-aluminum compared to the use of water as a coolant.

Karami and Rahimi [32] 2014 used different concentrations of boehmite ($AlOOH\_xH_2O$) in water for the cooling of two PV/T systems. The first contains a straight rectangular channel, while the second contains a helical channel. Experimental results showed that the use of nanofluid caused a clear reduction in the temperature of the photovoltaic cell. The highest observed temperature drop was 18.33 °C and 24.22 °C for both straight and helical channels, respectively, while the concentration of nanoparticles in water was only 0.1% by weight. The electrical efficiency of the two systems was increased by 20.57% and 37.67% for the straight and spiral channels, respectively.

Sardarabadi et al. [33] studied experimentally the effect of the addition of nano-silica to water to form a nanofluid acting as a coolant in the PV/T system. The experimental results showed that the addition of SiO_2 nanoparticles (1% by weight) to water increased the thermal and total efficiency by 7.6% and 3.6%, respectively. When using a nanofluid containing 3% by weight of nano-silica, the thermal and total efficiency increased by 12.8% and 7.9%, respectively.

Hussein et al. [34] studied the effect of using nanofluid consisting of 0.1–0.5% by weight of nano-Al_2O_3 and water with forcible cooling fluid flow. The research team compared the use of nanofluid with the use of water as a cooling fluid and the use of an individual photovoltaic cell under the same conditions of temperature, solar radiation intensity, and so on. The results of the study showed that the photovoltaic panel was at 79.1 °C when water was used, while the use of nanofluid (0.3% Al_2O_3 + water) reduced the PV panel temperature to 42.2 °C.

Jing et al. [35] studied mixing silica nanoparticles by different sizes with water and using them to cool a PV/T system. The results of the use of 5 nm silica particles by a 2% volumetric concentration in water resulted in a 20% thermal conductivity increase compared with distilled water.

Gadiri et al. [36] conducted an experimental study using a nanofluid, in which its basic fluid was water and the selected nanoparticle was ferro-(Fe_3O_4) with concentrations of 1% and 3% by weight. In the practical experiments, the nanofluids were exposed to a fixed and alternated magnetic field to show their effect on the overall efficiency of the PV/T system used in the study. The results of the study showed the total system efficiency was at 52% with distilled water. For the case when nanofluid was used, the total system efficiency reached 76% when the frequency was 50 Hz. When the alternating magnetic field was used, the total efficiency was increased by about 4–5%.

Shamani et al. [37] examined the impact of three types of nanomaterials (SiO_2, TiO_2, and SiC) when they were added to water to form nanofluids that been used in cooling a PV/T system operating in the tropical climate of Malaysia. Figure 6.3 shows a diagram of the experimental structure used in the tests. The results of the study were promising, with researchers achieving a total efficiency of 81.73% and electrical efficiency of 13.52% when using SiC nanofluid.

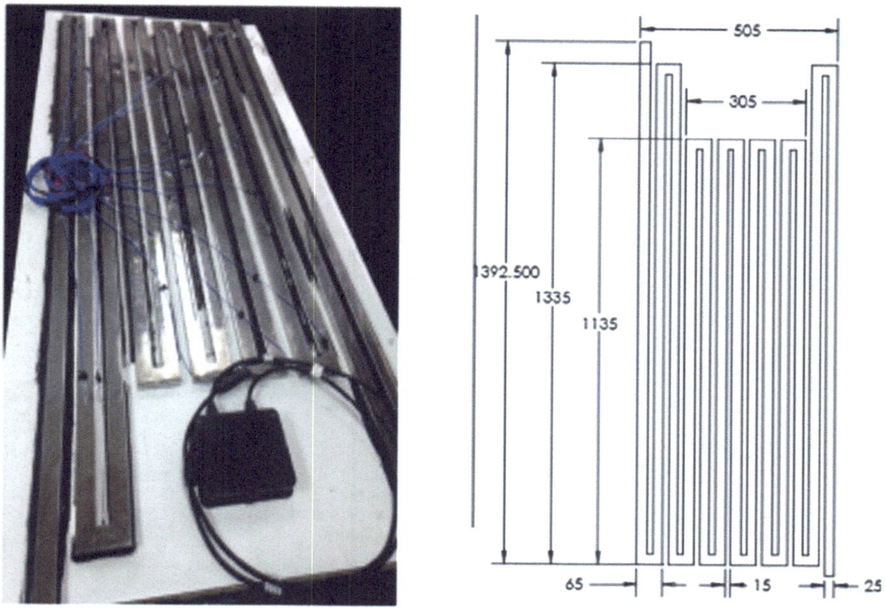

Fig. 6.3 PV/T design and collector used in [37]

Al-Waeli et al. [38] compared between three types of nanoparticles were added to water to form nanofluids (Al_2O_3, CuO, and SiC) used as coolants in a PV/T system. The researchers examined the physical thermal properties of the produced nanoparticles with their stability tests and adopted this stability for trade-off between the used nanofluids. The deposition of nanoparticles in the system will make nanofluid lose the good thermal conductivity property and thus become useless. The results of the study showed that SiC nanofluid stability was better with greater thermal conduction ability compared with Al_2O_3 and CuO nanofluids.

Al-Waeli et al. [39] used SiC-water nanofluid to evaluate the performance of a PV/T system. The study focused on the use of nano-silicon carbide as a high thermal conductivity nanoparticle and has great stability when mixed with water. The results of the study showed a significant improvement in electrical and thermal efficiency up to 24.1% and 100.19%, respectively, compared to using water as a coolant. The researchers concluded that the studied system, which uses SiC-water as a cooling liquid, has a total efficiency of 88.9% compared to the use of a single PV panel (Figs. 6.4, 6.5, and 6.6).

Al-Waeli et al. [40] added nano-silicon carbide (SiC) to the water and used the resulting nanofluid to cool the PV/T system inside the laboratory and compared the results obtained with experiments of the same system outside the laboratory. The researchers made several assumptions and used a special method to represent the solar radiation change laboratory to get closer to what actually happens. The results

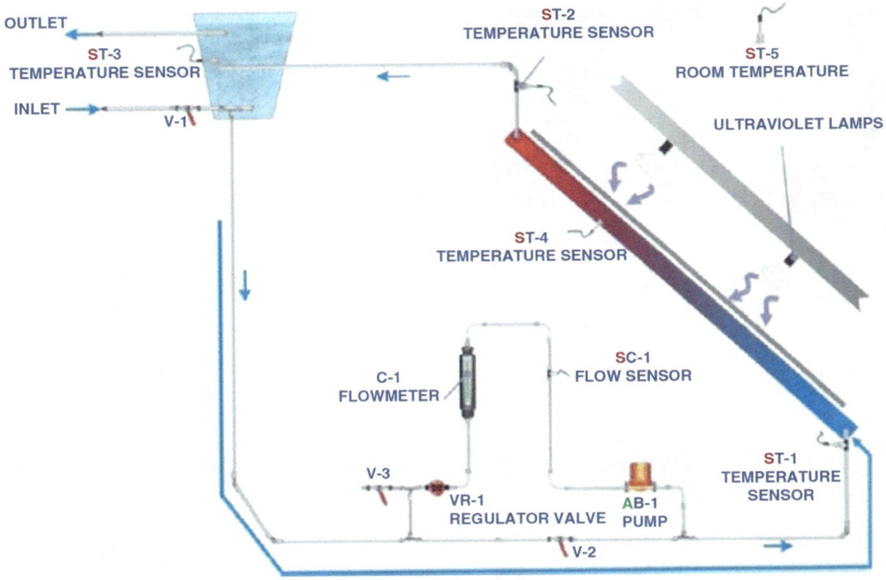

Fig. 6.4 A schematic diagram of the indoor solar simulator used by Ref. [38]

Fig. 6.5 SEM image of the (Al_2O_3, CuO, and SiC) nanoparticles used by Ref. [38]

showed a clear improvement in the outputs of the system. The researchers stressed the change in the thermal physical properties of the produced nanoparticles and found that to maintain steady stability of the fluid, an ultrasonic vibration should be subjected every 3 months to prevent the accumulation of nanoparticles and prevent their deposition on the walls of the system (Figs. 6.7 and 6.8).

Aberoumand et al. [41] investigated the efficiencies of the PV/T system cooled by nanofluid. In this study, a nanofluid contained water (base fluid) and nano-Ag were used. The researchers studied electrical, thermal, and exergy efficiency to evaluate the performance of the PV/T system. Also, they studied the effect of flow type (flux, transient, and turbulent flow) on these competencies. The results showed that the use of nanofluids in the PV/T system as a coolant instead of water enhances all the studied efficiencies generated by the system. This positive effect is increased by increasing the concentration of nanoparticles in nanofluids and by

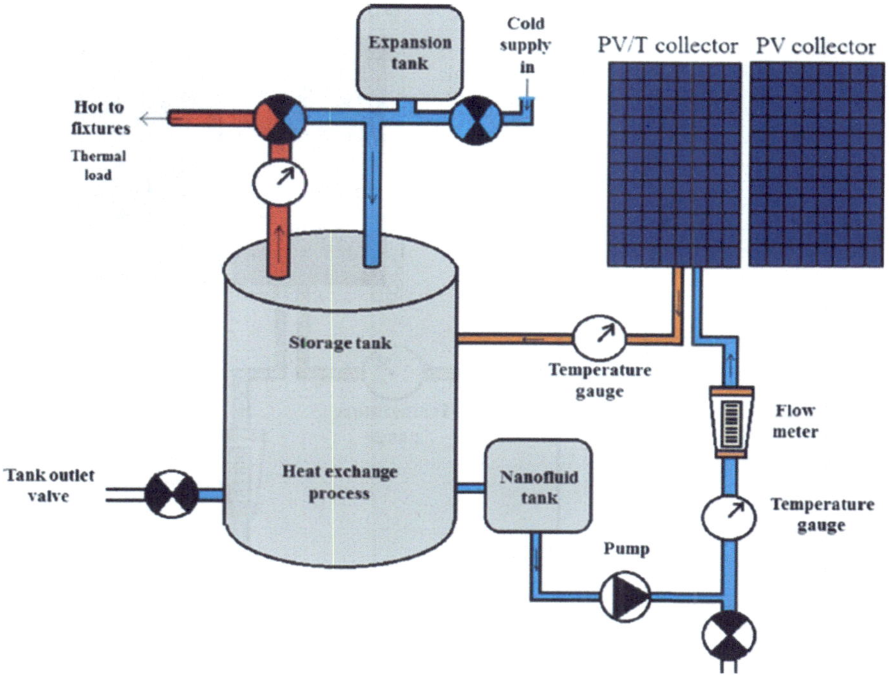

Fig. 6.6 Schematic diagram of the experimental rig used by Ref. [39]

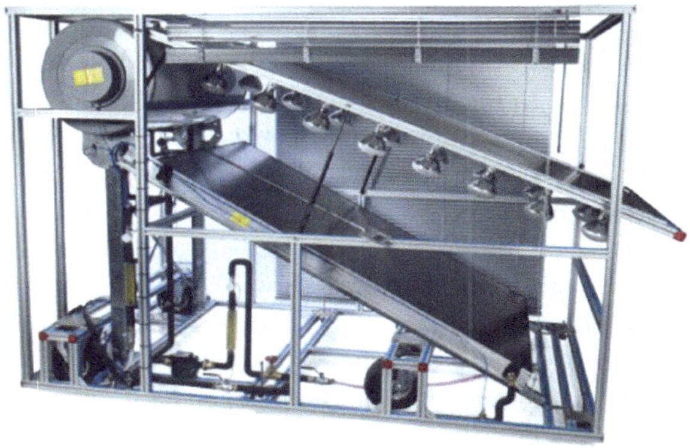

Fig. 6.7 A photo of the indoor solar simulator used in the tests conducted by Ref. [40]

working with a turbulent flow (i.e., by increasing the velocity of the nanofluid flow). Measurements showed a 35% improvement when using 4% weight of nanoparticles with water and turbulent flow compared to the non-cooling case (single PV cell) and 10% enhancement when compared to water cooling case.

The basic fluid is determined by the basic thermal physical properties of any nanotubes. Thus, Al-Waeli et al. (2019) [42] have studied three common types of liquid, namely, water, ethylene glycol, and propylene glycol, with a small percentage

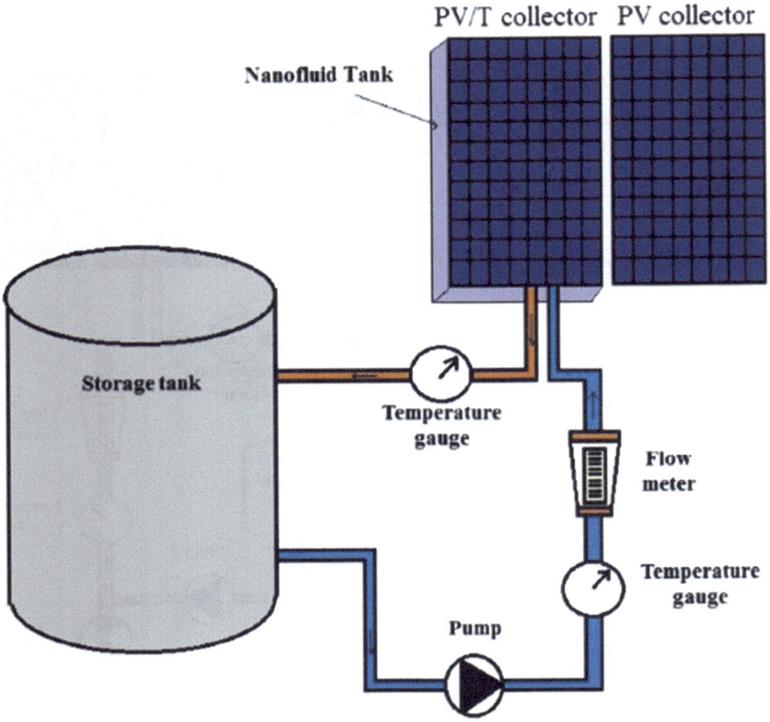

Fig. 6.8 A schematic diagram shows the outdoor system details used by Ref. [40]

of cetyl-trichromyl ammonium bromide surfactant. Ethylene glycol and propylene were added to the water by 35% each individually to create the best base fluid to work with nanoparticles for silicon carbide (nano-SiC). The density and viscosity of the basic fluids increased with the addition of SiC nanoparticles, but the most increased measured was by ethylene glycol. The results showed that the water-EG-nano-SiC solution was higher viscosity and density than the rest of the nanofluids. The thermal conductivity was close for the studied nanofluids, but its stability varied according to the basic fluid. Glycol solutions showed greater stability compared to water after using ultrasonic shaking for at least 4–6 hours.

The properties of nanofluid depend on its constituents, and because most researchers prefer to use surfactant to increase the stability of suspense produced from the basic fluid and the added nanoparticles, Al-Waeli et al. [43] studied this matter through the use of a basic nanofluid composed of water as a base liquid and silicon carbide (SiC) nanoparticles as added material to enhance the thermal conductivity. In this study, five common types of substances with surfactant activity were used to slow the deposition of the suspended nanoparticles in the emulsifier and increase its stability. The study showed that the stability period of the nano-emulsion depends largely on the type of surfactant used in addition to ultrasonic shaking time. The researchers used five substances: tannic acid + ammonia, sodium dodecyl benzene sulphonate (SDBS), cetyltrimethylammonium bromide (CTAB), sodium dodecyl sulfate (SDS), and sodium deoxycholate. When using ammonium cetyl cetyl (CTAB), the maximum stability was reached for more than 60 days with ultrasonic vibration for 6 hours, while the added dose was 0.1 mL. The maximum settling time was reached using 0.5 mL of the same surfactant above, and the period was 88 days suggesting that increasing the amount of surfactant caused better suspension of nanoparticles in the solution.

6.4.5 PV/T Cooling by Phase Change Materials (PCM) and Nanofluids

All the materials in the world are substances that change its phase as long as they turn from solid to liquid to gas and vice versa. But, certain materials store large amounts of thermal energy when they change from liquid to solid or liquid to gas. These materials, especially the solid ones, have a melting point that varies according to the substance. Applications of solar energy have different temperatures depending on the specified application. In PV/T applications, the PCM material can be used to absorb the heat out of the photovoltaic cell and store it inside it through the changing phase period. This aspect has taken a great deal of research and development during the past 10 years in particular. Some of these studies are presented in the following paragraphs.

Hassan et al. [44] used five variable phase change materials with a melting point of 25 ± 4 °C and 140–213 kJ/kg storing energy in four PV-PCM systems, two of which have a heat conductive container and the other two were isolated and one PV panel was used for comparison purposes. The results of the study showed that control of the temperature of the PV cell for the four systems depends on both the amount of PCM in the tanks and thermal conductivity of the used PCM. Of the four PV-PCM systems, an aluminum container with a thermal conductivity of 237 watts/m K has a 5 mm thickness and a 5 cm width showed the highest reduction in the photovoltaic temperature and gave better thermal regulation. The use of $CaCl_2$ salt hydrates enabled the extraction of the highest temperatures under the studied radiation levels. The temperature was controlled for a period of up to 5 hours reached when using this material with solar radiation intensity less than 1000 W/m^2.

The low thermal conductivity of the appropriate phase change materials in the PV/T applications is the main obstacle to its use in such application, as it hinders the dissipation of the heat obtained from the photovoltaic cell. Huang [45] tried to reduce the impact of this point using two types of designs for the container containing the PCM. The first design was triangular shaped, while the second was half circle. The researcher used five types of PCM in both designs. The aim of the research was to reach the best design with the PCM to keep the cell temperature at 25 °C. The results of the study showed that two studied PCMs were able to maintain the temperature of the photovoltaic cell at a temperature very close to the target value (25 °C). It is noted that the triangular reservoirs in the PV/T systems studied gave greater ease in size and control of heat for a longer period. The two provable PCMs were RT27 and RT21 as they showed the most suitable abilities to absorb the most heat from the photovoltaic cell.

Huang et al. [46] studied some properties of great importance when using PCM in PV/T systems, the thermal effect and crystal separation that occur in PCM on the performance of PV-PCM systems. The researchers studied the types of convection in the PV-PCM system and found that it arises from the effect of three basic variables: forced convection is generated due to pressure gradients induced by electrical and magnetic forces. The second factor is the natural convection forces caused by gravitational action and density gradation. The third variable is the contraction of the solidification. The process of the crystal is carried out in the material. The researchers found that the temperature control of the PV-PCM system (hardening, melting, and hardening) resulted in cavities or porosity in PCMs which in turn caused heat transfer resistance. The authors suggested as a solution to this problem using fins in a PCM container. They experimented with fins and PCM container. The results showed that the operation of the system without fin allowed achieving the temperature of 42 °C for the photovoltaic panel in 250 minutes. As for the use of fins, the rate of heat increase was decreased, and the crystal separation was interrupted during the same phase.

Maiti et al. [47] studied the performance of three PV/T systems in the laboratory and in the open air: the first system does not contain PCM, the second contained PCM only, and the third contained PCM attached to aluminum (1 mm width) pieces to achieve the best heat conduction for all PCM in the container. The indoor test results showed that the temperature of the PV module without PCM reached 90 °C in 15 minutes. In the second case PV-PCM (with PCM only), the photovoltaic module temperature reached 84 °C in less than 50 minutes; in the third case PV-PCM system (with metal parts within paraffin wax), the PV cell temperature reached 65–68 °C for about 3 hours due to the retention of the PCM heat. In the case of outdoor experiments, the researchers used the third case in these experiments and found that the maximum temperature possible for the photovoltaic module was 78–80 °C. These results showed the possibility of obtaining energy gains up to 55%.

Hassan et al. [48] installed five PV/T systems in the UAE, which is characterized by warm atmosphere most of the year. Four of these systems were different PV-PCM cells, and their impact on the system's performance was examined to make sure that they were suitable for use in the PV-PCM system. For the case of a PCM-free system, the researchers found that the peak heat of the PV cell at peak time was above 45 °C on a cloudy day and 58 °C on a sunny day. For systems containing PCM, the temperature of their photovoltaic cells dropped to 44 °C and 47 °C, respectively, under the same weather conditions. The maximum reduction in peak temperature obtained was 5 °C on cloudy day and 11 °C in clear sky conditions. The results showed that this decrease in temperature caused an increase in the output voltage difference from 1.3 to 1.7 V. The use of PCM especially after the afternoon (when the temperature is above 40 °C) was effective in regulating the PV module temperature.

Al-Waeli et al. [49] studied the use of the SiC-water-cooled PV/T system with a reservoir containing a PCM (paraffin wax), and to improve the conductivity of this material, it was mixed with nano-SiC. Experiments were conducted in the outdoor atmosphere of the city of Selangor, Malaysia. The results of the study showed an improvement in electrical efficiency (13.7%) and a high thermal efficiency increase (72%) compared to the case of a single photovoltaic cell. The system was able to reduce the temperature of the solar cell within 30 °C during the afternoon from 12:30 to 1:30 (peak period), thus improving the output of the cell as the open cell voltage increased from 11–13 to 20–21 V and the resulting power from 61.1 watts up to 120.7 watts (Figs. 6.9, 6.10, and 6.11).

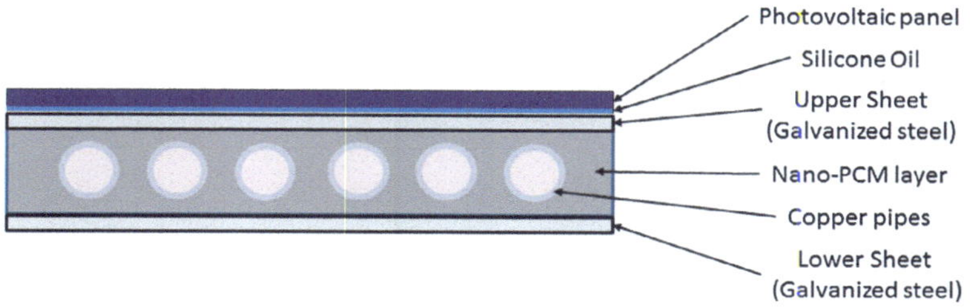

Fig. 6.9 Schematic cross-sectional diagram of the system used by Ref. [49]

Fig. 6.10 Large-scale production of SiC-paraffin for PCM container conducted by Ref. [49]

Fig. 6.11 FESEM of SiC nanoparticles and SiC-Paraffin mixture [49]

6.5 Theoretical Research

6.5.1 PV/T Cooling by Air

Hendry [50] developed a mathematical model for a flat-air PV/T system and tested the thermal and electrical performance of the system when it was cooled by air and water. The results of the study showed that the thermal efficiency of the system when the photovoltaic cells do not work is 42.5% and 40% for air and liquid, respectively. When photovoltaic cells work, thermal efficiency is reduced to 40.4% and 32.9% for both air and liquid cooling systems, respectively. The maximum electrical efficiency measured was 6.8% for the studied cells.

Cox and Raguraman [51] used a computer program to simulate flat collector air-cooled PV/T systems and analyzed their effectiveness and performance. Simulation results: Air-cooled PV/T systems are usually less efficient than liquid-cooled systems. The reason for this result can be returned to the low air heat capacity to absorb solar energy and due to low convection heat transfer coefficient of air. The study suggested a system that could be considered optimal when using air-cooled PV/T units. This system consists of PV network cells and nonselective absorption with high-flow/low-flow coating installed over photovoltaic cells.

Grag and Agarwal [52] studied (using a simulation model) the influence of design factors and the mode of operation on the performance of a PV/T system cooled by air, and this air then was used in heating. The authors also studied the effect of the use of single and double glass covers in the mentioned system and compared them to the fact that such covers are not used. The simulation demonstrated the system's dependence on its design temperatures. The addition of the glass cover will increase the heat transfer losses. For a double glass case, more heat can be stored in the double glass. The results showed that increasing the length of the compound caused an increase in the efficiency of the system, as well as with increased cell density and mass flow rate. The system efficiency was decreased by increasing the depth of the air channel. The researchers emphasized that the choice of the optimal design of the system should be focused on its cost. The addition of glass covers, increasing the length of the collectors, and the density of the cell, as well as the mass flow rate, necessitate increasing the cost of the system. So, the researchers saw that the best way to reduce the cost of the system is by reducing the cost of the life cycle of the system.

Bergene and Lovvik [21] worked on developing a mathematical model to predict the performance of an air-cooled PV/T system. The model depends on the analysis of the conversion of energy (solar to electric or thermal) and introduced the forms of heat transfer (conductivity, convection, and radiation) in the model. Results from the model showed that the overall efficiency of the PV/T system is within 60–80%.

Sopian et al. [14] developed a mathematical model that studies the stable state and analyzes the performance of air-cooled PV/T complexes using single or double channels. Mathematical results and analysis showed that the performance of the double-pass PV/T system results in better performance compared to single-pass system. In addition, the use of dual-lane cooling caused an increase in thermal, electrical, and overall efficiency.

Gu et al. (2018) [53] developed an analytical model for the usefulness of the use of the PV/T plant with an area of 10.37 m^2 at a total cost and capital cost of 448.2–537.8 €/m^2. The results of the study showed that investing in the proposed PV/T system is a profitable investment; expected profits exceed expected cost. The study also shows that the economic performance of the PV/T system in Sweden is proportional (most of the year) to the solar radiation intensity in this cold country.

Alfegi et al. [54] suggested PV/T system that has a single-pass air collector and is designed using a mathematical model to pass the PV/T cooling air through only one channel with concentrators and fins. In this model, emphasis was placed on the upper and lower sides and the edges of the heat absorbing collector to predict the thermal performance of the proposed PV/T system. Figure 6.12 shows the use of air as a cooling fluid that flows between the upper glass of the absorber collector and the bottom plates of it. The results of the study showed that when solar radiation up to 400 W/m^2 and the air mass flow rates ranging from 0.0316 to 0.09 kg/s, the thermal efficiency of the proposed system increased from 26.6% to 39.13%.

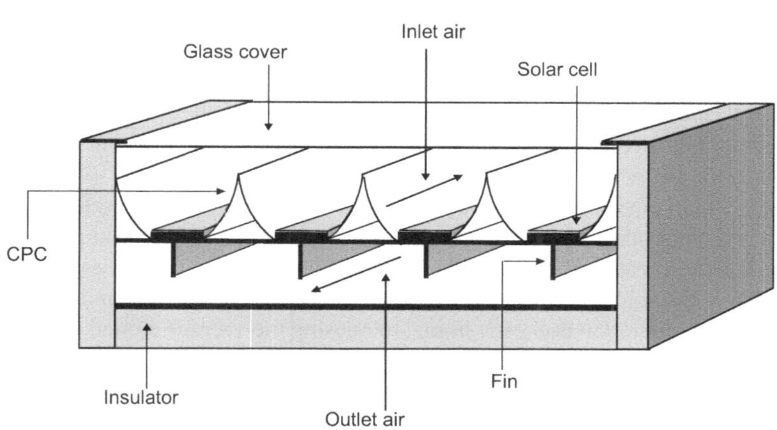

Fig. 6.12 Cross-sectional diagram of the single glass collector used in the study of Ref. [54]

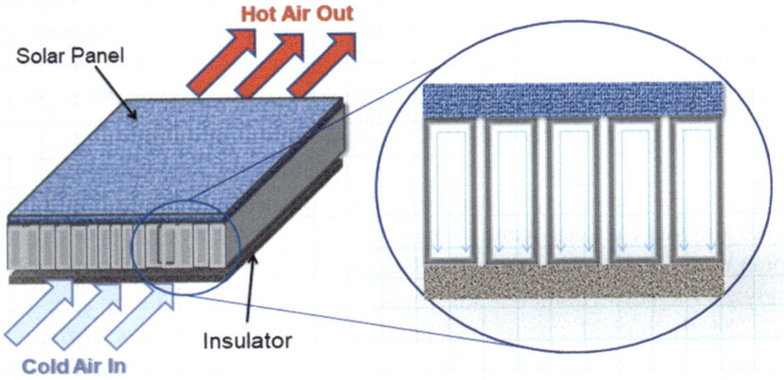

Fig. 6.13 Cross-sectional diagram of the rectangular-shaped air collector used in the study of Ref. [55]

Ibrahim et al. [55] compared the effect of air flow rates in a PV/T using air-cooling collector on thermal and electrical efficiencies. Figure 6.13 illustrates the proposed system, which consisted of a single air pass that has a rectangular tunnel collector designed and compared with a collector designed as spiral flow absorber. The rectangular tunnel is used to gain heat from the system and generate hot air and electricity. The spiral flow collector was designed to generate hot water and electricity. Absorption collectors for both systems were installed under one flat PV module. The results of the study showed that the water-cooled PV/T system with a spiral tunnel had a total efficiency of 64% and an electrical efficiency of 11% with maximum power output of 25.35 W. As for the air-cooled PV/T system with one rectangular path, the total efficiency resulting (at the same conditions) was 55%, the electrical efficiency was 10%, and the maximum generating power was 22.45 W.

6.5.2 PV/T Cooling by Water

Kalogirou [22] implemented modeling and simulation article for the performance of mixed solar photovoltaic water system using TRNSYS, adopting a transient simulation program for meteorological conditions (TMY) of Nicosia, Cyprus. The photovoltaic system consisted of a series of PV modules, battery, and inverter, while the thermal system consists of a hot water storage cylinder, a differential pump, and a differential thermostat. The results show that the hybrid system increased the annual efficiency of the solar PV system efficiency rate from 2.8% to 7.7% and covers 49% of the hot water needs at domestic home, increasing the average annual efficiency to 31.7%. Life cycle savings for the system were calculated at 790.00 Euros and the repayment time was 4.6 years.

Huang et al. [56] studied the energy balance of each component of the PV/T system; an analytical expression of the photovoltaic and water temperature was derived. The simulation predicted that total energy efficiency is around 58%, which is well consistent with the experimental value (61.3%) obtained by practical tests.

Chow [57] developed an explicit dynamic model containing seven nodes of the flat-water collector PV/T system has a single glass panel suitable for use in dynamic simulation of systems. This model, derived from the size variation formula specified in the control size and integrated with a secondary transfer program, can provide information on transient performance, including thermal/electrical current gains and efficiency and thermal conditions of many components. In addition to extending the nodal scheme to include multidimensional thermal conductivity on PV modules and adsorption, this model was able to conduct a complete energy analysis on the hybrid collector.

Dupeyrata et al. [58] evaluated the performance of a PV/T system cooled by water and the system outcomes; the authors compared their results with a single solar thermal collector and an individual PV module, using TRNSYS simulation. The comparison was made assuming the same surface area and the same climatic conditions for both systems. The results showed that the use of the PV/T system is better in specific conditions (the space available to the solar collector area) than the use of standard photovoltaic cells and solar water heaters each individually.

Using computer simulations, Tse et al. 2016 [59] studied the possibility of using a liquid-cooled PV/T system to equip an office building with its electricity and hot water needs in Hong Kong. The researchers also conducted an economic analysis to assess the possibility of installing a PV/T system instead of using PV cells and solar water heating collectors individually. The results of the study were encouraging in the case of the use of the PV/T system with a low recovery period (which included maintenance costs and spare parts). Net worth was positive for most of the year. The study demonstrated a very acceptable possibility of integrating PV/T systems into office buildings in semitropical climates.

Khelifa et al. [60] introduced a mathematical model for a water-cooled PV/T system through a sheet- and tube-type heat exchanger installed under the surface of the photovoltaic cell. In this work, the focus was on the study of fluid flow and heat transfer in the system, and they used ANSYS14 and FLUENT for this purpose. The results of the study showed that the heat extracted from the system is sufficient to heat water for domestic purposes and the electrical efficiency generated by reducing the photovoltaic cell temperature.

Bigorajski and Chwieduk [61] studied the possibility of using PV/T systems in moderately experimental climatic conditions in Patras, Greece. The researchers used a mathematical model to determine the performance of the system. The supposed system produced both electricity and heat, which is used to produce hot water enough to be used by a one-family house. The results of the study showed that the operation of the system is at its best during the months of December and January, as the electricity generated more than the thermal energy produced from the PV/T system. During the summer, when the cooling fluid temperatures rise to a high level, it caused low electrical efficiency.

6.5.3 PV/T Cooling by Heat Pipes

Guoying et al. [62] developed the PV/T system by adding a heat pump to the solar cell as shown in Fig. 6.14. For this purpose, the researchers used flat aluminum tubes with multiple outlets instead of conventional round copper tubes. This system was used to generate electricity and heat water for household purposes and heating the rooms. The simulation results of the system showed that it achieved higher COP rate by about 7% when it is used to heat the water in the summer and has a higher thermal efficiency of up to 6% compared to the traditional PV/T-HP system. The simulation results showed that the proposed system could generate electricity more efficiently with 150 L of water heating up to 50 °C throughout the year in the Chinese cities of Nanjing and Hong Kong. The control of the pump speed and fixing it at three different levels according to the different seasons caused the system COP to be improved significantly in the summer months, and it increased the heat produced, which almost supplied most of the heating load required in winter.

Ying et al. [63] suggested a PV/T system that uses a heat tube to absorb excess heat from a photovoltaic cell evenly over each photovoltaic solar cell space. The researchers built a theoretical model for heat transfer analysis to predict the thermal and electrical performance of the proposed system. In this model, the temperature of the water inside, the flow rate of the water mass, the factor of filling the solar cells, and the coefficient of heat loss were considered based on the first and second laws of thermodynamics. The results showed that the thermal and electrical efficiency of the studied system was 63.65% and 8.45%, respectively. The study confirmed the possibility of applying the PV/T hybrid tube system.

Gang et al. [64] investigated a new PV/T system that can provide both thermal and electrical powers capable of working in cold areas without freezing, conversely to the systems that use water as a coolant. The researchers built a dynamic model of

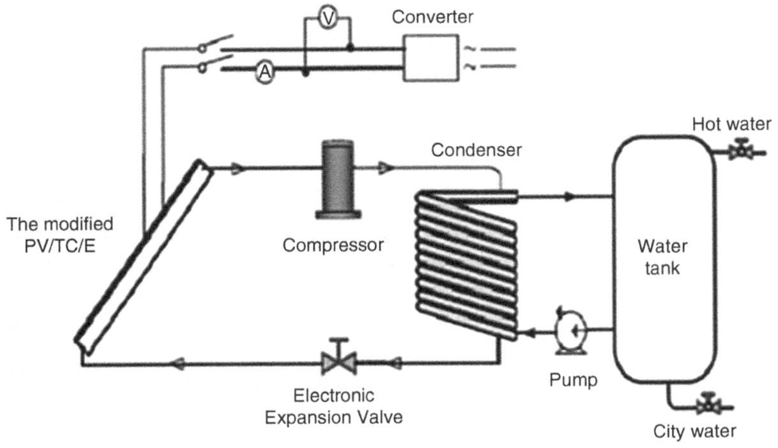

Fig. 6.14 The proposed PV/T-HP schematic diagram [62]

the heat pipe PV/T system with the creation of a test platform. Practical experiments were conducted to validate the theoretical simulation results. The researchers studied the proposed system performance in multiple parametric conditions, such as different water flow rates, the photovoltaic fill factor for the collector, heat pipe, and absorbent coatings. The results of the studies showed promising results for this type of system.

Zhang et al. [65] studied the effect of reservoir size and tilt angle PV/T system composed of photovoltaic cells cooled by heat pipes and calculated the generated electrical energy and hot water produced. The researchers used TRNSYS to simulate the proposed PV/T system and compared the theoretical results with experimental data. The results of the study showed that the generated electricity and hot water are increased by increasing the size of the circular reservoir. The highest overall efficiency of the proposed system was 67.5% with an 80 L tank size.

Hui et al. [66] have built an integrated system of photovoltaic cells and heat pipes (BiHP-PVT) to be used in electricity generation and water heating. The researchers used a dynamic model of the HP-PV/T system and compared experimental results to evaluate how the system's annual performance was evaluated under the conditions of the Hong Kong climate. The results of the theoretical study showed that the annual efficiency of water heating was 35% while for generating electricity efficiency was up to 10%. The total electrical power saved was 315 kWh/year per unit surface area. The studied system showed a high ability to adapt to the climate of the studied area in addition to its acceptable performance.

6.5.4 PV/T Cooling by Nanofluids

Many researchers studied the potential of enhancing the thermal conductivity of PV/T systems by cooling them with nanofluids and using mathematical calculations and computer simulations for this purpose. From the latest of these studies, we have chosen the following.

Khandari et al. [67] studied the effect of using nanofluid rather than water in cooling a PV/T system. Nanofluid enhances thermal conductivity compared to water. Researchers used computational fluid dynamics (CFD) to simulate the use of three types of liquids: pure water, nanofluid (Ag-water), and nanofluid (alumina-water). During the study, they focused on the effect of the concentration of nanoparticles added in volumetric proportions to water on the performance of the studied PV/T system. The results of the study found that the increase of nanoparticle proportion in the base fluid increases the efficiency and heat transfer. The heat transfer coefficient of alumina-water nanofluid has increased by 12%, whereas for Ag-water, it has increased to 43% compared to water. The produced electricity has increased by 8–10% using alumina-water compared to pure water. The increase in Ag-water case was up to 28–45% compared to pure water.

Rajab et al. [68] developed a numerical model in FORTRAN to study the effect of using nanofluids as a heat transfer agent in a PV/T system and compared the numerical results using a pilot empirical model. Practical experiments have been carried out by changing the type of nanoparticles used to produce nanoparticles and the effect of this change in the type of nanoparticle on the performance of the electrical and thermal system. To assess the impact of climatic conditions on the designed PV/T system, experiments were conducted in three different locations in Lyon-France, Mashhad-Iran, and Monastir-Tunisia. The results of the study showed that the use of pure water as a base fluid caused the best performance compared to using ethylene glycol. The best thermal and electrical performance of the system was achieved when using copper/water as a nanofluid. Also, increasing the proportion of nanoparticles in suspension from 0% to 4% enhanced thermal conductivity. The study showed that the best electrical and thermal performance was achieved in the city of Monastir-Tunisia (its climate is a model of semiarid cold areas).

Adriana [69] investigated the thermophysical properties of three types of oxide-based nanoparticles. The researcher used the numerical study to study the properties of Al_2O_3, TiO_2, and SiO_2 added to water. The results of the study showed a change in all thermophysical properties of all studied nanoparticles due to addition of nanoparticles, and thermal conductivity was increased by at least 12%. The increased thermal conductivity of nanofluids resulted from the increased convection heat transfer coefficient, increasing the Reynolds number, which improved significantly by increasing the concentration of nanoparticles.

6.5.5 PV/T Cooling Using PCM and Nanofluids

The researchers used numerical analysis and computer simulation to arrive at confirmation of some practical results on the use of BMS in the BFT systems. Most of these studies showed promising results and showed a significant improvement in the cooling of the photovoltaic cell and an apparent rise in its performance. Here are some of these studies.

Huang et al. [70] were considered one of the first efforts to develop a numerical model of the PV-PCM system (heat-simulating system use (PV-PCM)), which presented a number of models, including a three-dimensional digital model. The results of this model were compared with the results of the 2D model to verify the numerical results. The researchers compared the methods of heat transfer from the PV-PCM system, from the adiabatic edges, and when fins were introduced inside the PCM container. They also used solar radiation of 750 W/m^2 and the atmospheric temperature of 20 °C in the theoretical simulation. In the practical side of the study, the researchers selected the paraffin type RT25 because it is nontoxic and does not react chemically with most substances. The results of the study showed the importance of the heat transfer effect of the side edges of the system as well as the importance of using pin fins to promote heat transfer to PCM but at the same time restricted the heat natural convection. The researchers found that this dilemma can be reduced through the use of a suitable design for the direction of the fins.

Cellura et al. [71] studied the effect of adding a tank containing PCM to an accompanying photovoltaic module and the effect of this addition to the heat storage system in hot weather. In this study, COMSOL Multiphysics was used to achieve a theoretical analysis of the behavior of PV-PCM thermal system. The results of the theoretical study showed that the average efficiency of converting solar radiation into electricity for PV system without the use of PCM was approximately 12%, while for PV-PCM system status, this value increased to 26%.

Hendrix and Sark [72] studied the ability to enhance the performance of the PV/T-PCM system for a full year. Authors used for this purpose numerical modeling using MathWorks MATLAB 2009 and conducted a practical study in two different locations: Utrecht-Netherlands (selected as reference for mild climate) and Malaga-Spain (selected as reference for hot climate). The aim of choosing two different locations is to monitor the behavior of the PV-PCM system under two different climate conditions for a full year. The results of the numerical and analytical study showed that the energy gain of the PV-PCM system during summer for both the moderate and hot sites is the same, but in the winter the studied system in the hot zone (Malaga) shows a relatively high energy. Numerical results (assuming that the heat storage capacity of PCM is unlimited) found that the energy gain of the PV-PCM system reached 162 kWh. The actual capacity of the Malaga and Utrecht systems was 3.3 and 1.8 kWh, respectively. The researchers found that this difference can be traced back to the fact that the amount of heat storage in practice is very limited. However, the use of PCM was able to reduce the rise in the temperature of the photovoltaic module during the peak hour of the day to ensure a constant temperature throughout the day. The researchers also studied the effect of another factor on energy gain, the thickness of the PCM layer used in the system. The study showed that heat absorption affects the energy gain of the photovoltaic cell in the PVM-PCM system.

Brano et al. [73] investigated the utility of the use of PV/T system containing PCM numerically and experimentally. The researchers used the specific difference method to build an energy balance algorithm in the supposed PV/T system. This algorithm was used to analyze four cases of phase changing states, when there was no change in phase, phase change status, just starting phase change condition, and phase change phase termination condition. To validate the theoretical results, the researchers constructed a PV/T system and performed experiments using paraffin wax as PCM. The results of the numerical study of the four cases showed a good agreement with the experimental results. The researchers suggested that the proposed system would discharge excess heat overnight in the PV-PCM system. The numerical and practical results of the study showed poor performance of the paraffins (according to authors) as a thermal storage medium. The reason for this weakness is that its thermal spread (thermal conductivity) for the paraffins reduces its ability to get rid of excess heat at night.

Brano et al. [74] conducted a practical and theoretical study in external conditions using a PV/T system employing a PCM type RT-27 with low melting point of 26 °C. Practical experiments were conducted to confirm the numerical study results.

The authors have come to the conclusion that the model prepared by them is capable of predicting a high degree of accuracy at PV temperature in the PV-PCM system as well as other performance parameters for both sunny and foggy days.

Machniewicz et al. [75] used a dynamic simulation to study the effect of the PCM transition temperature on the electrical and thermal efficiency of the PV/T system. The researchers used ESP-r program for this purpose. The researchers assumed specific measurements of the structure of the PCM container. They considered this container structure of aluminum and its shape parallel to rectangles with a thickness of 2 cm, and inside it there are aluminum sheets. They also assumed the use of four types of paraffin wax as PCMs characterized by different melting points. The results of the study showed that paraffin wax with a very low melting point remains melted during most of the operation period of the day, while for the other types of paraffins (with high melting points), the change in phase did not exist, and it remained solid during the day. The researchers found that there is no specified temperature that could not be determined as an optimal melting point for the entire year, but the best and most successful results were when using a PCM with a melting point between 18 and 25 °C.

Al-Waeli et al. [76] developed a mathematical model that simulated a PV/T system for different configurations of air, water, nanofluid, and phase change materials (PCM) and a mixture of cooling methods and studied their respective effects on improving electrical and thermal efficiency. The researchers also conducted practical experiments to validate the results of the mathematical model. The results of the comparison showed a clear convergence in the results of the proposed mathematical model and its potential to simulate and match the experimental results. The convergence between the results of the experimental and numerical studies can be ascertained by reviewing the results when PV/T-PCM cooled by nanofluid system such as electrical efficiency, which were 13.7% and 13.2%, and thermal efficiency 72% and 71.3% for both theoretical and experimental results, respectively, indicating a good consistency and agreement.

6.6 Current Practical Applications of PV/T

Although PV/T technology is a very modern technology and is still in the R&D process, some of its applications have been introduced to the market as commercial products, and there are some engineering projects based on PV/T applications. From these commercial applications, we will mention what we currently have.

Germany's Grammer Solar GmbH has developed an air-cooled PV/T system called TWINSOLAR. The hot air gained by the system is used to heat the ventilation air in buildings. The system has an area of absorption ranging from 1.3 to 12.5 m^2 and according to the desired design. The parts of the units can be installed vertically or horizontally on the surface. The system is constructed facing the south, southeast, or southwest as long as the location is the northern half of the Earth. It was noticed that the temperature of air outside the system rose to 40°C when the maximum solar irradiance reached (700 W/m^2), which means the system converted up to 70% of the solar energy; the cooling air is heated and then transferred into the building, as shown in Figs. 6.15 and 6.16 [77].

SolarVenti PV/T units (Fig. 6.17) are currently being used in Denmark to provide necessary supplementary assistance in air ventilation and heating as well as in dehumidification. Large-capacity variants of SolarVenti can produce heat energy suitable for processing a large amount of hot air depending on buoyancy strength. The system absorbs heat energy directly from solar radiation and is currently used as a supplementary part of heating systems in buildings, whether residential or commercial [78].

The Canadian "Conserval Engineering Inc." has introduced SolarWall and Solar Duct products to the surface to heat the solar air and then push the air with ventilating air to warm the building. These systems can also rotate hot air through walls or ceilings to heat buildings or take advantage of including in the drying operations of agricultural products. The company's studies show that SolarWall, which has a PV/T system, has a much lower recovery period than the PV system, which means

Fig. 6.15 TWINSOLAR air heater produced by [77]

Fig. 6.16 TWINSOLAR PV/T air cooling system produced by [77]

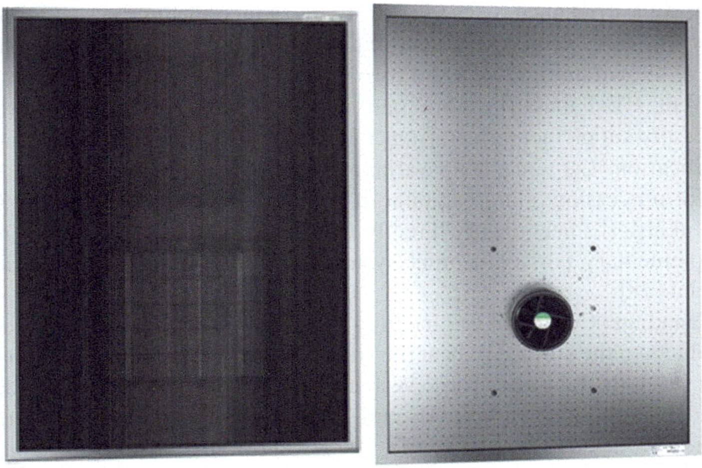

Fig. 6.17 SolarVenti PV/T air-cooled units [78]

Fig. 6.18 A diagram represents how the SolarWall PV/T system works [79]

it is more cost-effective and can produce up to 400% thermal and electric energy that can be used compared to using the PV system only (Figs. 6.18, 6.19, 6.20, 6.21, and 6.22). The PV/T system manufactured by the abovementioned company is installed on the surface and has a total operating efficiency of more than 50%. The PV shelving system is used to maximize the heat extracted [79].

"Millennium Electric Co." has installed 26 megawatt PV stations in different parts of the world. The company has the ability to manufacture highly efficient mono/poly PV panels. This company has manufactured many PV/T technologies by combining photovoltaic cells with solar thermal heaters to produce comprehensive renewable energy systems (Figs. 6.23 and 6.24). The production systems of this company are concerned with the management of electricity and storage of thermal energy and can work with the network or outside the network. The figure shows one of the water-cooled PV/T systems produced by the company above [80].

Fig. 6.19 A photo of one of the SolarWall systems [79]

Fig. 6.20 A photo of a practical SolarWall system used in heating spaces [79]

Fig. 6.21 A schematic diagram represents the facilities of SolarWall PV/T system when used in poultry fields [79]

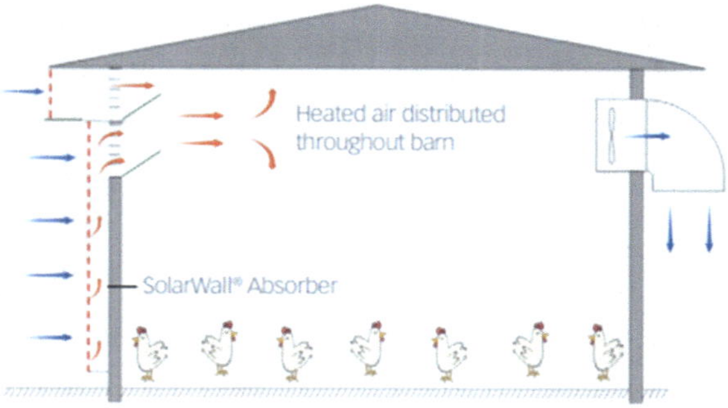

The Swedish company "Absolicon" is engaged in the development and manufacture of solar energy systems, including PV/T systems (Figs. 6.25 and 6.26). The company also provides solar condensers to increase the electrical and thermal energy produced, and its systems provide steam from the system itself. The solar condensate systems produced by this company generate higher temperatures than

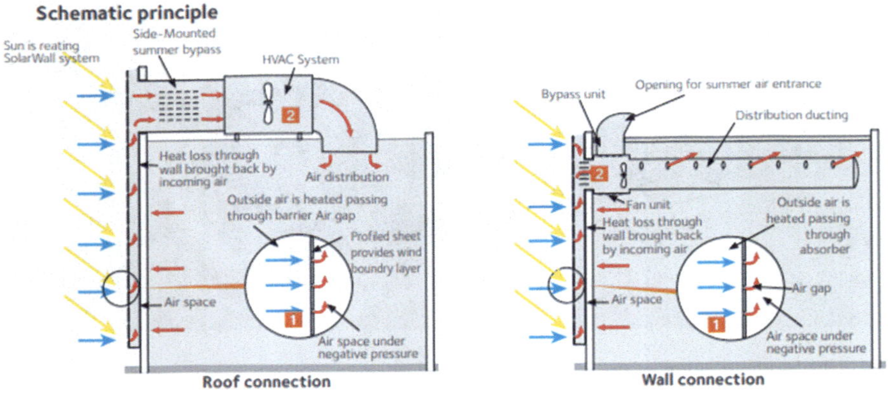

Fig. 6.22 Schematic diagrams represent the SolarWall PV/T systems instillation methods [79]

Fig. 6.23 Water-cooled PV/T system produced by "Millennium Electric Co." [80]

normal solar heating systems with a clear reduction in the size of storage tanks when working at high temperatures, so the requirements of the system of installation pipes are lower. The collector of solar energy is constantly directed toward the Sun, so the distribution of heat is done more evenly throughout the year than the normal solar collector. The systems of this company allow stopping the operation of

Fig. 6.24 Water-cooled PV/T system produced by "Millennium Electric Co." [80]

Fig. 6.25 The PV/T system manufactured by "Absolicon Co." and established at Hemab Energy Park

solar collectors if not necessary through a control system. Solar collectors are often installed on the roof of the building and are protected from wind. These solar systems produce heat energy even during the cold winter [81].

The International Energy Agency (IEA) established the Solar Heating and Cooling Cooperation Program (SHC TCP) in 1977, which is complemented by international efforts of experts from member countries of the Agency and the European Union. This program was established to accelerate and develop solar energy technologies in the field of heating and cooling, strengthening national research and development programs, and to save money and time. There are many obstacles that reduce the global public's desire to use solar energy. To overcome these obstacles and spread through the global heating and cooling market, IEA has

Fig. 6.26 The PV/T system manufactured and installed by "Absolicon Co."

Fig. 6.27 The newly installed Solarus energy system on the roof of the 15 on Orange Hotel, in Cape Town

chosen to focus globally on three axes: awareness and enlightenment of users and decision-makers, expanding the solar thermal market, hardware, materials, and designs. Until recently, solar thermal energy is often excluded from official renewable energy statistics. The SHC program has been developed to compare solar thermal energy with other renewable sources of energy (Figs. 6.27, 6.28, 6.29, and 6.30). Solar energy comes second only to wind energy in meeting global energy demands. The diagram generated by the program shows the annual energy output of TWh and cumulative capacity (GW) by the end of 2005 (Fig. 6.26) [82].

Fig. 6.28 64 PV/T collectors providing heat and electricity for 162 people in a home for the elderly in Málaga, Spain

Fig. 6.29 PV/T combined in a single product easy to integrate in roofs or facades and even under concentration

PV/T systems are also used in crop drying application. Food is one of the basic needs of man for his life and is an integral part of human energy, which he needs to accomplish almost every activity. The need for food security and sustainability is essential to guarantee the future of any people or country, and meeting the demand for food for the citizens is the basis of the rule of any party. Food security is ensured by increasing crop production or reducing postharvest losses or both. Modern civilization has been able to provide distinctive methods to increase crop productivity and improve the agricultural land needed. It remains very important to reduce postharvest losses during crop production, harvest, or storage to a minimum. Dehydration of crops is one of the traditional and effective ways to reduce this type of loss. During this process, removing excess moisture from the crops is done, thus increasing the safe storage period for a longer period while preserving the quality of the products [83].

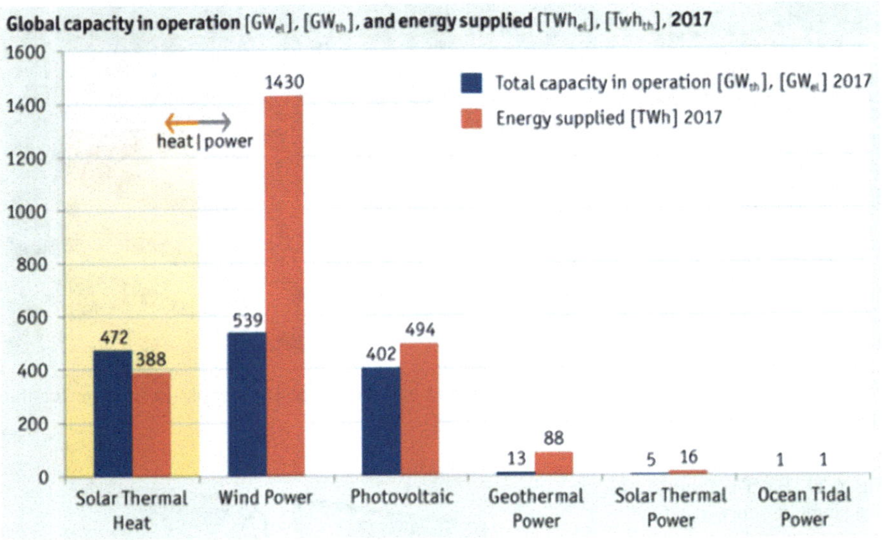

Fig. 6.30 The annual energy output in TWh for renewable energies as supplied by IEA SHC Solar Heat Worldwide 2018

In recent years, the R&D that began on air-cooled PV/T systems, and the possibility of increasing the utilization of the conversion of solar radiation to electricity and heat of by these systems with construction it on the roofs of buildings and the possibilities of this application to provide hot air or hot water at high temperatures made researchers study multiple aspects of this application [84, 85]. One of these successful applications is the drying of crops. Some researchers have studied the use of PV/T systems in drying with forced convection. Farkas et al. [86] used a PV panel to operate an air-cooling fan; the fan was installed on the front side of the drying system at an adjustable angle that is fitted with the solar radiation azimuth angle at different periods of the year. Mumba [87] enables the drying of grains by using a photovoltaic panel running a DC fan and reducing the drying time by up to 70% compared to the case of leaving the crops to dry under the Sun. Ref. [88] found that the PV/T systems can be used for space heating applications, water heating, lighting, electricity processing, etc.

Tiwari and M. S. Sodha [89] attempted to produce a dryer using a combined PV/T system with greenhouse structures for use in crop drying applications. The DC power unit provided a DC fan to generate the required heat and pump it into the greenhouse construction, which helps accelerate the drying of crops. The authors concluded that this kind of solar dryers is very practical in regions where there are no power grids.

The SERI (Solar Energy Research Institute at UKM-Malaysia) has conducted several successful experiments to produce air dryers used in tropical and subtropical climates, which are based primarily on PV/T systems. Figure 6.31 shows a schematic diagram of one of these systems. In the structure of the PV/T module used, the design made was the V-Groove (Fig. 6.32), rectangular tunnel made of aluminum (Fig. 6.33), and honey comb collectors (Fig. 6.34) (Table 6.1).

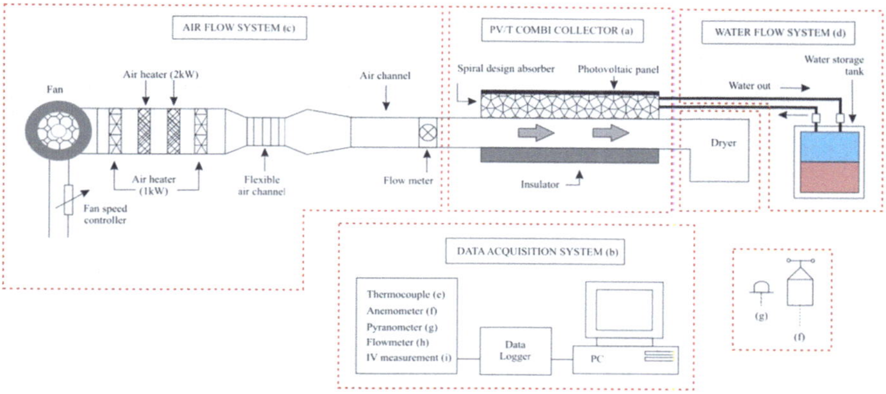

Fig. 6.31 Schematic diagram of an air dryer system using a PV/T system to generate hot air

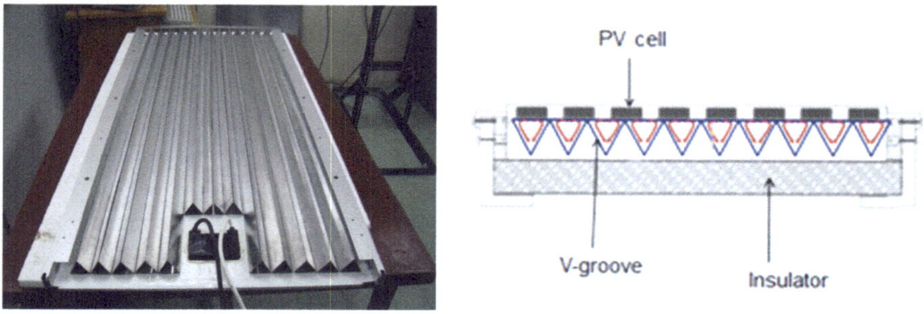

Fig. 6.32 SERI V-Groove PV/T system air dryer [90]

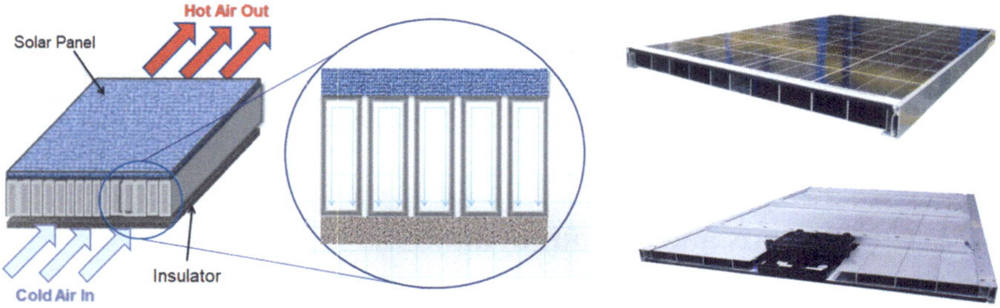

Fig. 6.33 SERI rectangular tunnel made of aluminum PV/T system air dryer [90]

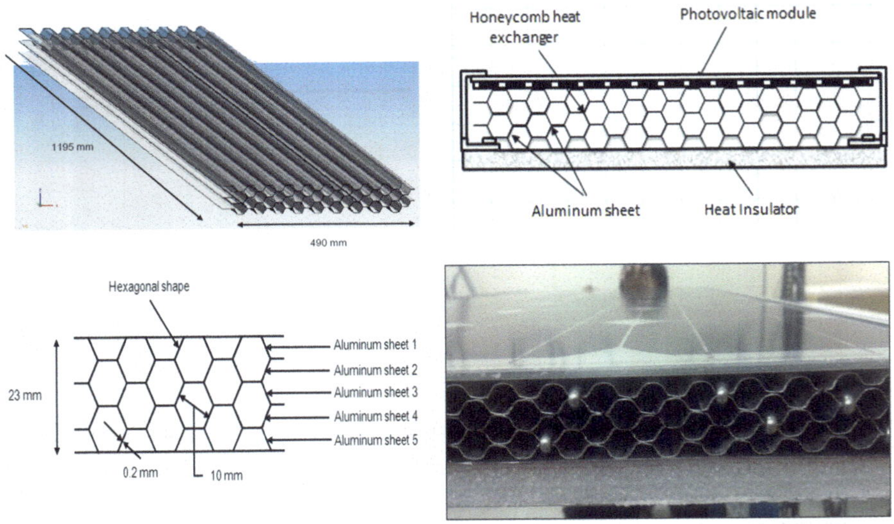

Fig. 6.34 SERI honey comb PV/T system air dryer [90]

Table 6.1 PV/T companies and projects worldwide [91]

Company name	Project/product name	Type of PV/T technology	Location of installation(s)	Size (kW or kWh/m^2)
Millennium electric	Multi solar system	Water-based	Israel	1.5 kWh/m^2
SolarWerk	Spectrum	Water-based PV/T	Germany	–
SolarWatt	MultiSolar	Water-based PV/T	Germany	–
Aidt Miljø A/S	SolarVenti	Air-based PV/T & BIPV/T	Denmark	–
Conserval engineering Inc.	PV SolarWall	Air-based BIPV/T	Canada	–
Grammer solar	TWINSOLAR compact	Air-based PV/T	Germany	50–250 kW

6.7 Conclusions and Recommendations

This chapter revisits the classifications of working fluid-based PV/T systems with systematic review based on the type of study, practical and theoretical studies. Moreover, the chapter discussed current practical applications of PV/T systems in the field.

Conclusions

1. PV/T systems are a suitable energy source for maximized energy generation under limited areas such as residential rooftops.
2. Research in PV/T systems show their viability as grid-connected and building-integrated configurations, in addition to offering a uniform appearance, unlike separate PV arrays and solar thermal collectors.

Recommendations

1. To consider the market potential of PV/T systems across several the service and industrial sectors
2. To perform case studies for different industrial-based PV/T-assisted systems, e.g., water desalination, with respect to novel designs and their cost-effectiveness

References

1. Energy – Consumption "Consumption by fuel, 1965–2008" (XLS), Statistical Review of World Energy 2009, BP. 31 July 2006. Accessed 24 Oct 2009
2. *Global Energy Review* (Enerdata Publication, 2009)
3. Global warming (2011), http://en.wikipedia.org/wiki/Globalwarming. Accessed 2 Nov 2011
4. "Renewables in global energy supply: an IEA facts sheet", IEA/OECD; 2007.
5. Solar heating and cooling for a sustainable energy future in Europe (Revised), European solar thermal technology platform (ESTTP) (2009), http://www.estif.org/fileadmin/estif/content/projects/downloads/ESTTPSRARevisedVersion.pdf
6. X. Zhang, X. Zhao, S. Smith, J. Xu, X. Yu, Review of R&D progress and practical application of the solar photovoltaic/thermal (PV/T) technologies. Renew. Sustain. Energy Rev. **16**, 599–617 (2012)
7. Solar thermal action plan for Europe: heating and cooling from the Sun, European Solar Thermal Industry Federation (ESTIF), http://www.estif.org/fileadmin/estif/content/policies/STAP/SolarThermalActionPlan2007A4.pdf (2007)
8. Technology Roadmap-Solar photovoltaic energy, International Energy Agency, http://www.iea-pvps.org (2010)
9. H.M.S. Al-Maamary, H.A. Kazem, M.T. Chaichan, Changing the energy profile of the GCC states: A review. Int. J. Appl. Eng. Res. **11**(3), 1980–1988 (2016)
10. A.H.A. Al-Waeli, K. Sopian, H.A. Kazem, M.T. Chaichan, PV/T (photovoltaic/thermal): status and future prospects. Renew. Sustain. Energy Rev. **77**, 109–130 (2017)
11. R. Komp, T. Reeser, Design, construction and operation of a PV/Hot air hybrid energy system, in *ISES Solar World Congress*, 1987
12. B. Moshtegh, M. Sandbergy, Investigation of fluid flow and heat transfer in a vertical channel heated from one side by PV elements, Part I–numerical study. Renew. Energy **8**, 248–253 (1996)
13. M. Sandbergy, B. Moshtegh, Investigation of fluid flow and heat transfer in a vertical channel heated from one side by PV elements, Part II – experimental study. Renew. Energy **8**, 254–258 (1996)
14. K.S. Sopian, H.T. Yigit, H.T. Liu, S. Kakac, T.N. Veziroglu, Performance analysis of photovoltaic/thermal air heaters. Energ. Conver. Manage. **37**(11), 1657–1670 (1996)
15. J. Ji, T.-T. Chow, W. He, Dynamic performance of hybrid photovoltaic/thermal collector wall in Hong Kong. Build. Environ. **38**, 1327–1334 (2003)

16. J.K. Tonui, Y. Tripanagnostopoulos, Performance improvement of PV/T solar collectors with natural air flow operation. Sol. Energy **82**, 1–12 (2008)
17. A. Shahsavar, M. Ameri, Experimental investigation and modeling of a direct coupled PV/T air collector. Sol. Energy **84**, 1938–1958 (2010)
18. B.J. Huang, Performance rating method of thermosyphon solar water heaters. Sol. Energy **50**, 435–440 (1993)
19. R.K. Agarwal, H.P. Garg, Study of a photovoltaic–thermal system – thermosyphonic solar water heater combined with solar cells. Energ. Conver. Manage. **35**(7), 605–620 (1994)
20. H.P. Grag, R.K. Agarwal, Some aspects of a PV/T collector/forced circulation flat-plat solar water heater with solar cells. Energ. Conver. Manage. **36**, 87–99 (1995)
21. T. Bergene, O.M. Lovvik, Model calculations on a flat plate solar heat collector with integrated solar cells. Sol. Energy **55**(6), 453–462 (1995)
22. S.A. Kalogirou, Use of TRNSYS for modelling and simulation of a hybrid PV–thermal solar system for Cyprus. Renew. Energy **23**, 247–260 (2001)
23. M.D. Bazilian, H. Kamalanathan, D.K. Prasad, Thermographic analysis of a building integrated photovoltaic system. Renew. Energy **26**, 449–461 (2002)
24. T.T. Chow, W. He, J. Ji, Hybrid photovoltaic-thermosyphon water heating system for residential application. Sol. Energy **80**, 298–306 (2006)
25. T.T. Chow, W. He, J. Ji, An experimental study of facade-integrated photovoltaic/water-heating system. Appl. Therm. Eng. **27**, 37–45 (2007)
26. S. Dubey, A.A.O. Tay, Testing of two different types of photovoltaic–thermal (PVT) modules with heat flow pattern under tropical climatic conditions. Energy Sustain. Dev. **17**, 1–12 (2013)
27. H. Jouhara, M. Szulgowska-Zgrzywa, M.A. Sayegh, J. Milko, J. Danielewicz, T.K. Nannou, S.P. Lester, The performance of a heat pipe based solar PV/T roof collector and its potential contribution in district heating applications. Energy **136**, 117–125 (2017)
28. M. Nishikawa, T. Sone, S. Ito, A heat pump using solar hybrid panels as the evaporator, in *ISES Solar World Congress*, 1993
29. X. Zhao, X. Zhang, S.B. Riffat, X. Su, Theoretical investigation of a novel PV/e roof module for heat pump operation. Energ. Conver. Manage. **52**, 603–614 (2011)
30. M. Moradgholi, S.M. Nowee, I. Abrishamchi, Application of heat pipe in an experimental investigation on a novel photovoltaic/thermal (PV/T) system. Sol. Energy **107**, 82–88 (2014)
31. T. Yousefi, F. Veysi, E. Shojaeizadeh, S. Zinadini, An experimental investigation on the effect of Al_2O_3–H_2O nanofluid on the efficiency of flat-plate solar collectors. Renew. Energy **39**, 293–298 (2012)
32. N. Karami, M. Rahimi, Heat transfer enhancement in a PV cell using Boehmite nanofluid. Energ. Conver. Manage. **86**, 275–285 (2014)
33. M. Sardarabadi, M. Passandideh-Fard, S.Z. Heris, Experimental investigation of the effects of silica/water nanofluid on PV/T (photovoltaic thermal units). Energy **66**, 264–272 (2014)
34. S. Hassani, R. Saidur, S. Mekhilef, R.A. Taylor, Environmental and exergy benefit of nanofluid-based hybrid PV/T systems. Energ. Conver. Manage. **123**(1), 431–444 (2016)
35. D. Jing, Y. Hu, M. Liu, J. Wei, L. Guo, Preparation of highly dispersed nanofluid and CFD study of its utilization in a concentrating PV/T system. Sol. Energy **112**, 30–40 (2015)
36. M. Ghadiri, M. Sardarabadi, M. Pasandideh-fard, A.J. Moghadam, Experimental investigation of a PVT system performance using nano ferrofluids. Energ. Conver. Manage. **103**, 468–476 (2015)
37. A.N. Al-Shamani, K. Sopian, S. Mat, H.A. Hasan, A.M. Abed, M.H. Ruslan, Experimental studies of rectangular tube absorber PV thermal collector with various types of nanofluids under the tropical climate conditions. Energ. Conver. Manage. **124**, 528–542 (2016)
38. A.H.A. Al-Waeli, M.T. Chaichan, H.A. Kazem, K. Sopian, Comparative study to use nano-(Al_2O_3, CuO, and SiC) with water to enhance photovoltaic thermal PV/T collectors. Energ. Conver. Manage. **148**(15), 963–973 (2017)
39. A.H.A. Al-Waeli, K. Sopian, M.T. Chaichan, et al., An experimental investigation of SiC nanofluid as a base-fluid for a photovoltaic thermal PV/T system. Energ. Conver. Manage. **142**, 547–558 (2017)

40. A.H.A. Al-Waeli, M.T. Chaichan, H.A. Kazem, K. Sopian, A. Ibrahim, S. Mat, M.H. Ruslan, Comparison study of indoor/outdoor experiments of a photovoltaic thermal PV/T system containing SiC nanofluid as a coolant. Energy **151**, 33e44 (2018)
41. S. Aberoumand, S. Ghamari, B. Shabani, Energy and exergy analysis of a photovoltaic thermal (PV/T) system using nanofluids: An experimental study. Sol. Energy **165**, 167–177 (2018)
42. A.H.A. Al-Waeli, M.T. Chaichan, K. Sopian, H.A. Kazem, Influence of the base fluid on the thermo-physical properties of PV/T nanofluids with surfactant. Case Stud. Therm. Eng. **13**, 100340 (2019)
43. A.H.A. Al-Waeli, M.T. Chaichan, H.A. Kazem, K. Sopian, Evaluation and analysis of nanofluid and surfactant impact on photovoltaic-thermal systems. Case Stud. Therm. Eng. **13**, 100392 (2019)
44. A. Hasan, S.J. McCormack, M.J. Huang, B. Norton, Evaluation of phase change materials for thermal regulation enhancement of building integrated photovoltaics. Sol. Energy **84**, 1601–1612 (2010)
45. M.J. Huang, The effect of using two PCMs on the thermal regulation performance of BIPV systems. Sol Energy Mater Solar Cells **95**, 957–963 (2011)
46. M.J. Huang, P.C. Eames, B. Norton, N.J. Hewitt, Natural convection in an internally finned phase change material heat sink for the thermal management of photovoltaics. Sol. Energy Mater. Sol. Cells **95**, 1598–1603 (2011)
47. S. Maiti, S. Banerjee, K. Vyas, P. Patel, P.K. Ghosh, Self-regulation of photovoltaic module temperature in V-trough using a metal–wax composite phase change matrix. Sol. Energy **85**, 1805–1816 (2011)
48. A. Hassan, H. Nouman, A. Assi, B. Norton, Temperature regulation and thermal energy storage potential of phase change materials layer contained at the back of a building integrated photovoltaic panel, in *Proceedings of the 30th International Plea Conference*, 2014, pp. 16–18
49. A.H.A. Al-Waeli, K. Sopian, M.T. Chaichan, H.A. Kazem, A. Ibrahim, S. Mat, M.H. Ruslan, Evaluation of the nanofluid and nano-PCM based photovoltaic thermal (PVT) system: An experimental study. Energ. Conver. Manage. **151**, 693–708 (2017)
50. S.D. Hendrie, Evaluation of combined photovoltaic/thermal collectors, in *ISES International Congress and Silver Jubilee*, 1980, p. 1865–1869.
51. C.H. Cox, P. Raghuraman, Design considerations for flat-plate photovoltaic/thermal collectors. Sol. Energy **35**(3), 227–241 (1985)
52. H.P. Grag, R.K. Agarwal, Some aspects of a PV/T collector/forced circulation flat-plat solar water heater with solar cells. Energy Convers. Manage. **36**, 87–99 (1995)
53. Y. Gu, X. Zhang, J.A. Myhren, M. Han, X. Chen, Y. Yu, Techno-economic analysis of a solar photovoltaic/thermal (PV/T) concentrator for building application in Sweden using Monte Carlo method. Energ. Conver. Manage. **165**, 8–24 (2018)
54. M.E.A. Alfegi, K. Sopian, M.Y.H. Othman, B.B. Yatim, Transient mathematical model of both side single pass photovoltaic thermal air collector. ARPN J. Eng. Appl. Sci. **2**, 22–26 (2007)
55. A. Ibrahim, G.L. Jin, R. Daghigh, M.H.M. Salleh, M.Y. Othman, M.H. Ruslan, et al., Hybrid photovoltaic thermal (PV/T) air and water based solar collectors suitable for building integrated applications. Am. J. Environ. Sci. **5**, 618–624 (2009)
56. B.J. Huang, T.H. Liu, W.C. Hung, F.S. Sun, Performance evaluation of solar photovoltaic/thermal systems. Sol. Energy **70**, 443–448 (2001)
57. T.T. Chow, Performance analysis of photovoltaic–thermal collector by explicit dynamic model. Sol. Energy **75**(2), 143–152 (2003)
58. P. Dupeyrata, C. Ménézo, S. Fortuin, Study of the thermal and electrical performances of PVT solar hot water system. Energ. Buildings **68**, 751–755 (2014)
59. K.-K. Tse, T.-T. Chow, Y. Su, Performance evaluation and economic analysis of a full scale water-based photovoltaic/thermal (PV/T) system in an office building. Energ. Buildings **122**, 42–52 (2016)
60. A. Khelifa, K. Touafek, H. Ben Moussa, I. Tabet, Modeling and detailed study of hybrid photovoltaic thermal (PV/T) solar collector. Sol. Energy **135**, 169–176 (2016)

61. J. Bigorajski, D. Chwieduk, Analysis of a micro photovoltaic/thermal (PV/T) system operation in moderate climate. Renew. Energy **137**, 1–10 (2018). https://doi.org/10.1016/j.renene.2018.01.116
62. X.U. Guoying, S. Deng, X. Zhang, L. Yang, Y. Zhang, Simulation of a photovoltaic/thermal heat pump system having modified collector/evaporator. Sol. Energy **83**, 1967–1976 (2009)
63. S.-Y. Wu, Q.-L. Zhang, L. Xiao, F.-H. Guo, A heat pipe photovoltaic/thermal (PV/T) hybrid system and its performance evaluation. Energ. Buildings **43**, 3558–3567 (2011)
64. P. Gang, F. Huide, Z. Huijuan, J. Jie, Performance study and parametric analysis of a novel heat pipe PV/T system. Energy **37**, 384–395 (2012)
65. B. Zhang, J. Lv, H. Yang, T. Li, S. Ren, Performance analysis of a heat pipe PV/T system with different circulation tank capacities. Appl. Therm. Eng. **87**, 89–97 (2015)
66. H. Long, T.-T. Chow, J. Ji, Building-integrated heat pipe photovoltaic/thermal system for use in Hong Kong. Sol. Energy **155**, 1084–1091 (2017)
67. Y. Khanjari, F. Pourfayaz, A.B. Kasaeian, Numerical investigation on using of nanofluid in a water-cooled PV thermal system. Energ. Conver. Manage. **122**, 263–278 (2016)
68. O. Rejeb, M. Sardarabadi, C. Ménézo, M. Passandideh-Fard, M.H. Dhaou, A. Jemni, Numerical and model validation of uncovered nanofluid sheet and tube type photovoltaic thermal solar system. Energ. Conver. Manage. **110**, 367–377 (2016)
69. M.A. Adriana, Hybrid nanofluids based on Al_2O_3, TiO_2 and SiO_2: Numerical evaluation of different approaches. Int. J. Heat Mass Transfer **104**, 852–860 (2017)
70. M. Huang, P. Eames, B. Norton, Comparison of predictions made using a new 3D phase change material thermal control model with experimental measurements and predictions made using a validated 2D model. Heat Transfer Eng. **28**, 31–37 (2007)
71. M. Cellura, V.L. Brano, A. Marvuglia A Heat Transfer Eng. Photovoltaic panel coupled with a phase changing material heat storage system in hot climates, in *PLEA 2008 – 25th conference on passive and low energy architecture*, Dublin, Ireland
72. J.H.C. Hendricks, W.G.J.H.M. Sark, Annual performance enhancement of building integrated photovoltaic modules by applying phase change materials. Progr. Photovolt. **21**, 620–630 (2013)
73. V.L. Brano, G. Ciulla, A. Piacentino, F. Cardona, On the efficacy of PCM to shave peak temperature of crystalline photovoltaic panels: an FDM model and field validation. Energies **6**, 6188–6210 (2013)
74. V.L. Brano, G. Ciulla, A. Piacentino, F. Cardona, Finite difference thermal model of a latent heat storage system coupled with a photovoltaic device: description and experimental validation. Renew. Energy **68**, 181–193 (2014)
75. A. Machniewicz, D. Knera, D. Heim, Effect of transition temperature on efficiency of PV/PCM panels. Energy Procedia **78**, 1684–1689 (2015)
76. A.H.A. Al-Waeli, M.T. Chaichan, K. Sopian, H.A. Kazem, H.B. Mahood, A.A. Khadom, Modeling and experimental validation of a PVT system using nanofluid coolant and nano-PCM. Sol. Energy **177**, 178–191 (2019)
77. TWINSOLAR (2010), http://www.grammer-solar.com/en/products/twinsolar/index.shtml. Accessed 20 Dec 2010
78. SolarVenti (2010), http://www.solarventi.com. Accessed 20 Dec 2010
79. Solar Wall (2010), http://solarwall.com/en/home.php. Accessed 20 Dec 2010
80. MULTI SOLAR PV/T System (2010), http://www.millenniumsolar.com. Accessed 20 Dec 2010
81. Solar Collector X10 (2010), http://www.absolicon.com. Accessed 20 Dec 2010
82. PV/T examples (2010), http://www.iea-shc.org/task35/examples.htm. Accessed 11 Oct 2010
83. P. Barnwal, G.N. Tiwari, Life cycle energy metrics and CO2 credit analysis of a hybrid photovoltaic/thermal greenhouse dryer. Int. J. Low Carbon Technol **3**(3), 203–220 (2019)
84. T.T. Chow, J.W. Hand, P.A. Strachan, Building-integrated photovoltaic and thermal applications in a subtropical hotel building. Appl. Therm. Eng. **23**, 2035–2049 (2003)
85. S.A. Kalogirou, Y. Tripanagnostopoulos, Industrial application of PV/T solar energy systems. Appl. Therm. Eng. **27**, 1259–1270 (2007)

86. I. Farkas, I. Seres, C.S. Meszaros, Analytical and experimental study of a modular solar dryer'. Renew. Energy **16**, 773–778 (1999)
87. J. Mumba, Development of a photovoltaic powered forced circulation grain dryer for use in the tropics. Renew. Energy **6**, 855–862 (1995)
88. J. Mumba, Design and development of a solar grain dryer incorporating photovoltaic powered air circulation. Energ. Conver. Manage. **37**, 615–621 (1996)
89. A. Tiwari, M.S. Sodha, Parametric study of various configurations of hybrid PV/thermal air collector: experimental validation of theoretical model. Sol. Energy Mater. Sol. Cells **91**, 17–28 (2006)
90. Solar Energy Research Institute, Malaysia, http://www.ukm.edu.my/seri. Accessed 1 Feb 2019
91. M. Bosanac, B. Sorensen, K. Ivan, H. Sorensen, N. Bruno, B. Jamal, Photovoltaic/thermal solar collectors and their potential in Denmark (2003). Final Report, EFP Project, www.solenergi.dk/rapporter/pvtpotentialindenmark

Chapter 7
Research Opportunities and Future Work

7.1 Background

In this book, fundamentals and concepts of photovoltaic thermal (PV/T) technology were introduced, and significant literature review efforts were devoted to establishing a clear map of research and development (R&D) in the field. Chronology of PV/T development and impact of nanofluids as energy carriers for PV/T systems were presented in detail. The review mainly discussed the performance evaluation of PV/T system with different types of thermal collectors and under various operating conditions. Photovoltaic thermal (PV/T) systems can offer a high-performance conversion rate and requires lesser area than separate photovoltaic and solar thermal systems. Implementation of such hybrid technologies can lead to establishing an alternative energy source, energy autonomy and dependency, and cost-competitiveness with fossil fuel energy resources, in addition to being more environmentally friendly. This chapter presents the conclusions and recommendations of this book and further expectations of PV/T development.

7.2 Conclusions

1. Photovoltaic thermal (PV/T) collectors provided simultaneous electrical and thermal energy yields. They allow for increased photovoltaic yield compared to standard photovoltaic under the same conditions.
2. Photovoltaic thermal (PV/T) collectors can be designed with bias to thermal or electrical generation.
3. Photovoltaic thermal (PV/T) collectors can be classified according to the working fluid into air-based, water-based, air- and water-based, refrigerant-based, and nanofluid-based. Another form of classification is by the type of thermal absorber such as sheet and tube.

A. H. A. Al-Waeli et al., *Photovoltaic/Thermal (PV/T) Systems*,
https://doi.org/10.1007/978-3-030-27824-3_7

4. There are different configurations of photovoltaic thermal (PV/T) collectors according to their passage flow shape and absorber tube shape. Passage flow shapes include direct, serpentine, web, honeycomb, etc. Absorber tube shapes include square, circular, elliptical, rectangular, hexagonal, etc.
5. Increase of solar irradiance is generally correlated with increase in energy yield and efficiency of PV/T. However, insufficient cooling would lead to reduction of electrical efficiency due to increased PV cell temperature.
6. Increase of mass flow rate is generally correlated with increased electrical and thermal efficiencies. However, power requirements must be considered for high mass flow rates.
7. Increase of air's relative humidity leads to degradation of the photovoltaic module.
8. Increase of wind speed although leads to improvement in solar cell efficiency for PV systems causes decrease in system efficiency for PV/T systems, depending on the type of PV/T system.
9. Dust accumulation on PV/T surface causes degradation in system efficiency.
10. Increased contact area is correlated with enhanced PV/T efficiency due to more heat transfer between photovoltaic module and attached thermal absorber.
11. The optical efficiency has a great effect on the exergy efficiency of PV/T collector. Optical losses cause degradation in overall efficiency of PV/T.
12. Exergy analysis provides critical details to quantify inefficiencies and magnitudes.
13. Majority of studies implement the two-step method for preparation of nanofluids for PV/T application. Nanofluids are found to enhance the thermal conductivity and consequently overall heat transfer between the PV module and working fluid within absorber.
14. It is commonly observed throughout the literature that aluminum and copper are suitable channel materials.
15. The use of nanofluids leads to enhanced thermal conductivity, thermal diffusivity, viscosity, convective heat transfer coefficient, absorption coefficient, and optical absorption. Moreover, because of the increase of heat transfer, the radiative and convective losses are minimized.
16. Increase of nanofluid volume fraction is commonly observed to lead to better cooling and hence increase of energy yield. However, it is also found to lead to cost increase due to increase of the number of nanoparticles needed. Above a certain point, increase of volume fraction leads to increase in viscous forces and hence decrease in heat transfer rate.
17. Increase of nanofluid volume fraction means that the nanofluid becomes heavier, and hence more pumping power per unit length is needed.
18. The use of surfactant is highly useful for maintaining the stability of nanofluids and exhibiting higher thermal conductivities.
19. Increase of PV/T power and energy yields leads to decrease in the payback period (PBP) of the system.
20. The life cycle costs (LCC) of cooled PV systems are lower than conventional PV systems, specifically, for passive cooled PV systems, due to less need for maintenance and replacement or energy use.

21. The energy yield of a nanofluid-based PV/T system, installed in a small area, is found to be equivalent to that of a standard PV system installed in a larger area. This is attributed to the cooling achieved, which allows for the maintenance of PV/T's electrical yield and producing thermal yield, which is not possible for a standard PV system.
22. Jet impingement is an effective method of cooling photovoltaic modules and producing thermal energy. Nanofluids have been found to work as excellent heat transfer fluids for jet impingement.
23. Application of organic phase change material (PCM) for thermal regulation and increase of power generation of PV module is highly effective and exhibits massive potential for storage of thermal energy. Higher performance is achieved using nano-enhanced PCM or "nano-PCM."
24. Different PV/T systems can be compared in terms of equal Reynolds number basis and/or equal pumping power requirement basis.
25. Genetic algorithm (GA) and artificial neural networks (ANN) are useful methods for prediction of PV/T performance for long term and for implementation of PV and PV/T systems in remote areas where there is shortage of data to assess their feasibility.

7.3 Recommendations and Future Work

This section provides recommendations and future work for research in the field of photovoltaic thermal (PV/T) collectors. The recommendations for future studies are classified into five types: (i) original work, (ii) case studies, (iii) replication studies, (iv) literature review or review articles, and (v) collaboration studies.

(i) The recommendations for original work involve presenting ideas of advanced designs of PV/T collector and/or systems to be fabricated, tested, and evaluated. These ideas can be implemented both numerically and experimentally. Further recommendations are made with regard to elements to include in these studies, which are in many cases neglected.
(ii) The recommendation for case studies is to consider well-established PV/T designs for performance evaluation and/or cost analysis under different settings, either in different climate conditions or for different applications.
(iii) The recommendations for replication studies are of quite importance for a more solid understanding of PV/T performance. This is due to inconsistency in some of the findings in the literature, particularly in the nanofluid aspect of PV/T systems.
(iv) The recommendations for literature review involve proposing ideas for different approaches to summarize, discuss, and provide critical review of the research conducted in the field.
(v) The recommendations for collaborations provide ideas for publications through mutual concepts. It is of high importance to conduct research collaborations

among scientists in the field to produce valuable research and establish solid research connections. This is highly useful to bridge the gap in the literature and share research methods.

7.3.1 Original Work

1. To design, fabricate, and test the performance of a PCM-based PV/T collector with partial-honeycomb absorber.
2. To design, fabricate, and test the performance of a nano-PCM-based solar still with nanofluid as working fluid.
3. To assess conventional types of absorber flow configurations, e.g., direct flow, parallel flow, etc., using rough or twisted pipes to investigate their performance.
4. To design, fabricate, and test the performance of a double-pass v-groove back-contact PV/T collector. In addition to investigation of back-contact performance in PV/T systems.
5. To design a passive cooling nanofluid-based PV/T system through elevation of head.
6. To design, fabricate, and test nanofluid-based bifacial PV/T collector.
7. To implement nano-coating methods for enhanced heat transfer between photovoltaic module and thermal absorber.
8. To investigate the utility of absorber plate in PV/T collectors by comparing their performance with PV/Ts of different designs without absorber plates.
9. To utilize nano-PCM material in building façade of BIPV/T systems.
10. Further investigation of nano-PCM material for PV/T systems is required. Consideration of PCM type, shell material, number of heat and cooling cycles, and carrier fluid along with investigation of charging and discharging period.
11. Higher enhancement in PV/T efficiency is observed when using carbon nanotubes and carbon nanohorns. It is recommended to test these nanofluids and other nanofluids, such as magnetic nanofluids, boehmite nanofluids, etc., for the various established designs of PV/T systems.
12. To study the effect and limit of cascaded PV/T collectors on their individual energy and exergy efficiencies.
13. To investigate design considerations for PV/T arrays and study the absorber shape, the mismatch effects, etc., when using a PV/T array.
14. To assess the stability of nanofluids for continuous heating and cooling cycles. This could be done by conducting thermophysical properties and stability tests prior to entering the PV/T tubes and after exiting those tubes.
15. To assess different types of base fluids, such as ethylene glycol and other oils, for nanofluid preparation and their feasibility for use in different types of PV/T systems.
16. To determine optimum pipe angle between different types of absorber flow configurations in terms of pressure drop.

17. To conduct overall loss coefficient for accurate calculation of exergy efficiency of nano-PCM-based PV/T systems.
18. To use artificial neural networks (ANN) to predict the stability of nanofluids depending on different input parameters such as volume fraction, physical properties of particle, etc.
19. To conduct economic analysis for novel designs of PV/T systems is crucial to assess their potential for future commercialization and industrial consideration.
20. In order to evaluate the high-grade exergy and environmental impacts of advanced PV/T hybrid systems, it is recommended to conduct life cycle exergetic analysis. The analysis should also extend to conventional PV and PV/T systems for comparison purposes. This will help in determining the cumulative exergies and pollutant emissions across different stages of PV/T lifetime, for example, the carbon dioxide reduction.
21. To compare between double-pass PV/T and two-channel (separate) PV/T utilizing (i) nanofluids of the same type and (ii) different types of nanofluids.
22. To study a nanofluid cooled concentrated photovoltaic thermal collector with a nanofluid-based liquid optical filter.
23. To conduct study for PV/T-assisted heat pump for solar drying purposes. To assess the quality of dried products and status of continuous drying.
24. To conduct techno-economic and thermo-economic studies on PV/T systems. There is a lack in calculating levelized cost of energy (LCOE) and levelized cost of heat (LCOH) in the literature.
25. To investigate water-floating PV/T systems with flowing channel above PV layer using either air, water, or selective liquids.
26. To investigate nano-coated fins for passive cooling of photovoltaic modules.
27. To present mathematical models for long-term evaluation of PV/T.
28. To study nanofluid-based spectral filters for different types of PV cells.
29. Awareness is an important issue in the field of solar energy. Quantitative studies with the use of questionnaires are recommended to further assess these issues and create methods for establishing awareness of solar energy, in general, and PV/T technology, specifically.
30. Studies to investigate optimum criteria for PV/T and PV plant installation location with consideration for resources, geomorphological parameters, hydrological parameters, transportation parameters, and land use parameters. Solar energy system can be installed in land, on water bodies, and on vehicles.

7.3.2 Review Articles

Although comprehensive reviews are established across the literature, the following review ideas can be addressed:

1. Review, and revisit, each type of working fluid used for PV/T systems. As provided in Chap. 2 of the book, the designs of conventional and novel working

fluid-based PV/T are continuously developing. Hence, it is necessary to continuously review the development according to type of working fluid.
2. Review different types of nanofluids used in PV/T systems in terms of preparation, volume fractions, thermophysical properties, heat transfer enhancement, and stability (if present).
3. Review PV/T systems according to methodology, experimental setup, and findings. Moreover, reviewing these studies with respect to time of publication. This is useful to view the development of theory and concepts across time, providing a historical prospective.
4. Review these systems with respect to performance under varying mass flow rates. This point is critical to understand the range of mass flow rates present in the literature. Comparison of achieved efficiencies under similar mass flow rates can provide an understanding of efficiency enhancement and effect of absorber type.
5. Review the mathematical models presented in the literature in terms of precision and accuracy of results. These reviews can include experimental works which are used to validate these models.
6. Review the use of artificial neural network (ANN) and genetic algorithm (GA) approaches for performance prediction of PV/T collectors.
7. Review the different types of modeling software, for example, TRNSYS software, ANSYS Fluent, MATLAB Simulink, standard Excel sheets, etc., for PV/T performance evaluations, and assess the advantages/disadvantage of each type according to the outcome of the studies.

7.3.3 Case Studies

The case studies allow for understanding the technologies' varying performance and feasibility according to solar energy policies and regulations depending on the country of installation and effect of different climatic conditions.

1. Photovoltaic thermal (PV/T) systems for supply of office building demands (commercial scale). This concept can be linked to building-integrated PV/T (BIPV/T).
2. Photovoltaic thermal (PV/T) systems for autonomous supply of water and electricity in typical residential setting. This concept can be linked to building-integrated PV/T (BIPV/T).
3. Life cycle cost analysis of a PV/T system according to the price of the systems components.
4. To investigate the energy and cost savings of nanofluid-based PV/T systems for energy support in palm oil mills or other industrial facilities, e.g., water desalination.
5. To study the viability of PV/T systems as reliable solutions for electrification of remote off-grid regions. The study must account for cost and lack of resources.

6. To investigate the utility of nano-PCM-based PV/T in cold climate conditions, where it is useful to store thermal energy for later use.
7. To investigate the performance of PV/T-assisted systems. There is a clear lack in this area in the field.
8. To investigate the use of advanced passive cooling PV/T systems' assisted water pumping system.

7.3.4 Replications Studies

1. To replicate the experimental studies done to test the performance of different PV/T systems indoors. This particular work is possible due to controlled environment of the laboratory. However, for comparison sake, the replication of indoor studies is only possible if the published article describing the work provides comprehensive information of the methodology, material, and experimental procedures carried out by the researchers. This also calls for the utility of submitting data-in-brief to journals and/or providing the data in appendices.
2. To replicate the experimental studies done to test the performance of different PV/T systems in outdoors environment. This duty is possible to be carried out by members of the same institute, given their proximity to the actual testing location. However, the study must account for margins of error associated with different environmental conditions.
3. To replicate the published numerical studies in the literature. In addition to assessing the assumptions of those studies and changing accordingly. For example, assessing the material specification. Further optimization of the outcome by employing up-to-date material. Essentially, revisiting those studies with an update prospective. This aspect could also be included as an opportunity for research collaboration.

7.3.5 Collaborations

1. To replicate numerical studies using input data from different locations around the world. For instance, a numerical model is provided in a study to assess the behavior and performance of PV/T for a set location, e.g., city (A) in country (A). Collaboration can be done to view the outcomes under different locations, e.g., city (A) in country (B) and/or city (B) in country (A).
2. To conduct international collaborations to establish a global standard for PV/T system design, testing, installation, and commissioning.

7.4 Risks, Problems, and Hurdles in the Field of PV/T Systems

7.4.1 Technical Challenges

Different issues are associated with conventional working fluids depending on their type. Air-based PV/T exhibits the lowest efficiency and is considered the least effective due to relatively lower heat transfer. Water-based PV/T exhibits higher efficiencies. However, there is a risk of freezing when implementing in extremely cold climates and issues associated with leakage and possibly damage. Water- and air-based type exhibits and causes higher cooling of PV module; however, it is more complicated in terms of design and difficult to establish in passive mode. Novel working fluids face issues of increase complexity, fluid costs, and maintenance requirements. Refrigerant-based PV/T requires having heat exchangers which introduce other points of concern such as the heat exchanger efficiency, initial costs, and maintenance costs. Similarly, nanofluid-based PV/T exhibits those issues, in addition to issues of nanofluids' stability and effectiveness over long term. As for PCM and nano-PCM-based PV/T, this technology is very recent and with that comes issues such as lack of standard for preparation of nano-PCMs and discrepancies of measuring methods for the different parameters to assess the behavior of nano-PCMs. The limitation of repeated use for nano-PCMs and the difficulty of their replacement on the end consumer are major issues that boost both operation and maintenance costs and reduce favorability of this technique for the end consumer. Finally, the lack of an international standard for the testing, monitoring, and performance evaluation of PV/T systems is a major issue due to the dependency of the field on different views which can negatively affect the development of this technology, and hence focused attention to establishing an international PV/T standard is critical.

7.4.2 Economic Challenges

Despite the efforts mentioned in Chap. 6, there is still a lack of commercialization and acceptance in the market of PV/T systems, as appose to separate photovoltaic (PV) and solar thermal (ST) technologies, which are described in Chap. 1.

It is certainly of great importance to bridge the gap between academic research and industrial development of this technology. The research provides the concepts and fundamentals, while the industry can introduce scaling and commercialization of this technology.

7.4.3 Social Awareness Challenges

The lack of social awareness imposes many difficulties in expanding PV/T technology research and industrial acceptance. To simply put it, demand is of highest priority to motivate supply. Hence, it is important to establish methods of spreading awareness of PV/T systems and their value to end consumers in terms of autonomy and usefulness for residential installation in cities and remote areas. Assessing these issues is possible through public surveys and polls, and solutions such as awareness campaigns in digital forms can help in gaining the public's acceptance of this technology.

7.4.4 Policy and Regulation

Policy and regulation are highly important aspects to speeding up the development and market integration of this technology. Hence, it is also recommended to bridge the gap between the research and decision-makers through continuous social pressure through the public and awareness campaigns. Positive outcome includes governmental subsidies and programs for end consumers, tax cuts for investors, awareness initiatives, and research grants.

7.5 Points of Contention/Debate

1. What is the most accurate method for assessing the performance of PV/T collectors?
2. Comparing PV/T systems under equal Reynolds number basis or under equal pumping power basis or using different criteria.
3. To favor advanced designs of PV/T systems in the research or focus on pushing to commercialization of simpler designs of PV/T systems.
4. Nanofluids are effective long-term strategies as base fluids for PV/T systems.
5. Although nanofluids were found to be effective as heat transfer fluids for jet impingement PV/T, it is reasonable to question their stability using such cooling mechanism.
6. Many research studies use steady-state analysis of PV/T collectors. However, some authors claim its ineffectiveness for performance analysis as an approach and to replace it with transient-state analysis.

7.6 Future Expectations of PV/T Technology Development (Opinion)

In our view, with the increase of research and development along with scientific awareness of PV/T technology, more businesses will probably be attracted to this field. Given the promising output and ecological impact this technology holds, we (the authors) predict it will be more acceptable and attractive to end consumers and consequently investors. The main hurdle in this field is the lack of an international standard to follow for, at the very least minimum requirements, design, installation, testing, monitoring and performance evaluation for PV/T systems. Further developments will occur once the technology has gained enough reputation and established manufacturing methods. In case such stage is reached, it is safe to assume that more demand will be established in the market. Meeting the demand is simultaneous when full economies of scale come into effect; subsequently drop in price of PV/T collectors will make it an attractive option to different classes of end consumers. This interest would be optimal for commercial development of PV/T systems and their prominence as cost-effective, viable, and competitive sources of energy.

Index

A. H. A. Al-Waeli et al., *Photovoltaic/Thermal (PV/T) Systems*,
https://doi.org/10.1007/978-3-030-27824-3

The manufacturer's authorised representative in the EU is Springer Nature Customer Service Centre GmbH, Europaplatz 3, 69115 Heidelberg, Germany. If you have any concerns regarding our products, please contact ProductSafety@springernature.com

Printed and bound by CPI Group (UK) Ltd, Croydon, CR0 4YY
05/10/2026
02228130-0007